L$_p$-Theory for Incompressible Newtonian Flows

T0238443

Matthias Köhne

L_p-Theory for Incompressible Newtonian Flows

Energy Preserving Boundary Conditions, Weakly Singular Domains

 Springer Spektrum

Matthias Köhne
Darmstadt, Deutschland

Dissertation, Technische Universität
Darmstadt, 2012, D 17

Referenten:
Prof. Dr. Dieter Bothe
Prof. Dr. Jan Prüß
Prof. Dr. Yoshihiro Shibata

ISBN 978-3-658-01051-5   ISBN 978-3-658-01052-2 (eBook)
DOI 10.1007/978-3-658-01052-2

The Deutsche Nationalbibliothek lists this publication in the Deutsche Nationalbibliografie; detailed bibliographic data are available in the Internet at http://dnb.d-nb.de.

Library of Congress Control Number: 2012953925

Springer Spektrum
© Springer Fachmedien Wiesbaden 2013
This work is subject to copyright. All rights are reserved by the Publisher, whether the whole or part of the material is concerned, specifically the rights of translation, reprinting, reuse of illustrations, recitation, broadcasting, reproduction on microfilms or in any other physical way, and transmission or information storage and retrieval, electronic adaptation, computer software, or by similar or dissimilar methodology now known or hereafter developed. Exempted from this legal reservation are brief excerpts in connection with reviews or scholarly analysis or material supplied specifically for the purpose of being entered and executed on a computer system, for exclusive use by the purchaser of the work. Duplication of this publication or parts thereof is permitted only under the provisions of the Copyright Law of the Publisher's location, in its current version, and permission for use must always be obtained from Springer. Permissions for use may be obtained through RightsLink at the Copyright Clearance Center. Violations are liable to prosecution under the respective Copyright Law.The use of general descriptive names, registered names, trademarks, service marks, etc. in this publication does not imply, even in the absence of a specific statement, that such names are exempt from the relevant protective laws and regulations and therefore free for general use. While the advice and information in this book are believed to be true and accurate at the date of publication, neither the authors nor the editors nor the publisher can accept any legal responsibility for any errors or omissions that may be made. The publisher makes no warranty, express or implied, with respect to the material contained herein.

Printed on acid-free paper

Springer Spektrum is a brand of Springer DE.
Springer DE is part of Springer Science+Business Media.
www.springer-spektrum.de

To my Family

Acknowledgements

First of all, I would like to thank PROF. DR. DIETER BOTHE for his invaluable support and encouragement over many years. He was always prepared to promote my research in numerous fruitful and inspiring discussions. On the other hand, he gave me the freedom to develop my own ideas. Last, but not least, he made it possible for me to become a member of the mathematical community by establishing many contacts, which lead to fruitful and lasting collaborations.

Moreover, I would like to thank PROF. DR. JAN PRÜSS, who was always prepared to give me thoughtful advice and support, not only on the topic of this thesis. I absolutely enjoyed each of the numerous research visits at the *Department of Mathematics of the Martin-Luther-Universität Halle-Wittenberg* and each of the plenty illuminative discussions with him and the members of his research group. In particular, the collaboration with DR. MATHIAS WILKE, who was always open for the exchange of ideas, lead to various insights and valuable perspectives.

Furthermore, I express my greatest thanks to PROF. DR. YOSHIHIRO SHIBATA for the kind hospitality and many stimulating discussions during my research stay at *Waseda University, Tokyo*. It was really a pleasure for me to work with him and the members of his research group and the time in Japan was an invaluable experience not only concerning mathematics.

Last, but not least, I am greatly indebted to PROF. DR. MATTHIAS HIEBER for giving me the opportunity to join the *International Research Training Group "Mathematical Fluid Dynamics"* at *TU Darmstadt* and to participate in an exchange program, which made it possible to visit *Waseda University, Tokyo*. He never avoided any effort to support me and the members of the research group, which lead to numerous fruitful discussions with guest researchers and many new contacts within the mathematical community.

Moreover, I would like to express my greatest gratitude to PROF. DR. JÜRGEN SAAL and PROF. DR. GIERI SIMONETT for their interest in my research and many useful advices and hints, which certainly improved this thesis.

Furthermore, I would like to convey special thanks to the publisher, *Springer Spektrum*, in particular to MRS. SABINE SCHÖLLER and MRS. UTE WRASMANN for their kind support, and to my colleagues MR. ANDRÉ FISCHER and MR. SIEGFRIED MAIER for many helpful comments on the manuscript.

Finally, I express my greatest thanks to my former and present colleagues at the *RWTH Aachen University*, the *Center of Smart Interfaces at TU Darmstadt* and the *Department of Mathematics at TU Darmstadt* for the always pleasant ambiance and plenty hints and suggestions.

Contents

Bounded Weakly Singular Domains

Introduction

There exist various ways to describe the motion of a fluid, most of which result in a mathematical model for the evolution of the functional properties of the fluid. Of course, the physics of such a flow heavily depend on the particular properties of the fluid under consideration and, hence, a particular model may only describe the behaviour of a certain class of fluids, which share their reaction to environmental influences, e. g. driving forces, to a sufficiently large extent.

The isothermal flow of a special class of fluids that constitutes of the *incompressible Newtonian fluids* may be described by the *Navier-Stokes equations*

$$\partial_t(\rho u) + \operatorname{div}(\rho u \otimes u - S) = \rho f, \qquad \operatorname{div} u = 0,$$

which postulate the conservation of mass and momentum in this special case. Indeed, the density $\rho > 0$ of an incompressible fluid is constant in the isothermal case, which implies the second of the above equations postulating the conservation of volume to be equivalent to the conservation of mass. Here u denotes the velocity field of the flow and, hence, ρu equals the mass flux density. On the other hand, ρu also equals the momentum density and, therefore, the first of the above equations encodes the conservation of momentum, where the momentum flux density is composed of a convective part $\rho u \otimes u$ and a stress part, which is modelled via

$$S = 2\mu D - p.$$

The stress part combines the viscous stress $2\mu D$, which is proportional to the rate of deformation tensor

$$D = \tfrac{1}{2}(\nabla u + \nabla u^{\mathsf{T}}),$$

and the pressure p, which serves as a force density potential for the conservative forces resulting from externally applied forces and the constraint on the velocity field to be solenoidal. This model for the stress results from the assumption that the viscous stress depends linearly on the rate of deformation, which is the definition of a Newtonian behaviour of the fluid. The fact that the constitutive equation for the viscous stress has to be independent of the particular choice of the coordinate frame together with the constraint on the velocity field to be solenoidal imply that a linear dependence of the viscous stress on the rate of deformation can only be achieved, if the viscous stress is proportional to the rate of deformation. The arising proportionality constant $\mu > 0$ denotes the (dynamic) viscosity of the fluid. Finally, ρf denotes the density of external forces, which serve as momentum sources and sinks. A continuum mechanical derivation of the Navier-Stokes equations will be presented in Chapter 1. For a more detailed overview of fluid dynamics and the Navier-Stokes equations we refer to [Ari90], [Bat00], [CM00], [RT00] and [Tem05]. An interesting presentation of the history of fluid dynamics is provided in [And98].

Since the Navier-Stokes equations constitute a system of evolution equations, they have to be complemented by suitable initial conditions. However, since the density is assumed to be constant, the conservation of mass reduces to the constraint on the velocity field to be solenoidal and, thus, only the conservation of momentum exhibits an evolutionary character. As a rule of thumb, the number of initial conditions should therefore equal the number of equations that describe the conservation of momentum, and the prescription of the initial velocity field via

$$u(0) = u_0$$

is the canonical choice to achieve this.

On the other hand, if the fluid does not occupy the whole space but e. g. a bounded domain $\Omega \subset \mathbb{R}^3$, the Navier-Stokes equations have further to be complemented by suitable boundary conditions. Again, as a rule of thumb, the number of boundary conditions should equal the number of equations that describe the conservation of momentum. However, there exist various possibilities to model the behaviour of the flow on the boundary $\Gamma = \partial\Omega$ of the domain Ω and in many situations no canonical choice of boundary conditions seems to be available.

If the flow is surrounded by rigid walls, the interaction of the fluid and these walls has to be modelled. Here one possibility is to assume that the fluid may not slip along the wall, which leads to the boundary condition

$$u|_\Gamma = 0.$$

This is called the *no-slip condition* and was already proposed by G. G. STOKES in 1845 along with his derivation of the Navier-Stokes equations [Sto45]. Another possibility is to allow the fluid to slip along the wall while being stressed in tangential directions due to a friction between the fluid and the rigid wall, i. e.

$$u|_\Gamma \cdot \nu_\Gamma = 0 \quad \text{and} \quad P_\Gamma u|_\Gamma + \lambda P_\Gamma S|_\Gamma \nu_\Gamma = 0.$$

This is known as the *Navier condition*, since it was first proposed by C. L. M. H. NAVIER in 1823 along with his derivation of the Navier-Stokes equations [Nav23]. Here, ν_Γ denotes the outer unit normal field of Ω and $P_\Gamma = P_\Gamma(y)$ denotes the projection onto the tangent space $T_y\Gamma$ of Γ at a point $y \in \Gamma$. The constant $\lambda > 0$ is called the *slip length*. Thus, the Navier condition prevents the fluid from leaving the domain through the rigid wall while a movement tangentially to the wall is allowed. However, the stress in tangential directions equals the friction between the fluid and the rigid wall, which is assumed to be proportional to the velocity in tangential directions. As a variant one could assume that the fluid may slip along the wall without being stressed in tangential directions. This assumption, which formally corresponds to the limit case $\lambda \to \infty$ for the Navier condition, leads to

$$u|_\Gamma \cdot \nu_\Gamma = 0 \quad \text{and} \quad P_\Gamma S|_\Gamma \nu_\Gamma = 0.$$

This is known as the *free slip* or *perfect slip condition*. On the other hand, the slip length $\lambda = 0$ results in the no-slip condition. All these three conditions are well-known and well-accepted in both, the engineering literature as well as the mathematical literature. For a more detailed overview we refer to the survey articles [LS03] and [BLS07] and the references therein.

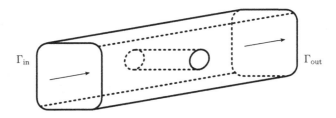

– A Tube with Cylindrical Obstacle –

If the flow is surrounded by a *free boundary*, i. e. $\Omega = \Omega(t)$ and $\Gamma = \Gamma(t) = \partial\Omega(t)$ depend on time, since the boundary is allowed to be moved and deformed by the fluid, the behaviour on the boundary is usually modelled via a *free boundary condition*

$$S|_{\Gamma(t)}\nu_{\Gamma(t)} = H(\Sigma, \Gamma(t)),$$

where the function H models the stress exerted to the fluid via the boundary. If the free boundary is assumed to be elastic, e. g. as in the case of a balloon filled with a liquid or a gas, this stress depends on the deviation of an actual configuration $\Gamma(t)$ of the free boundary from a reference configuration Σ. In other situations the free boundary may be the *free surface* of a liquid, e. g. as in the case of an ocean, where the evolution of the atmosphere is neglected in the model. In these cases the stress $H = H(\Gamma(t))$ depends only on the actual configuration $\Gamma(t)$ of the free boundary and its physical properties, e. g. *surface tension* or *surface viscosity*. A similar condition occurs in the case of *two-phase flows*, where the two fluids are assumed to be immiscible and the free boundary $\Gamma(t)$ is assumed to be a *sharp interface*, which separates the fluid contained in $\Omega(t)$ from another fluid outside of $\Omega(t)$. A typical example would be an ocean, where the evolution of the atmosphere is not neglected in the model. In these cases the free boundary condition has to be substituted by a similar *transmission condition*, which models the exchange of momentum between the two fluids via the interface $\Gamma(t)$. However, nearly all known mathematical approaches to obtain a solution to such a model require the reduction to a fixed domain Ω with a fixed boundary $\Gamma = \partial\Omega$, where an inhomogeneous boundary condition

$$S|_{\Gamma}\nu_{\Gamma} = h$$

is imposed. This is called the *Neumann condition* for the Navier-Stokes equations. Again, this condition is well-known and well-accepted.

More delicate issues arise for *artificial boundaries*. These appear for example in models for numerical simulations, since it is often convenient to keep the domain as small as possible to capture the most interesting part of the flow without wasting computation time. As an example one might consider a tube, where the flow has to pass a cylindrical obstacle. In this case the most interesting part of the flow surely takes place in a neighbourhood of the obstacle. Moreover, for a numerical simulation it is impossible to assume the tube to be of infinite length and, hence, the tube has to be cut somewhere before and somewhere after the obstacle. For such a geometry there appear two artificial boundaries, an inflow boundary Γ_{in} and an outflow boundary Γ_{out}. Of course, the shell of the tube and the surface of the obstacle may be treated as rigid walls and one may impose a Navier type boundary condition. However, a reasonable choice of boundary conditions for the two artificial boundaries is not at all obvious. For the inflow boundary Γ_{in} one may for example assume that the flow enters the tube as a fully developed steady profile, since this would be an admissible approximation, if the distance between Γ_{in} and the obstacle is chosen to be sufficiently large. In that case an inhomogeneous *Dirichlet condition*

$$u\big|_{\Gamma_{in}} = u_{in}$$

with a given velocity profile u_{in} would be appropriate. On the other hand, the behaviour of the flow at the outflow boundary Γ_{out} heavily depends on the effect of the obstacle on the flow. Therefore, one would have to choose a boundary condition on Γ_{out}, which, on one hand, is able to fix a unique solution to the model problem, and, on the other hand, allows the flow to freely develop around the obstacle as if the artificial boundary would not be present. An ideal boundary condition for such an outflow boundary would always ensure the unique solution to coincide with the solution to the model problem of an infinitely extended tube. Unfortunately, no such so-called *transparent boundary condition* is available, if the computation should be restricted to the tube of finite length.

As a result, a vast number of *artificial boundary conditions* has been developed in the numerical literature, each of which is adapted to a particular class of geometries and particular flow characteristics, cf. [Gre95]. A popular approach is to combine an inhomogeneous boundary condition for the tangential velocity respectively tangential vorticity, i. e.

$$u\big|_\Gamma \times \nu_\Gamma = u_{out} \times \nu_\Gamma \quad \text{respectively} \quad (\mathrm{rot}\, u)\big|_\Gamma \times \nu_\Gamma = v_{out} \times \nu_\Gamma,$$

with an inhomogeneous boundary condition for the normal velocity respectively the pressure, i. e.

$$u\big|_\Gamma \cdot \nu_\Gamma = u_{out} \cdot \nu_\Gamma \quad \text{respectively} \quad p\big|_\Gamma = p_{ext}.$$

This way one obtains a complete set of boundary conditions, if the tangential velocity $u_{out} \times \nu_\Gamma$, the tangential vorticity $v_{out} \times \nu_\Gamma$, the normal velocity $u_{out} \cdot \nu_\Gamma$ or the external pressure p_{ext} at an artificial outflow boundary is known in a sufficiently accurate approximation. On one hand, for the field of computational fluid dynamics these conditions deliver quite satisfactory results. On the other hand, a rational derivation of such conditions is often missing in the literature. Even more severe, almost no analytical results concerning the strong well-posedness of the resulting models are available.

Two of the main objectives of this thesis are therefore

- to provide a rational derivation of a large class of boundary conditions, which contains all of the above mentioned standard and artificial boundary conditions, and

- to develop an L_p-theory for the resulting models, which in particular provides strong well-posedness results.

To achieve the first objective, we introduce the concept of *energy preserving boundary conditions*, which provides a large class of boundary conditions with distinct physical properties. The rational derivation of these boundary conditions thereby employs the same methods as the continuum mechanical derivation of the Navier-Stokes equations.

The L_p-theory for the resulting models is then developed based on the establishment of *maximal L_p-regularity* for the Stokes equations subject to one of the derived energy preserving boundary conditions. Since this class of boundary conditions also includes the above mentioned standard boundary conditions such as the Dirichlet condition and the Neumann condition, we thereby recover most of the known results on maximal L_p-regularity for the Stokes equations. However, we present a generic approach, which is not restricted to the energy preserving boundary conditions considered in this thesis. As a result, we obtain a rather complete picture concerning the L_p-theory for incompressible Newtonian flows that has been developed up to now. On the other hand, the overall approach may inspire further developments.

Last, but not least, there is a further serious issue concerning the strong well-posedness of the models, which are in use for numerical simulations. Revisiting the example of a tube as illustrated in the figure on page 3 we have to note that the boundary of this tube is *not smooth*. Indeed, there exist edges joining the in- and outflow boundaries and the shell as well as edges resulting from the placement of the obstacle in the interior of the tube. Moreover, as has been explained above, a reasonable model should impose different boundary conditions on these different parts of the boundary, e. g. an inflow condition on Γ_{in}, a no-slip or Navier type condition on the shell and on the surface of the obstacle and, finally, an outflow condition on Γ_{out}. Unfortunately, the theory of strong solutions to the Stokes and Navier-Stokes equations developed up to now is only applicable, if the boundary is sufficiently smooth. As a result, geometries like the tube are not covered by the existing theory. Thus, even an L_p-theory that covers all the energy preserving boundary conditions derived in this thesis may not deliver any strong well-posedness results for most of the models that are in use for numerical simulations, if it is restricted to smooth domains. To make a first step forward in the analysis of incompressible Newtonian flows in non-smooth domains we will consider a restrictive class of non-smooth domains, which we call *weakly singular domains*, and establish a first L_p-theory for the Stokes and Navier-Stokes equations subject to different energy preserving boundary conditions on different parts of the boundary of such a weakly singular domain. Here the notion *weakly singular* is motivated by the fact that we allow for the presence of edges joining two smooth components of the boundary, where, however, we require such two components to meet perpendicularly at their joining edge. Therefore, the boundary of a weakly singular domain locally exhibits a rather simple geometry, since it is either approximately a plane, which is the case in the neighbourhood of points in the interior of one of the smooth components of the boundary, or it is approximately a rectangular wedge, which is the case in the neighbourhood of points on one of the edges. However, even if these requirements seem to be quite restrictive, geometries like the tube as illustrated in the figure on page 3 belong to the class of weakly singular domains. The third and last objective of this thesis is therefore

- to develop an L_p-theory for incompressible Newtonian flows in weakly singular domains subject to one of the energy preserving boundary conditions on each smooth component of the boundary, which in particular provides strong well-posedness results.

To achieve this we give a precise definition of weakly singular domains and then apply the approach developed for smooth domains to boil down the problem to the analysis of the Stokes equations in a rectangular wedge subject to two possibly different energy preserving boundary conditions.

References

[And98] J. D. ANDERSON, JR.: *Some Reflections on the History of Fluid Dynamics.* In: The Handbook of Fluid Dynamics (R. W. JOHNSON, ed.), chap. 2, Springer, 1998.

[Ari90] R. ARIS: Vectors, Tensors, and the Basic Equations of Fluid Mechanics. Dover Publications, 1990.

[Bat00] G. K. BATCHELOR: An Introduction to Fluid Dynamics. Cambridge University Press, 3rd ed., 2000.

[BLS07] M. P. BRENNER, E. LAUGA, and H. A. STONE: *Microfluidics: The No-Slip Boundary Condition.* In: Handbook of Experimental Fluid Dynamics (J. FOSS, C. TROPEA, and A. YARIN, eds.), (1219–1240), Springer, 2007.

[CM00] A. J. CHORIN and J. E. MARSDEN: A Mathematical Introduction to Fluid Mechanics. Springer, 3rd ed., 2000.

[Gre95] P. M. GRESHO: *Incompressible Fluid Mechanics: Some Fundamental Formulation Issues.* Annual Review of Fluid Mechanics, 23, 413–453, 1995.

[LS03] E. LAUGA and H. A. STONE: *Effective Slip in Pressure Driven Stokes Flow.* J. Fluid Mech., 489, 55–77, 2003.

[Nav23] C. L. M. H. NAVIER: *Mémoir sur les Lois du Mouvement des Fluides.* Mem. Acad. Inst. Sci. Fr., 6, 389–440, 1823.

[RT00] K. R. RAJAGOPAL and C. TRUESDELL: An Introduction to the Mechanics of Fluids. Birkhäuser, 2000.

[Sto45] G. G. STOKES: *On the Theories of the Internal Friction of Fluids in Motion, and of the Equilibrium and Motion of Elastic Solids.* Trans. Cambridge Phil. Soc., 8, 287–319, 1845.

[Tem05] R. TEMAM: Mathematical Modeling in Continuum Mechanics. Cambridge University Press, 2005.

Abstract and Notation

This thesis is organised as follows. The first part, which constitutes of the first two chapters, is devoted to the mathematical modelling. We start with a rational derivation of the Navier-Stokes equations based on the principles of continuum mechanics in Chapter 1. Afterwards, in Chapter 2, we give an overview of the modelling of boundary conditions for the different types of boundaries like e. g. impermeable walls, free surfaces, interfaces and artificial boundaries, which may occur in reality and numerical simulations, followed by a rational derivation of a class of energy preserving boundary conditions. These energy preserving boundary conditions are then complemented by two artificial boundary conditions, which are related to the prior due to their structure but do not share their physical properties. However, since these two boundary conditions are in use for numerical simulations and nicely fit into the later developed framework, we include them in our analysis.

The second part, which constitutes of the Chapters 3 to 7, is devoted to the development of an L_p-theory for incompressible Newtonian flows in a bounded, smooth domain subject to one of the derived boundary conditions. In Chapter 3 we give a precise formulation of the Stokes and Navier-Stokes equations in this setting and collect the known results related to this problems. Afterwards, we derive the necessary regularity and compatibility conditions and formulate in Theorem 3.30 our first main result, which is the maximal L_p-regularity of the Stokes equations subject to one of the derived boundary conditions. Due to its complexity, the proof of this theorem is delayed to the following chapters. However, we collect some immediate consequences, which include the corresponding results on the associated Stokes operators in the L_p-setting, cf. Corollaries 3.40 and 3.44, as well as results on the well-posedness of the Navier-Stokes equations subject to one of the derived boundary conditions, cf. Theorems 3.46 and 3.48. Since all employed tools are stated at the point they are needed and all collected consequences are proved immediately, Chapter 3 is in a sense self-contained and delivers a rather complete picture concerning the L_p-theory for incompressible Newtonian flows, which has been developed up to now, and, which is extended to the derived boundary conditions by Theorem 3.30. Note that we rediscover maximal L_p-regularity results for some well-known boundary conditions. However, most of the considered boundary conditions, in particular those involving the prescription of the vorticity or the pressure, have not been treated in the literature yet and, thus, Theorem 3.30 provides a whole set of new results. Of course, the above mentioned immediate consequences concerning the associated Stokes operators and the strong solvability of the Navier-Stokes equations constitute new results, too. In particular, we obtain a rather abstract method to derive the precise mapping properties of the associated Stokes operators including density results of their domains, which are not at all obvious. To prepare the proof of Theorem 3.30 we collect the essential tools and methods for the proof in Chapter 4. This includes results on related elliptic and parabolic problems in an L_p-setting as well as an introduction of the theory of sectorial operators, which plays a key role for the analysis of the Stokes equations in a halfspace. The proof of Theorem 3.30 is then carried out by a localisation procedure, which starts with an analysis of the Stokes equations in a halfspace, cf. Chapter 5, and then transfers the results via bent halfspaces, which are treated in Chapter 6, to bounded, smooth domains. This final step is presented in Chapter 7. To clarify the subsequent steps of the proof each of the Chapters 5 to 7 starts with a brief description of the strategy followed by the establishment of maximal L_p-regularity of the Stokes equations in the particular geometry. Disregarding the tools and methods collected in Chapter 4, Chapters 5 to 7 are in a sense self-contained and deliver a complete analysis of the Stokes equations in the particular geometry.

The third part, which constitutes of the Chapter 8, is devoted to the development of an L_p-theory for incompressible Newtonian flows in weakly singular domains subject to possibly different boundary conditions each of which belongs to the class derived in Chapter 2. The precise notion of a weakly singular domain is introduced in Chapter 8 and we collect some basic facts and derive a localisation procedure, which adapts the corresponding procedure for bounded smooth domains. This way most of the auxiliary results obtained in Chapters 3 to 7 may be reused. In particular, the splitting scheme presented in Theorem 4.60 for smooth domains, which forms one of the cornerstones for the proof of Theorem 3.30, admits a suitable generalisation to weakly singular domains. Finally, we study the Stokes equations in an acute wedge, which forms the prototype domain for the localisation procedure for weakly singular domains. The maximal L_p-regularity results for the Stokes equations in a weakly singular domain subject to fully inhomogeneous boundary conditions that may be obtained by our approach are collected in Theorem 8.24. To the best of our knowledge such problems have not been treated in the literature yet.

The style and organisation of this thesis is adapted from the *Academic Press* series *Pure and Applied Mathematics*. Each of the chapters starts with a short introduction. Starting with the first paragraph of the first section in a chapter the paragraphs are numbered to simplify cross-references. Definitions, theorems, propositions and corollaries are thereby treated as a single paragraph each. Moreover, each chapter closes with a collection of remarks, which may deliver details, which have been omitted in the foregoing sections to increase the readability, or, which may shortly discuss related topics or related results from the literature. Finally, each chapter contains a bibliography, which collects the references used in the particular chapter. A complete bibliography containing all referenced sources may be found at the end of the thesis. This way the separate chapters become more or less self-contained and a reader who is interested in the topics of a particular chapter only may omit the preceding chapters without suffering a loss of readability.

The thesis is supplemented by a table of symbols, which collects the most important notations, conventions, function spaces and equations. Finally, an index delivers references to the most important notions.

The Model

Chapter 1

The Navier-Stokes Equations

As mentioned in the introduction, the Navier-Stokes equations constitute the conservation of mass and momentum for incompressible Newtonian fluids. Of course, these basic equations of fluid dynamics as well as their derivation can be found in many popular and classical books, see e. g. [Lam32] or [Bat00]. However, we want to keep this thesis as self-contained as possible and present a short derivation of the Navier-Stokes equations based on the principles of continuum mechanics. Last, but not least, the derivation of a class of energy preserving boundary conditions as presented in Chapter 2 relies on the principles of continuum mechanics and follows the spirit of this presentation.

To keep the presentation of the governing equations more structured we strictly distinguish between the derivation of balance equations for the relevant extensive quantities, which are valid due to the choice of the continuum mechanical framework, and the imposition of constitutive laws, which act as closure relations and model a specific material behaviour. This modern approach, which is presented e. g. in [NT03] and [BFG$^+$12], allows for a short derivation of the desired equations without hiding their particular origin.

To describe the behaviour of incompressible Newtonian fluids it is sufficient to consider the extensive quantities mass and momentum. Of course, the resulting model could be modified and complemented by balance equations for other extensive quantities, e. g. energy, and closure relations, which model a different and more complex material behaviour. However, the Navier-Stokes equations already constitute one of the main mathematical challenges in the field of fluid dynamics as some of the most fundamental problems related to these equations are still open, cf. [Fef07]. Moreover, it is not necessary to work in the framework of continuum mechanics in order to derive a mathematical model for the behaviour of a particular class of fluids. However, this approach is appropriate to describe the behaviour of a fluid, which is assumed to be continuously distributed in the occupied region of space. For more detailed derivations of models based on the principles of continuum mechanics as well as alternative approaches we refer to [Lam32], [Ari90], [Bat00], [CM00], [RT00], [NT03], [Tem05], [BFG$^+$12] and the references therein.

1.1 The Principles of Continuum Mechanics

1.1 Let $\Omega \subseteq \mathbb{R}^3$ be a region in space. A physical quantity Φ is called *extensive*, if every volume $V \subseteq \Omega$ may be assigned a value $\Phi(V)$ for this quantity, and, if Φ behaves additive w. r. t. disjoint volumes, i. e. $\Phi(U \cup V) = \Phi(U) + \Phi(V)$, whenever $U, V \subseteq \Omega$ with $U \cap V = \varnothing$. In the framework of *continuum mechanics* we assume every extensive quantity to be determined via a volume specific *density*, i. e. there exists an integrable function $\phi : \Omega \longrightarrow \mathbb{R}$ such that

$$\Phi(V) = \int\limits_V \phi \, \mathrm{d}\mathcal{H}^3,$$

whenever $V \subseteq \Omega$ is Lebesgue measurable. We will always use $\mathcal{H}^n$ to denote the n-dimensional Hausdorff measure, which in the above case coincides with the three-dimensional Lebesgue measure.

1.2 To describe the evolution of an extensive quantity Φ we assume its density ϕ to depend on time, i. e. to be a continuous function $\phi : J \times \Omega \longrightarrow \mathbb{R}$, where $J = (0, a)$ denotes the considered time interval, which w. o. l. g. starts at $t = 0$ and ends at a certain point $t = a > 0$. If the quantity Φ is conserved, the *master balance equation*

$$\begin{array}{ccc} \text{rate of change} \\ \text{of } \Phi(V) \end{array} = - \begin{array}{c} \text{rate of outflow} \\ \text{of } \Phi \text{ across } \partial V \end{array} + \begin{array}{c} \text{rate of change of } \Phi(V) \\ \text{due to sources and sinks} \end{array}$$

is valid. In the framework of continuum mechanics we assume the existence of a continuous *flux density* $q : J \times \Omega \longrightarrow \mathbb{R}^3$ and a continuous *density of sources and sinks* $f : J \times \Omega \longrightarrow \mathbb{R}$ such that the master balance equation may be formulated as

$$\frac{\mathrm{d}}{\mathrm{d}t} \Phi(t, V) = \frac{\mathrm{d}}{\mathrm{d}t} \int\limits_V \phi(t, \cdot) \, \mathrm{d}\mathcal{H}^3 = - \int\limits_{\partial V} q(t, \cdot) \cdot \nu_{\partial V} \, \mathrm{d}\mathcal{H}^2 + \int\limits_V f(t, \cdot) \, \mathrm{d}\mathcal{H}^3$$

for every Lebesgue measurable volume $V \subseteq \Omega$ with sufficiently smooth boundary ∂V. Above, $\nu_{\partial V} : \partial V \longrightarrow \mathbb{R}^3$ denotes the outer unit normal field of V.

1.3 To pass to a local formulation of the above master balance equation we fix a point $x \in \Omega$ and consider $V = B_r(x) \subseteq \Omega$. Using the Gaussian theorem we obtain

$$\frac{\mathrm{d}}{\mathrm{d}t} \int\limits_{B_r(x)} \phi(t, \cdot) \, \mathrm{d}\mathcal{H}^3 = \int\limits_{B_r(x)} \left\{ f(t, \cdot) - \operatorname{div} q(t, \cdot) \right\} \mathrm{d}\mathcal{H}^3.$$

In the framework of continuum mechanics we assume ϕ to be sufficiently smooth to allow for an interchange of differentiation and integration on the left hand side. Thus, we infer

$$\frac{1}{\mathcal{H}^3(B_r(x))} \int\limits_{B_r(x)} \left\{ \partial_t \phi + \operatorname{div} q - f \right\} \mathrm{d}\mathcal{H}^3 = 0$$

and a passage to the limit $r \to 0+$ yields

$$(1.1) \qquad\qquad \partial_t \phi + \operatorname{div} q = f \qquad \text{in } J \times \Omega.$$

1.4 Based on the principles of continuum mechanics as stated in 1.1 – 1.3 the master balance equation (1.1) holds for every extensive quantity Φ. However, to obtain a complete model for the evolution of such a quantity, it still remains to determine the density ϕ, the flux density q and the density of sources and sinks f. In general this is only possible by the imposition of one or more *constitutive laws*.

1.5 To simplify the determination of the flux density q it is often convenient to employ the decomposition $q = \phi u + j$ of the flux density into a *convective flux density* ϕu, where $u : J \times \Omega \longrightarrow \mathbb{R}^3$ denotes the velocity field, which describes the motion of the fluid particles, and an additional flux density $j : J \times \Omega \longrightarrow \mathbb{R}^3$, which models all fluxes that are not a result of convection, e. g. *diffusive fluxes* as introduced in Section 1.3. The master balance equation (1.1) may then be rewritten as

$$(1.2) \qquad \partial_t \phi + \operatorname{div}(\phi u) + \operatorname{div} j = f \qquad \text{in } J \times \Omega.$$

The motivation of this decomposition stems from the fact that it is sometimes more convenient to consider balance equations based on advected volumes instead of fixed ones, which have been used in the derivation of (1.1). Indeed, once a velocity field u is present to describe the motion of the fluid particles, a family of flow maps

$$\Psi_s^t : \Omega \longrightarrow \Omega, \qquad s,\, t \in J,\ |t - s| < \varepsilon$$

is available, where $\Psi_s^t(x)$ denotes the position of a particle that starts at a time $s \in J$ in $x \in \Omega$ at a time $t \in J$. Note that we do not want to impose any restrictions on the velocity field u, which would prevent such a particle to leave the region Ω in finite time. Therefore, the final time t has to be chosen sufficiently close to s, which is the reason for the restriction $|t - s| < \varepsilon$, where $\varepsilon = \varepsilon(s) > 0$ has to be determined based on $u(s, \cdot)$. In fact, the flow map Ψ_s^t is uniquely defined for every $x \in \Omega$ as $\Psi_s^t(x) = \psi(t \,|\, s,\, x)$, where $\psi(\,\cdot\,|\, s,\, x)$ solves the equation of motion

$$\psi'(t \,|\, s,\, x) = u(t,\, \psi(t \,|\, s,\, x)), \qquad t \in J,\ |t - s| < \varepsilon, \qquad \psi(s \,|\, s,\, x) = x,$$

which admits a unique local solution, provided u satisfies some mild regularity conditions. Hence, a volume $V = V(s) \subseteq \Omega$ is advected according to

$$V(t) = \Psi_s^t(V(s)) = \Big\{ \Psi_s^t(x) \ :\ x \in V(s) \Big\}, \qquad t \in J,\ |t - s| < \varepsilon$$

and the master balance equation for a conserved extensive quantity Φ reads

$$\begin{array}{ccc} \text{rate of change} \\ \text{of } \Phi(V(t)) \end{array} = - \begin{array}{c} \text{rate of non-convective} \\ \text{flux of } \Phi \text{ across } \partial V(t) \end{array} + \begin{array}{c} \text{rate of change of } \Phi(V(t)) \\ \text{due to sources and sinks.} \end{array}$$

Note that the convective flux of the quantity Φ does not enter this balance equation, since the set of particles that form $V(t)$ is independent of t, since the volume is advected by the flow. This equation may be formulated as

$$(1.3) \qquad \frac{\mathrm{d}}{\mathrm{d}t}\Phi(t,\, V(t)) = \frac{\mathrm{d}}{\mathrm{d}t}\int_{V(t)} \phi(t,\, \cdot)\,\mathrm{d}\mathcal{H}^3 = -\int_{\partial V(t)} j(t,\, \cdot) \cdot \nu_{\partial V(t)}\,\mathrm{d}\mathcal{H}^2 + \int_{V(s)} f(t,\, \cdot)\,\mathrm{d}\mathcal{H}^3$$

and differentiation and integration on the left hand side may be interchanged using *Reynolds' transport theorem*

$$\frac{\mathrm{d}}{\mathrm{d}t}\int_{V(t)} \phi(t,\, \cdot)\,\mathrm{d}\mathcal{H}^3 = \int_{V(t)} \Big\{ \partial_t\phi(t,\, \cdot) + \operatorname{div}(\phi(t,\, \cdot)u(t,\, \cdot)) \Big\}\,\mathrm{d}\mathcal{H}^3.$$

Hence, we may apply the Gaussian theorem and use the localisation procedure as described in 1.3 to recover (1.2). In summary, the non-convective flux density j may be considered as the rate of outflow of the quantity Φ across the boundary of an advected volume. This will prove useful in the derivation of the momentum balance equation in Section 1.3.

1.2 Conservation of Mass

1.6 We consider mass as an extensive quantity with volume specific density

$$\rho : J \times \Omega \longrightarrow (0, \infty).$$

As in the previous section $J = (0, a)$ denotes the considered time interval and $\Omega \subseteq \mathbb{R}^3$ denotes the considered region in space. Note that we assume the mass density ρ to be strictly positive in time and space. Moreover, we assume that the mass flux results from convection only, i. e. the mass flux density is given by ρu, where $u : J \times \Omega \longrightarrow \mathbb{R}^3$ denotes the velocity field, which describes the motion of the fluid particles, cf. 1.5. Furthermore, we assume that there do not exist any sources or sinks for mass, i. e. mass is completely conserved. Thus, the master balance equation (1.1) provides

(1.4) $$\partial_t \rho + \operatorname{div}(\rho u) = 0 \qquad \text{in } J \times \Omega,$$

the so-called *continuity equation*.

1.7 A fluid is called *incompressible*, if the mass density ρ is independent of the pressure, which will be introduced in the next section. Since we do not want to include an energy balance in our model and, therefore, do not want to introduce the temperature as an intensive quantity, we focus on *isothermal* flows. However, for an isothermal flow of an incompressible fluid the mass density is constant in time and space. In this case the continuity equation (1.4) simplifies to

(1.5) $$\operatorname{div} u = 0 \qquad \text{in } J \times \Omega.$$

1.8 We can also consider volume as an extensive quantity with volume specific density $\phi \equiv 1$. The volume flux results from convection only, i. e. the flux density is given by u. Now, if we assume the absence of any sources and sinks for volume, i. e. volume is completely conserved, equation (1.5) is nothing but the master balance equation (1.1) for the extensive quantity volume. If volume is a conserved quantity, the flow is called *isochoric*. Note that by the definitions in 1.7 the isothermal flow of an incompressible fluid is always isochoric.

1.3 Conservation of Momentum

1.9 Based on the principles of classical mechanics the momentum density is given as ρu, where $\rho : J \times \Omega \longrightarrow (0, \infty)$ denotes the density of the fluid as introduced in 1.6 and $u : J \times \Omega \longrightarrow \mathbb{R}^3$ denotes the velocity field of the flow as introduced in 1.5. As in the previous sections $J = (0, a)$ denotes the considered time interval and $\Omega \subseteq \mathbb{R}^3$ denotes the considered region in space. Now, Newton's second law states that the rate of change of momentum in an advected volume $V = V(t) \subseteq \Omega$ equals the sum of all forces that act on $V(t)$. According to Cauchy one has to distinguish between two different types of forces that act in a continuum:

- *contact forces*, which are caused by the adjacent material and act as surface forces via the boundary of the volume;

- *body forces*, which are caused by external fields, e. g. gravitation or electromagnetic fields, and act as volume forces over long distances.

The contact forces, also called *stresses*, do not only depend on time and space but also on the direction of the boundary they act on. Thus, we describe them by a family of functions

$$\Sigma(\,\cdot\,, \cdot\,, n\,) : J \times \Omega \longrightarrow \mathbb{R}^3,$$

where $n \in \mathbb{R}^3$ denotes an arbitrary unit vector. Moreover, we assume the body forces to be determined by a mass specific density

$$f : J \times \Omega \longrightarrow \mathbb{R}^3$$

and obtain the momentum balance equation

$$\frac{\mathrm{d}}{\mathrm{d}t} \int_{V(t)} \rho(t, \,\cdot\,) u(t, \,\cdot\,) \, \mathrm{d}\mathcal{H}^3 = \int_{\partial V(t)} \Sigma(t, \,\cdot\,, \nu_{\partial V(t)}) \, \mathrm{d}\mathcal{H}^2 + \int_{V(t)} \rho(t, \,\cdot\,) f(t, \,\cdot\,) \, \mathrm{d}\mathcal{H}^3.$$

Note the similarity to (1.3), which was one reason for its introduction.

1.10 To recast the momentum balance equation into a form, which allows the determination of the non-convective momentum fluxes, we need a more precise knowledge about the stresses Σ. For a fluid at rest, these stresses are in equilibrium and we have

$$\Sigma(t, x, n) + \Sigma(t, x, -n) = 0, \qquad t \in J, \ x \in \Omega, \ n \in \mathbb{R}^3, \ |n| = 1.$$

Fortunately, this relation transfers to moving fluids, too. Indeed, if we assume the stresses to be continuous functions of time, space and direction, then there exists a second rank tensor

$$S : J \times \Omega \longrightarrow \mathbb{R}^{3 \times 3}$$

such that

$$\Sigma(t, x, n) = S(t, x)n, \qquad t \in J, \ x \in \Omega, \ n \in \mathbb{R}^3, \ |n| = 1.$$

This is known as *Cauchy's theorem*, cf. [BFG⁺12] for an elegant and self-contained geometrical proof or [FV89] for a variational proof. The tensor S is called the *stress tensor* and the above relation allows for a determination of the non-convective momentum fluxes via

$$\frac{\mathrm{d}}{\mathrm{d}t} \int_{V(t)} \rho(t, \,\cdot\,) u(t, \,\cdot\,) \, \mathrm{d}\mathcal{H}^3 = \int_{\partial V(t)} S(t, x) \, \nu_{\partial V(t)} \, \mathrm{d}\mathcal{H}^2 + \int_{V(t)} \rho(t, \,\cdot\,) f(t, \,\cdot\,) \, \mathrm{d}\mathcal{H}^3.$$

Now, we may either employ the Gaussian theorem and the localisation procedure as described in 1.3 or the alternative form (1.2) of the master balance equation to obtain the *Cauchy equations*

(1.6) $\partial_t(\rho u) + \operatorname{div}(\rho u \otimes u - S) = \rho f \qquad \text{in } J \times \Omega,$

a local momentum balance.

1.4 Constitutive Equations

1.11 In order to completely describe the flow of a fluid via (1.4) (respectively (1.5)) and (1.6) it remains to impose constitutive equations for the stress tensor S. For an *inviscid fluid* the stress exerted to the surface of an advected volume is solely given by a pressure, i. e.

$$S(t, x)n = -p(t, x)n, \qquad t \in J, \ x \in \Omega, \ n \in \mathbb{R}^3, \ |n| = 1$$

with a scalar function $p : J \times \Omega \longrightarrow \mathbb{R}$, the pressure. Hence, we have $S = -pI$, where $I \in \mathbb{R}^{3 \times 3}$ denotes the identity tensor, and we also write $S = -p$ for short. In general, the pressure may depend on the temperature and on the density. However, in a simplified model without energy balance and temperature the pressure is assumed to be a function of the density only. In the case of isothermal flows of incompressible fluids as introduced in 1.7 the density is constant and the pressure has to be regarded as a free variable. Equations (1.5) and (1.6) then become

$$\partial_t(\rho u) + \operatorname{div}(\rho u \otimes u + p) = \rho f, \qquad \operatorname{div} u = 0 \qquad \text{in } J \times \Omega$$

and through the exploitation of the divergence equation further simplify to the *Euler equations*

$$\rho \partial_t u + \rho(u \cdot \nabla)u + \nabla p = \rho f, \qquad \operatorname{div} u = 0 \qquad \text{in } J \times \Omega.$$

Since the velocity field u consists of three components, these are four equations for the four unknowns u and p. Complemented by suitable initial and boundary conditions the Euler equations thus form a complete system of partial differential equations. However, the mathematical treatment of these equations seems to be even more delicate than the treatment of the Navier-Stokes equations introduced below. The major obstacle is the quadratic nonlinearity combined with the constraint on the divergence of the velocity field, which requires the pressure to be determined as a kind of Lagrangian parameter and not based on an evolution equation. However, in contrast to the Navier-Stokes equations, the linearization of the Euler equations does not lead to a parabolic problem.

1.12 In case of a *viscous fluid* the stress does not only depend on the pressure but also on internal friction, which is modelled via the so-called *viscous stress tensor*

$$E : J \times \Omega \longrightarrow \mathbb{R}^{3 \times 3}.$$

Thus, $S = E - pI$ and we also write $S = E - p$ for short. Since internal friction is a result of the relative movement of nearby fluid particles, the viscous stress should be a function of the velocity gradient. Now,

$$\nabla u = \tfrac{1}{2}(\nabla u + \nabla u^\mathsf{T}) + \tfrac{1}{2}(\nabla u - \nabla u^\mathsf{T}) =: D + R$$

with the symmetric part D, called the *rate of deformation tensor*, and the antisymmetric part R, called the *rate of rotation tensor* or *spin tensor*. The reason for these notions is simple: Since D is symmetric, it will be diagonal after a suitable orthogonal transformation of the coordinate system and, therefore, causes a deformation of advected volumes. On the other hand, R is anti-symmetric and, therefore,

$$Rh = \tfrac{1}{2}\omega \times h, \qquad h \in \mathbb{R}^3$$

for some $\omega \in \mathbb{R}^3$. Thus, R represents a rigid body rotation around the axis of rotation ω. Note that a simple computation reveals $\omega = \operatorname{rot} u$, the *vorticity* of the flow. However, a rigid body rotation does not lead to internal friction. Therefore, we assume the viscous stress tensor E to be a function of the rate of deformation tensor D only, i.e. $E = E(D)$.

1.13 A special class of viscous fluids are the *Newtonian fluids*, where the viscous stress tensor E is assumed to depend linearly on the rate of deformation tensor D. In general, this would lead to

$$E_{ij} = \sum_{k,l=1}^{3} a_{ijkl} D_{kl}, \qquad i, j = 1, 2, 3,$$

a relation, which contains 81 parameters a_{ijkl}, which would still have to be determined. However, any constitutive equation must be consistent with the basic laws of physics and especially satisfy the *principle of material frame indifference*, which states that every two observers in two different material frames must identify the same functional dependencies. Now, suitable orthogonal transformations of the coordinate system reveal that this implies $E = \lambda(\operatorname{tr} D)I + 2\mu D$ or

$$(1.7) \qquad\qquad E = \lambda \operatorname{tr} D + 2\mu D,$$

for short, to be the only possible linear dependence that may occur. The two remaining parameters λ and μ are called the *dilatational* respectively *dynamic viscosity* of the fluid. For more details on the fundamental properties of constitutive equations as well as a detailed derivation of (1.7) we refer to [NT03] and [BFG+12].

1.14 In general the viscosities λ and μ depend on the density ρ and the temperature. However, if we consider an isothermal flow of an incompressible fluid, we may assume these viscosities to be constant and equations (1.5), (1.6) and (1.7) become

$$\partial_t(\rho u) + \operatorname{div}(\rho u \otimes u - S) = \rho f, \qquad S = 2\mu D - p, \qquad \operatorname{div} u = 0 \qquad \text{in } J \times \Omega,$$

the *Navier-Stokes equations*, which describe the motion of an *incompressible Newtonian fluid* with constant density $\rho > 0$ and constant (dynamic) viscosity $\mu > 0$. Note that the dilatational viscosity λ does not appear in these equations, since $\operatorname{tr} D = \operatorname{div} u = 0$. Based on the divergence equation and the definition of the rate of deformation tensor the Navier-Stokes equations may also be written as

$$\rho\partial_t u + \rho(u \cdot \nabla)u - \mu\Delta u + \nabla p = \rho f, \qquad \operatorname{div} u = 0 \qquad \text{in } J \times \Omega,$$

which is the preferred notation in the mathematical community.

1.15 If the viscosity μ of an incompressible Newtonian fluid is large as compared to its density ρ, the inertia term $\rho(u \cdot \nabla)u$ may be approximately neglected and the flow is essentially governed by the *Stokes equations*

$$\rho \partial_t u - \mu \Delta u + \nabla p = \rho f, \qquad \text{div}\, u = 0 \quad \text{in } J \times \Omega.$$

Note that these equations represent the principal linearization of the Navier-Stokes equations. Therefore, the Stokes equations may prove very useful for the mathematical treatment of the Navier-Stokes equations, since some results concerning the non-linear problem may be obtained by perturbation arguments based on results for the linear problem. This is also the approach of our analysis as presented in the following chapters.

Remarks

1.16 The equations of motion employed in 1.5 are ordinary differential equations and, hence, may be treated by classical theorems as Peano's existence theorem and the Picard-Lindelöf theorem. In particular, there exist locally unique solutions, if the velocity field $u \in C(J \times \Omega, \mathbb{R}^3)$ is Lipschitz continuous w. r. t. the space variable. If $u(t, \cdot) \in C^m(\Omega, \mathbb{R}^3)$ for all $t \in J$ and some $m \in \mathbb{N}$, the flow maps Φ_s^t are even m-times differentiable. For more details on ordinary differential equations we refer to the monograph [Ama90].

1.17 A proof of Reynolds' transport theorem, which we employed in 1.5, may be found e. g. in [BFG+12].

1.18 An interesting presentation of the history of fluid dynamics, which also compares the two totally different ways, in which C. L. M. H. NAVIER and G. G. Stokes derived the Navier-Stokes equations, is provided by J. D. ANDERSON, cf. [And98].

References

[Ama90]   H. AMANN: Ordinary Differential Equations. de Gruyter, 1990.

[And98]   J. D. ANDERSON, JR.: *Some Reflections on the History of Fluid Dynamics*. In: The Handbook of Fluid Dynamics (R. W. JOHNSON, ed.), chap. 2, Springer, 1998.

[Ari90]   R. ARIS: Vectors, Tensors, and the Basic Equations of Fluid Mechanics. Dover Publications, 1990.

[Bat00]   G. K. BATCHELOR: An Introduction to Fluid Dynamics. Cambridge University Press, 3rd ed., 2000.

[BFG+12] D. BOTHE, R. FARWIG, M. GEISSERT, H. HECK, M. HIEBER, W. STANNAT, C. TROPEA, and S. ULBRICH: Mathematical Fluid Dynamics. Preprint, 2012.

[CM00]   A. J. CHORIN and J. E. MARSDEN: A Mathematical Introduction to Fluid Mechanics. Springer, 3rd ed., 2000.

[Fef07]   C. L. FEFFERMAN: *Existence & Smoothness of the Navier-Stokes Equation*. Official Problem Description of the Clay Mathematics Institute, 2007.

[FV89]   R. FOSDICK and E. VIRGA: *A Variational Proof of the Stress Theorem of Cauchy*. Arch. Rational Mech. Anal., 105, 95–103, 1989.

[Lam32]   H. LAMB: Hydrodynamics. Cambridge University Press, 6th ed., 1932.

[NT03]   W. NOLL and C. TRUESDELL: The Non-Linear Field Theories of Mechanics. Springer, 2003.

[RT00]   K. R. RAJAGOPAL and C. TRUESDELL: An Introduction to the Mechanics of Fluids. Birkhäuser, 2000.

[Tem05]   R. TEMAM: Mathematical Modeling in Continuum Mechanics. Cambridge University Press, 2005.

Chapter 2

Energy Preserving Boundary Conditions

As has been pointed out in the introduction the Navier-Stokes equations

$$(\partial_t \rho u) + \operatorname{div}(\rho u \otimes u - S) = \rho f, \qquad S = 2\mu D - p, \qquad \operatorname{div} u = 0 \qquad \text{in } J \times \Omega$$

form a system of partial differential equations and have to be complemented by suitable initial and boundary conditions. We employ the notation of Chapter 1, i.e. $J = (0, a)$ denotes the considered time interval and $\Omega \subseteq \mathbb{R}^3$ denotes the region in space that is occupied by the fluid. Moreover, $\rho > 0$ denotes the constant density of the fluid and $\mu > 0$ denotes its constant viscosity.

Unfortunately, in many situations the choice of reasonable boundary conditions is by no means obvious. To point out the main difficulties, which arise for the modelling of reasonable boundary conditions, it is convenient to distinguish the most relevant properties of the boundary under consideration.

Whenever a physical boundary is present, it separates the fluid from another phase, which may be a solid or another fluid. In both situations the boundary may be moving or non-moving. On the other hand, such a boundary may arise as an interface between two immiscible phases, e.g. between a fluid and a solid or between two immiscible fluids, or it may arise artificially, if the model describes only a part of a larger system. The last type of boundary, which is also called an artificial boundary, often appears in models for numerical simulations, since it is often convenient to save computation time by focusing on the interesting part of the flow while neglecting other parts of the system. Hence, such an artificial boundary "separates" a fluid from itself and may be freely passed in both directions by this fluid. As a consequence, no useful information is available about the exchange of mass, momentum and energy at such a boundary, which turns the modelling of reasonable artificial boundary conditions into a formidable task.

Now, the aim of this chapter is to give a tentative introduction to the modelling of boundary conditions for the above mentioned situations. Concerning artificial boundaries we will present a class of boundary conditions, which is derived based on suitable energy balances. The resulting *energy preserving boundary conditions* will then be the main topic of the following chapters, where we will show that these conditions are not only physically reasonable but also lead to mathematically well-posed models.

2.1 Generic Transmission Conditions

2.1 As has been pointed out above, a boundary separates the fluid from another phase, which, however, may be the fluid itself as in the case of an artificial boundary. Nevertheless, the balance equations for extensive quantities, which encode their conservation, are always valid in the continuum mechanical setting. In Chapter 1 we used these balance equations for the extensive quantities mass and momentum to derive the Navier-Stokes equations in Ω. However, analogue arguments imply the balance equations

$$(2.1) \qquad [\![\rho(u - u_\Gamma)]\!]_\Gamma \cdot \nu_\Gamma = 0, \quad [\![\rho u \otimes (u - u_\Gamma) - S]\!]_\Gamma \nu_\Gamma = f_\Gamma, \qquad t > 0, \ y \in \Gamma(t)$$

to be always valid. Here, $[\![\,\cdot\,]\!]_\Gamma$ denotes the jump of a quantity, if the boundary Γ is passed from the outside of Ω into the domain, i.e. in the direction opposite to the outer unit normal field ν_Γ. Moreover, u_Γ denotes the velocity of the boundary itself, e.g. $u_\Gamma = 0$ for a non-moving boundary or $u_\Gamma = u|_\Gamma$ for an interface separating two immiscible fluids, if the absence of phase transitions and interfacial slip is assumed. Hence, the first of the above *transmission conditions* encodes the conservation of mass, where we assume that there do not exist any sources or sinks for mass on the boundary. Analogously, the second transmission condition encodes the conservation of momentum, where we allow for the presence of sources or sinks given by their density f_Γ, which may e.g. arise due to surface tension in the case of a phase separating interface. Note that the density ρ and the viscosity μ, which enters the transmission conditions via the stress tensor S, will in general be discontinuous across the boundary and, thus, a jump of these quantities also contributes to the above balance equations.

2.2 In order to exploit the transmission conditions (2.1) to derive boundary conditions for the flow in Ω it is convenient to rewrite them via

$$[\![\rho(u - u_\Gamma)]\!]_\Gamma \cdot \nu_\Gamma = \rho[\![u - u_\Gamma]\!]_\Gamma \cdot \nu_\Gamma + [\![\rho]\!]_\Gamma(\bar{u}|_\Gamma - u_\Gamma) \cdot \nu_\Gamma,$$

where $[\![\rho]\!]_\Gamma = \bar{\rho} - \rho$, if we assume the phase on the outside of Ω to have a constant density $\bar{\rho} > 0$, and where $\bar{u}$ denotes the velocity of the outer phase. Hence, we obtain

$$(2.2) \qquad \rho[\![u - u_\Gamma]\!]_\Gamma \cdot \nu_\Gamma + (\bar{\rho} - \rho)(\bar{u} - u_\Gamma) \cdot \nu_\Gamma = 0, \qquad t > 0, \ y \in \Gamma(t).$$

Moreover, we may simplify the left hand side of the momentum transmission condition using

$$\begin{aligned}
[\![\rho u \otimes (u - u_\Gamma)]\!]_\Gamma \nu_\Gamma &= (u|_\Gamma \otimes [\![\rho(u - u_\Gamma)]\!]_\Gamma)\nu_\Gamma + ([\![u]\!]_\Gamma \otimes \bar{\rho}(\bar{u}|_\Gamma - u_\Gamma))\nu_\Gamma \\
&= ([\![\rho(u - u_\Gamma)]\!]_\Gamma \cdot \nu_\Gamma) u|_\Gamma + ([\![u]\!]_\Gamma \otimes \bar{\rho}(\bar{u}|_\Gamma - u_\Gamma))\nu_\Gamma \\
&= ([\![u]\!]_\Gamma \otimes \bar{\rho}(\bar{u}|_\Gamma - u_\Gamma))\nu_\Gamma
\end{aligned}$$

to obtain

$$(2.3) \qquad ([\![u]\!]_\Gamma \otimes \bar{\rho}(\bar{u} - u_\Gamma))\nu_\Gamma - [\![S]\!]_\Gamma \nu_\Gamma = f_\Gamma, \qquad t > 0, \ y \in \Gamma(t).$$

For more details on the modelling of boundary conditions based on transmission conditions we refer to [BFG+12] and [BKP12].

2.2 Impermeable Walls

2.3 If the boundary Γ separates the fluid from a solid, i. e. Γ represents a solid wall, we necessarily have $u_\Gamma = \bar{u}|_\Gamma$, where $\bar{u}$ denotes the velocity field describing the motion of the solid. Now, the mass transmission condition (2.2) implies

$$u \cdot \nu_\Gamma = \bar{u} \cdot \nu_\Gamma, \qquad t > 0, \ y \in \Gamma(t).$$

Note that this condition encodes the fact, that the boundary is impermeable. In particular, for an impermeable, non-moving solid wall we have

(2.4) $$u \cdot \nu_\Gamma = 0 \qquad \text{on } J \times \Gamma.$$

2.4 To obtain a complete set of boundary conditions for solid walls, we would need to exploit the momentum transmission condition (2.3), which in this case simplifies to

$$-[\![S]\!]_\Gamma \, \nu_\Gamma = f_\Gamma, \qquad t > 0, \ y \in \Gamma(t).$$

However, this would require to model the stress for the solid via constitutive equations, which would depend on the material properties of the solid, e. g. its rigidity or elasticity. Since this goes deeply beyond the scope of this thesis, we focus on rigid solids and assume the fluid to slip along the wall while being stressed in tangential directions due to friction. This leads to

(2.5) $$P_\Gamma(u - \bar{u}) + \lambda P_\Gamma S \nu_\Gamma = 0, \qquad t > 0, \ y \in \Gamma(t),$$

with a parameter $\lambda > 0$, which is also called the *slip length*. For an impermeable, non-moving, rigid wall the combination of (2.4) and (2.5) reads

(2.6) $$u \cdot \nu_\Gamma = 0, \qquad P_\Gamma u + \lambda P_\Gamma S \nu_\Gamma = 0 \qquad \text{on } J \times \Gamma,$$

which is known as the *Navier condition*, since it was first proposed by C. L. M. H. NAVIER in 1823 along with his derivation of the Navier-Stokes equations [Nav23].

2.5 The extremal case $\lambda = 0$ in (2.5) corresponds to the model assumption that the fluid is not allowed to slip along the wall and for an impermeable, non-moving, rigid wall the boundary condition becomes

(2.7) $$u = 0 \qquad \text{on } J \times \Gamma.$$

This is called the *no-slip condition* and was already proposed by G. G. STOKES in 1845 along with his derivation of the Navier-Stokes equations [Sto45].

2.6 The extremal case $\lambda \to \infty$ in (2.5) corresponds to the model assumption that the fluid may freely slip along the wall without being stressed due to friction and for an impermeable, non-moving, rigid wall the boundary condition becomes

(2.8) $$u \cdot \nu_\Gamma = 0, \qquad P_\Gamma S \nu_\Gamma = 0 \qquad \text{on } J \times \Gamma,$$

which is called the *free slip* or *perfect-slip condition*.

2.7 For more details on the family of boundary conditions (2.6), (2.7) and (2.8) we refer to [BLS07] and [LS03] and the references therein. A recent debate concerning the no-slip condition (2.7) may be found in [CD98].

2.3 Free Surfaces and Interfaces

2.8 If the boundary Γ separates two immiscible fluids, the situation is not as easy as for impermeable walls as described in 2.3. In fact, due to processes like phase transitions even the normal velocity might be discontinuous across the boundary and, moreover, the boundary velocity u_Γ may be totally different from the velocities of the fluids. However, if we exclude such phenomena and assume

$$[\![u]\!]_\Gamma \cdot \nu_\Gamma = 0, \qquad u_\Gamma \cdot \nu_\Gamma = u \cdot \nu_\Gamma, \qquad t > 0, \ y \in \Gamma(t),$$

the situation significantly simplifies and the boundary may be regarded as a *free interface* separating the two phases.

2.9 Now, if the interface carries no mass and no interfacial slip occurs, the tangential velocities may be neglected and we may assume

$$[\![u]\!]_\Gamma = 0, \qquad u_\Gamma = u, \qquad t > 0, \ y \in \Gamma(t).$$

Thus, the transmission conditions (2.2) and (2.3) simplify to

$$(2.9) \qquad [\![u]\!]_\Gamma = 0, \qquad -[\![S]\!]_\Gamma \nu_\Gamma = f_\Gamma, \qquad t > 0, \ y \in \Gamma(t),$$

where the right hand side f_Γ has to be modelled by constitutive equations to take care of effects like *surface tension* or *surface viscosity*. This way we may model e. g. two-phase flows of immiscible, incompressible Newtonian fluids or an ocean, where the evolution of the atmosphere is included in the model, while the atmosphere itself is also regarded as an incompressible Newtonian fluid.

2.10 On the other hand, we may simplify the above model of an ocean by assuming the atmosphere to be an inviscid gas. This way it is reasonable to substitute the two transmission conditions (2.9) by the boundary condition

$$(2.10) \qquad -S\nu_\Gamma = \bar{p}\nu_\Gamma + f_\Gamma, \qquad t > 0, \ y \in \Gamma(t),$$

where $\bar{p}$ denotes the pressure in the atmosphere and f_Γ has again to be modelled by constitutive equations to take care of effects like surface tension or surface viscosity. Hence, we may neglect the evolution of the atmosphere and exclude it from the model, which turns the boundary Γ into a *free surface*.

2.11 As may be expected, the mathematical treatment of problems including free surfaces and free interfaces is as difficult as their modelling. Nearly all approaches rely on a transformation of the problem to a fixed domain with a fixed boundary, which in case of the boundary condition (2.10) leads to model problems with an inhomogeneous *Neumann condition*

$$(2.11) \qquad -S\nu_\Gamma = h \qquad \text{on } J \times \Gamma.$$

2.4 Artificial Boundaries

2.12 A third kind of boundaries occur, if the model describes only a part of a larger system. In this case *artificial boundaries* like the inflow and outflow boundaries in the figure on page 3 appear. Now, the transmission conditions (2.1) do not contain any useful information. Indeed, the density, the viscosity, the velocity and the pressure are continuous across the boundary Γ, since it "separates" the fluid from itself. Therefore, it is not at all obvious how *artificial boundary conditions* for such situations should be modelled. In fact, this is an active field of research in mathematics as well as in engineering science.

2.13 For an inflow boundary one may for example assume that the fluid enters the domain with a velocity profile, which is assumed to be known. This leads to an inhomogeneous *Dirichlet Condition*

$$(2.12) \qquad u = h \qquad \text{on } J \times \Gamma,$$

where the right hand side h encodes the desired profile. Note that this condition is the inhomogeneous version of the no-slip condition (2.7).

2.14 In case of an outflow boundary as shown in the figure on page 3 it is completely unclear, how a reasonable boundary condition should be obtained. Ideally, such an outflow condition should, on one hand, lead to a well-posed problem and, on the other hand, ensure the unique solution to coincide with the solution to the model problem of an infinitely extended tube. Unfortunately, no such *transparent boundary condition* is available, if the model is restricted to the tube of finite length. This is one of the reasons, why we want to present a self-contained approach to derive a class of boundary conditions, which may especially be used at artificial boundaries. Their derivation is based on energy balances.

2.5 Energy Preserving Boundary Conditions

2.15 To derive the class of boundary conditions we have in mind we first observe that the *kinetic energy balance*

$$\frac{\mathrm{d}}{\mathrm{d}t} \int_\Omega \tfrac{1}{2}\rho|u|^2 \,\mathrm{d}\mathcal{H}^3 + 2\mu \int_\Omega |D|^2 \,\mathrm{d}\mathcal{H}^3 = \int_\Omega \rho\partial_t \tfrac{1}{2}(u \cdot u)\,\mathrm{d}\mathcal{H}^3 + \int_\Omega \nabla u : 2\mu D \,\mathrm{d}\mathcal{H}^3$$

$$= \int_\Omega u \cdot \rho\partial_t u \,\mathrm{d}\mathcal{H}^3 + \int_\Omega \nabla u : S \,\mathrm{d}\mathcal{H}^3$$

$$= \int_\Omega u \cdot (\rho\partial_t u + \operatorname{div}(\rho u \otimes u - S))\,\mathrm{d}\mathcal{H}^3$$

$$\qquad + \int_\Gamma (u \cdot \nu_\Gamma)\tfrac{1}{2}\rho|u|^2 \,\mathrm{d}\mathcal{H}^2 + \int_\Gamma u \cdot S\nu_\Gamma \,\mathrm{d}\mathcal{H}^2$$

$$= \int_\Gamma (u \cdot \nu_\Gamma)\tfrac{1}{2}\rho|u|^2 \,\mathrm{d}\mathcal{H}^2 + \int_\Gamma u \cdot S\nu_\Gamma \,\mathrm{d}\mathcal{H}^2$$

is valid for incompressible Newtonian flows, if we assume the absence of any driving forces, i. e. if we assume $f = 0$ in the Navier-Stokes equations. Note that we used

$$2\mu|D|^2 = 2\mu D : D = 2\mu(\tfrac{1}{2}(\nabla u + \nabla u^\mathsf{T}) : D) = 2\mu(\nabla u : D) = \nabla u : 2\mu D$$

based on the symmetry of D, which implies $\nabla u : D = \nabla u^\mathsf{T} : D$. Moreover, we used the incompressibility constraint, which implies $\nabla u : p = (\operatorname{div} u)\, p = 0$ as well as

$$\tfrac{1}{2}\operatorname{div}\big\{(\rho u \otimes u)u\big\} = \nabla u : (\rho u \otimes u).$$

The remaining equalities may then be obtained by partial integration as

$$\int_\Omega u \cdot \operatorname{div}(\rho u \otimes u)\, \mathrm{d}\mathcal{H}^3 = \int_\Gamma (u \cdot \nu_\Gamma)\rho|u|^2\, \mathrm{d}\mathcal{H}^2 - \int_\Omega \nabla u : (\rho u \otimes u)\, \mathrm{d}\mathcal{H}^3$$

$$= \int_\Gamma (u \cdot \nu_\Gamma)\rho|u|^2\, \mathrm{d}\mathcal{H}^2 - \tfrac{1}{2}\int_\Omega \operatorname{div}\big\{(\rho u \otimes u)u\big\}\, \mathrm{d}\mathcal{H}^3$$

$$= \int_\Gamma (u \cdot \nu_\Gamma)\tfrac{1}{2}\rho|u|^2\, \mathrm{d}\mathcal{H}^2$$

and

$$\int_\Omega \nabla u : S\, \mathrm{d}\mathcal{H}^3 = \int_\Gamma u \cdot S\nu_\Gamma\, \mathrm{d}\mathcal{H}^2 - \int_\Omega u \cdot \operatorname{div} S\, \mathrm{d}\mathcal{H}^3.$$

Of course, we assume the velocity field, the pressure and the domain to be sufficiently regular to allow for the above manipulations of the equations.

2.16 Now, according to the kinetic energy balance derived in 2.15, the rate of change of total kinetic energy plus the loss of kinetic energy due to internal friction are given as the power

$$\pi_{\mathrm{NF}} := \int_\Gamma \Big\{u \cdot S\nu_\Gamma - (u \cdot \nu_\Gamma)\tfrac{1}{2}\rho|u|^2\Big\}\, \mathrm{d}\mathcal{H}^2,$$

which changes the total amount of kinetic energy of the system via the boundary. A boundary condition that implies this contribution via the boundary to vanish may therefore be considered as an *energy preserving boundary condition* for incompressible Newtonian flows. In the case of a local boundary condition given by a linear operator $\mathcal{B}$ this contribution surely vanishes, if

(2.13a)
$$\mathcal{B}(u, p) = 0 \quad \text{on } J \times \Gamma$$
$$\Rightarrow\ u \cdot \nu_\Gamma = 0,\ u \cdot S\nu_\Gamma = 0 \quad \text{on } J \times \Gamma.$$

2.17 The energy preserving boundary conditions that are subject to the constraint (2.13a) always ensure the total kinetic energy to be monotonically decreasing as they imply the rate of its change to equal its loss due to internal friction. However, due to the incompressibility condition we have

$$\operatorname{div} D = \operatorname{div} R = \tfrac{1}{2}\Delta u, \qquad \operatorname{div} S = \operatorname{div} V = \mu\Delta u - \nabla p,$$

where $R = \frac{1}{2}(\nabla u - \nabla u^{\mathsf{T}})$ denotes the rate of rotation tensor and $V := 2\mu R - p$ denotes the antisymmetric counterpart of the stress tensor S. Hence, the *alternative form*

$$\frac{\mathrm{d}}{\mathrm{d}t}\int_{\Omega}\frac{1}{2}\rho|u|^2\,\mathrm{d}\mathcal{H}^3 + 2\mu\int_{\Omega}|R|^2\,\mathrm{d}\mathcal{H}^3 = \int_{\Gamma}(u\cdot\nu_{\Gamma})\frac{1}{2}\rho|u|^2\,\mathrm{d}\mathcal{H}^2 + \int_{\Gamma}u\cdot V\nu_{\Gamma}\,\mathrm{d}\mathcal{H}^2$$

of the kinetic energy balance is also available. Note that this alternative form may be obtained as in 2.15. Here we use

$$2\mu|R|^2 = 2\mu(\,R:R\,) = 2\mu(\tfrac{1}{2}(\nabla u - \nabla u^{\mathsf{T}}):R) = 2\mu(\nabla u:R) = \nabla u : 2\mu R$$

based on the antisymmetry of R, which implies $R:\nabla u = -R:\nabla u^{\mathsf{T}}$.

2.18 Therefore, the rate of change of total kinetic energy together with the loss of kinetic energy due to rotation are given as the power

$$\bar{\pi}_{\mathrm{NF}} := \int_{\Gamma}\left\{u\cdot V\nu_{\Gamma} - (u\cdot\nu_{\Gamma})\tfrac{1}{2}\rho|u|^2\right\}\mathrm{d}\mathcal{H}^2,$$

which changes the total amount of kinetic energy of the system via the boundary. Thus, the vanishing of this contribution also implies the total kinetic energy to be monotonically decreasing and we want to consider any boundary condition that ensures this contribution to vanish as an energy preserving boundary condition for incompressible Newtonian flows, too. In the case of a local boundary condition given by a linear operator $\mathcal{B}$ the contribution above surely vanishes, if

(2.13b)
$$\mathcal{B}(u,p) = 0 \quad \text{on } J\times\Gamma$$
$$\Rightarrow\ u\cdot\nu_{\Gamma} = 0,\ u\cdot V\nu_{\Gamma} = 0 \quad \text{on } J\times\Gamma.$$

As will be revealed below there are several boundary conditions satisfying the constraint (2.13b), which lead to physically reasonable and mathematically interesting models, and, which have been treated recently in an analytically rigorous way, see [BKP12].

2.19 To develop a first impression of how local, linear boundary conditions satisfying (2.13a) or (2.13b) may be constructed we decompose the inner product into a tangential and a normal part according to

$$u\cdot S\nu_{\Gamma} = P_{\Gamma}u\cdot P_{\Gamma}S\nu_{\Gamma} + (u\cdot\nu_{\Gamma})(S\nu_{\Gamma}\cdot\nu_{\Gamma}),$$

An analogous decomposition is valid for $u\cdot V\nu_{\Gamma}$. Due to

$$P_{\Gamma}S\nu_{\Gamma} = 2\mu P_{\Gamma}D\nu_{\Gamma} \quad \text{and} \quad P_{\Gamma}V\nu_{\Gamma} = 2\mu P_{\Gamma}R\nu_{\Gamma} = 2\mu R\nu_{\Gamma}$$

there are three directly arising energy preserving boundary conditions satisfying one of the constraints (2.13a) or (2.13b), which read

$(\mathrm{B}\,|\,a,\,\Omega)_0^{0,0}$
$$P_{\Gamma}u = 0 \quad \text{on } J\times\partial\Omega,$$
$$u\cdot\nu_{\Gamma} = 0 \quad \text{on } J\times\partial\Omega,$$

which equals the no-slip condition (2.7),

$$(B \,|\, a,\, \Omega)_0^{+1,0} \qquad \begin{aligned} 2\mu P_\Gamma D\nu_\Gamma &= 0 \quad \text{on } J \times \partial\Omega, \\ u \cdot \nu_\Gamma &= 0 \quad \text{on } J \times \partial\Omega, \end{aligned}$$

which equals the free slip condition (2.8), and, finally,

$$(B \,|\, a,\, \Omega)_0^{-1,0} \qquad \begin{aligned} 2\mu R\nu_\Gamma &= 0 \quad \text{on } J \times \partial\Omega, \\ u \cdot \nu_\Gamma &= 0 \quad \text{on } J \times \partial\Omega, \end{aligned}$$

which corresponds to the prescription of the vorticity and the normal velocity. Note that we use tags with two upper indices, since the following chapters will refer to these conditions as $(B \,|\, \cdot,\, \cdot)^{\alpha,\beta}$ with suitable parameters $\alpha,\, \beta \in \{\, -1,\, 0,\, +1 \,\}$. The lower index will be used to prescribe a suitable right hand side.

2.20 In the case of a Stokes flow, which is governed by the Stokes equations

$$\rho \partial_t u - \operatorname{div} S = \rho f, \qquad S = 2\mu D - p, \qquad \operatorname{div} u = 0 \quad \text{in } J \times \Omega,$$

analogous considerations as in 2.15 and 2.17 result in the kinetic energy balance

$$\frac{\mathrm{d}}{\mathrm{d}t} \int_\Omega \tfrac{1}{2}\rho |u|^2 \, \mathrm{d}\mathcal{H}^3 + 2\mu \int_\Omega |D|^2 \, \mathrm{d}\mathcal{H}^3 = \int_\Gamma u \cdot S\nu_\Gamma \, \mathrm{d}\mathcal{H}^2,$$

which implies the rate of change of total kinetic energy together with its loss due to internal friction to equal the power

$$\pi_{\mathrm{SF}} := \int_\Gamma u \cdot S\nu_\Gamma \, \mathrm{d}\mathcal{H}^2,$$

and the alternative form of the kinetic energy balance

$$\frac{\mathrm{d}}{\mathrm{d}t} \int_\Omega \tfrac{1}{2}\rho |u|^2 \, \mathrm{d}\mathcal{H}^3 + 2\mu \int_\Omega |R|^2 \, \mathrm{d}\mathcal{H}^3 = \int_\Gamma u \cdot V\nu_\Gamma \, \mathrm{d}\mathcal{H}^2,$$

which implies the rate of change of total kinetic energy together with its loss due to rotation to equal the power

$$\bar{\pi}_{\mathrm{SF}} := \int_\Gamma u \cdot V\nu_\Gamma \, \mathrm{d}\mathcal{H}^2.$$

2.21 In analogy to the discussion in 2.16 and 2.18 we will consider a boundary condition that ensures one of the contributions π_{SF} or $\bar{\pi}_{\mathrm{SF}}$ to vanish as an energy preserving boundary condition for incompressible Newtonian Stokes flows, since these boundary conditions imply the total kinetic energy to be monotonically decreasing. In the case of a local boundary condition given by a linear operator $\mathcal{B}$ we therefore require

$$(2.13c) \qquad \mathcal{B}(u,\, p) = 0 \quad \text{on } J \times \Gamma \qquad \Rightarrow \qquad \begin{aligned} u \cdot S\nu_\Gamma &= 0 \text{ on } J \times \Gamma \\ \text{or } u \cdot V\nu_\Gamma &= 0 \text{ on } J \times \Gamma, \end{aligned}$$

which is a weaker condition than (2.13a) respectively (2.13b).

2.22 To develop a first impression of how local, linear boundary conditions satisfying (2.13c) may be constructed we again decompose the inner product into a tangential and a normal part. Due to

$$S\nu_\Gamma \cdot \nu_\Gamma = 2\mu D\nu_\Gamma \cdot \nu_\Gamma - p = 2\mu\, \partial_\nu u \cdot \nu_\Gamma - p$$

$$\text{and} \quad V\nu_\Gamma \cdot \nu_\Gamma = 2\mu R\nu_\Gamma \cdot \nu_\Gamma - p = -p$$

there are, in addition to the energy preserving boundary conditions derived in 2.19, four directly arising energy preserving boundary conditions for Stokes flows satisfying the constraint (2.13c). These read

$$(\text{B} \,|\, a,\, \Omega)_0^{0,+1} \qquad \begin{aligned} P_\Gamma u &= 0 \quad \text{on } J \times \partial\Omega, \\ 2\mu\, \partial_\nu u \cdot \nu_\Gamma - p &= 0 \quad \text{on } J \times \partial\Omega, \end{aligned}$$

which corresponds to the prescription of tangential velocities and the normal component of normal stress,

$$(\text{B} \,|\, a,\, \Omega)_0^{+1,+1} \qquad \begin{aligned} 2\mu P_\Gamma D\nu_\Gamma &= 0 \quad \text{on } J \times \partial\Omega, \\ 2\mu\, \partial_\nu u \cdot \nu_\Gamma - p &= 0 \quad \text{on } J \times \partial\Omega, \end{aligned}$$

which equals the Neumann condition (2.11),

$$(\text{B} \,|\, a,\, \Omega)_0^{0,-1} \qquad \begin{aligned} P_\Gamma u &= 0 \quad \text{on } J \times \partial\Omega, \\ -p &= 0 \quad \text{on } J \times \partial\Omega, \end{aligned}$$

which corresponds to the prescription of tangential velocities and the external pressure, and, finally,

$$(\text{B} \,|\, a,\, \Omega)_0^{-1,-1} \qquad \begin{aligned} 2\mu R\nu_\Gamma &= 0 \quad \text{on } J \times \partial\Omega, \\ -p &= 0 \quad \text{on } J \times \partial\Omega, \end{aligned}$$

which corresponds to the prescription of the vorticity and the external pressure.

2.6 Vorticity and Pressure Boundary Conditions

2.23 The boundary conditions introduced in 2.19 and 2.22 are constructed based on the constraint that there should be no contribution to the kinetic energy of the system via the boundary. On the other hand, all those conditions have a similar structure, since they are all decomposed into a tangential part and a normal part based on the velocity u, the stress tensor S and its antisymmetric counterpart V. However, the two boundary conditions

$$(\text{B} \,|\, a,\, \Omega)_0^{+1,-1} \qquad \begin{aligned} 2\mu P_\Gamma D\nu_\Gamma &= 0 \quad \text{on } J \times \partial\Omega, \\ -p &= 0 \quad \text{on } J \times \partial\Omega, \end{aligned}$$

which corresponds to the prescription of the tangential part of the normal deformation rate and the external pressure, and

$$(B \mid a, \, \Omega)_0^{-1,+1} \qquad \begin{array}{ll} 2\mu R\nu_\Gamma = 0 & \text{on } J \times \partial\Omega, \\[2mm] 2\mu \, \partial_\nu u \cdot \nu_\Gamma - p = 0 & \text{on } J \times \partial\Omega, \end{array}$$

which corresponds to the prescription of the vorticity and the normal part of the normal stress, are not included in Section 2.5, because they do not directly arise from an energy balance.

2.24 As a consequence, their influence on the kinetic energy of the system may not be predicted. Both conditions may cause a contribution to or a consumption of kinetic energy via the boundary and may therefore not be considered as energy preserving boundary conditions, even in the case of a Stokes flow. However, we want to keep the well-posedness and maximal regularity results obtained in the next chapters as general as possible, since all of the conditions $(B \mid \cdot, \, \cdot)^{\alpha,\beta}$ with $\alpha, \, \beta \in \{-1, \, 0, \, +1\}$, may prove useful in several situations, e. g. as model problems or as artificial boundary conditions. Therefore, all of these conditions are covered by our analysis.

Remarks

2.25 The tags of the boundary conditions derived in 2.19, 2.22 and 2.23 are explained as follows: A boundary condition $(B \mid \cdot, \, \cdot)^{0,\cdot}$ always includes a tangential part of order zero, which prescribes the tangential velocity. A boundary condition $(B \mid \cdot, \, \cdot)^{\pm 1,\cdot}$ always includes a tangential part of order one, which either prescribes the tangential part of the normal deformation rate or the vorticity via the prescription of

$$\mu P_\Gamma (\nabla u \pm \nabla u^\mathsf{T}) \nu_\Gamma.$$

Analogously, a boundary condition $(B \mid \cdot, \, \cdot)^{\cdot,0}$ always includes a normal part of order zero, which prescribes the normal velocity. A boundary condition $(B \mid \cdot, \, \cdot)^{\cdot,\pm 1}$ always includes a normal part of order one, which either prescribes the normal part of the normal stress or the pressure via the prescription of

$$\mu (\nabla u \pm \nabla u^\mathsf{T}) \nu_\Gamma \cdot \nu_\Gamma - p.$$

Note that

$$\mu (\nabla u + \nabla u^\mathsf{T}) \nu_\Gamma \cdot \nu_\Gamma - p = 2\mu \, \partial_\nu u \cdot \nu_\Gamma - p$$

and

$$\mu (\nabla u - \nabla u^\mathsf{T}) \nu_\Gamma \cdot \nu_\Gamma - p = -p.$$

2.26 The no-slip condition (2.7), the perfect-slip condition (2.8), the Neumann condition (2.11) as well as their inhomogeneous versions like e. g. the Dirichlet condition (2.12) are included in the family of (energy preserving) boundary conditions $(B \mid a, \, \Omega)^{\alpha,\beta}$ via suitable choices of $\alpha, \, \beta \in \{-1, \, 0, \, +1\}$.

2.27 Another popular boundary condition for impermeable walls, which is not mentioned in Section 2.2, arises under the assumptions of 2.4, if we additionally assume the boundary Γ to be perfectly flat, i. e. to coincide with a two-dimensional plane in $\mathbb{R}^3$. In this case we have

$$P_\Gamma S \nu_\Gamma = \mu P_\Gamma \partial_\nu u$$

and the boundary condition becomes

(2.14) $$u \cdot \nu_\Gamma = 0, \qquad P_\Gamma u + \lambda P_\Gamma \partial_\nu u = 0 \qquad \text{on } J \times \Gamma,$$

which is known as the *Robin condition*. For this condition a complete L_p-theory including an $\mathcal{H}^\infty$-calculus due to J. SAAL is available, cf. [Saa06, Saa07], where the halfspace case is presented, and the thesis [Saa03], which additionally presents the corresponding results for bent halfspaces, perturbed halfspaces and bounded and exterior domains.

2.28 Again, the extremal case $\lambda = 0$ in (2.14) yields the no-slip condition (2.7), while the extremal case $\lambda \to \infty$ leads to

(2.15) $$u \cdot \nu_\Gamma = 0, \qquad \mu P_\Gamma \partial_\nu u = 0 \qquad \text{on } J \times \Gamma,$$

which may be regarded as a *do-nothing condition* for impermeable, non-moving, rigid walls. The popular do-nothing condition

(2.16) $$\mu \partial_\nu u - p \nu_\Gamma = 0 \qquad \text{on } J \times \Gamma$$

for artificial outflow boundaries may be derived as an energy preserving boundary condition, provided the incompressibility constraint is used to legitimate the simplification $\operatorname{div} S = \mu \Delta u - \nabla p$ before the procedure described in 2.15 is carried out. Instead of the Neumann condition (2.11) one then obtains (2.16). Analogously, one may obtain

(2.17) $$P_\Gamma u = 0, \qquad \mu \partial_\nu u \cdot \nu_\Gamma - p = 0 \qquad \text{on } J \times \Gamma,$$

which may also be regarded as a do-nothing condition, as an energy preserving boundary condition for Stokes flows. For more details on do-nothing conditions and their benefit for numerical simulations we refer to [HRT92, Gre95, Gri97] and the references therein.

2.29 Several of the derived (energy preserving) boundary conditions $(B \mid a, \Omega)_*^{\alpha,\beta}$ with $\alpha, \beta \in \{-1, 0, +1\}$ are well-known to be applicable for artificial outflow boundaries in numerical simulations. The vorticity conditions ($\alpha = -1$) and the pressure conditions ($\beta = -1$) have first been considered by C. CONCA, F. MURAT, C. PARES, O. PIRONNEAU and M. THIRIET, who constructed weak solutions to the steady Navier-Stokes equations and studied the applicability of these boundary conditions for numerical simulations, cf. [CMP94, CPPT95]. However, an analysis of the resulting initial boundary value problems is only recently available, see [BKP12].

References

[BFG+12] D. BOTHE, R. FARWIG, M. GEISSERT, H. HECK, M. HIEBER, W. STANNAT, C. TROPEA, and S. ULBRICH: Mathematical Fluid Dynamics. Preprint, 2012.

[BKP12]   D. BOTHE, M. KÖHNE, and J. PRÜSS: *On a Class of Energy Preserving Boundary Conditions for Incompressible Newtonian Flows.* Preprint at arXiv.org, URL http://arxiv.org/abs/1207.0707, 2012.

[BLS07]   M. P. BRENNER, E. LAUGA, and H. A. STONE: *Microfluidics: The No-Slip Boundary Condition.* In: Handbook of Experimental Fluid Dynamics (J. FOSS, C. TROPEA, and A. YARIN, eds.), (1219–1240), Springer, 2007.

[CD98]   S. CHEN and G. D. DOOLEN: *Lattice Boltzmann Method for Fluid Flows.* Annual Review of Fluid Mechanics, 30, 329–364, 1998.

[CMP94]   C. CONCA, F. MURAT, and O. PIRONNEAU: *The Stokes and Navier-Stokes Equations with Boundary Conditions Involing the Pressure.* Japan J. Math., 20 (2), 279–318, 1994.

[CPPT95]   C. CONCA, C. PARÈS, O. PIRONNEAU, and M. THIRIET: *Navier-Stokes Equations with Imposed Pressure and Velocity Fluxes.* Int. J. Numer. Meth. Fluids, 20, 267–287, 1995.

[Gre95]   P. M. GRESHO: *Incompressible Fluid Mechanics: Some Fundamental Formulation Issues.* Annual Review of Fluid Mechanics, 23, 413–453, 1995.

[Gri97]   D. F. GRIFFITHS: *The 'No Boundary Condition' Outflow Boundary Condition.* Int. J. Numer. Meth. Fluids, 24 (4), 393–411, 1997.

[HRT92]   J. G. HEYWOOD, R. RANACHER, and S. TUREK: *Artificial Boundaries and Flux and Pressure Conditions for Incompressible Navier-Stokes Equations.* Int. J. Numer. Meth. Fluids, 22, 325–352, 1992.

[LS03]   E. LAUGA and H. A. STONE: *Effective Slip in Pressure Driven Stokes Flow.* J. Fluid Mech., 489, 55–77, 2003.

[Nav23]   C. L. M. H. NAVIER: *Mémoir sur les Lois du Mouvement des Fluides.* Mem. Acad. Inst. Sci. Fr., 6, 389–440, 1823.

[Saa03]   J. SAAL: Robin Boundary Conditions and Bounded $\mathcal{H}^\infty$-Calculus for the Stokes Operator. Logos, 2003.

[Saa06]   J. SAAL: *Stokes and Navier-stokes Equations with Robin Boundary Conditions in a Half-Space.* J. Math. Fluid Mech., 8, 211–241, 2006.

[Saa07]   J. SAAL: *The Stokes Operator with Robin Boundary Conditions in Solenoidal Subspaces of* $L^1(\mathbb{R}_+^n)$ *and* $L^\infty(\mathbb{R}_+^n)$. Commun. Partial Differ. Equations, 32 (3), 343–373, 2007.

[Sto45]   G. G. STOKES: *On the Theories of the Internal Friction of Fluids in Motion, and of the Equilibrium and Motion of Elastic Solids.* Trans. Cambridge Phil. Soc., 8, 287–319, 1845.

Bounded Smooth Domains

Chapter 3

L_p-Theory for Incompressible Newtonian Flows

This chapter is devoted to the development of an L_p-Theory for incompressible Newtonian flows in bounded, smooth domains subject to one of the energy preserving respectively artificial boundary conditions introduced in Chapter 2. To be precise, we assume $a > 0$ and $\Omega \subseteq \mathbb{R}^n$ to be a bounded domain with sufficiently smooth boundary $\Gamma := \partial\Omega$. Given a constant density $\rho > 0$ and a constant viscosity $\mu > 0$, we will study the Navier-Stokes equations

$$\rho\partial_t u + \rho(u \cdot \nabla)u - \mu\Delta u + \nabla p = \rho f \quad \text{in } (0, a) \times \Omega,$$

$$\text{div } u = g \quad \text{in } (0, a) \times \Omega,$$

$$(\text{N} \,|\, a, \, \Omega)^{\alpha,\beta}_{f,g,h,u_0} \qquad \mathcal{B}^{\alpha,\beta}(u, \, p) = h \quad \text{on } (0, a) \times \partial\Omega,$$

$$u(0) = u_0 \quad \text{in } \Omega$$

as well as the Stokes equations

$$\rho\partial_t u - \mu\Delta u + \nabla p = \rho f \quad \text{in } (0, a) \times \Omega,$$

$$\text{div } u = g \quad \text{in } (0, a) \times \Omega,$$

$$(\text{S} \,|\, a, \, \Omega)^{\alpha,\beta}_{f,g,h,u_0} \qquad \mathcal{B}^{\alpha,\beta}(u, \, p) = h \quad \text{on } (0, a) \times \partial\Omega,$$

$$u(0) = u_0 \quad \text{in } \Omega$$

in a functional analytic framework based on L_p-spaces.

The boundary operators $\mathcal{B}^{\alpha,\beta}$ with parameters $\alpha, \, \beta \in \{ -1, \, 0, \, +1 \}$ realise in each case one of the energy preserving respectively artificial boundary conditions (B $\,|\, a, \, \Omega)^{\alpha,\beta}_h$ introduced in 2.19, 2.22 and 2.23. We decompose the boundary operators as

$$\mathcal{B}^{\alpha,\beta}(u, \, p) = P_\Gamma\mathcal{B}^{\alpha,\beta}(u, \, p) + Q_\Gamma\mathcal{B}^{\alpha,\beta}(u, \, p),$$

where $P_\Gamma = P_\Gamma(y)$ denotes the projection onto the tangent space $T_y\Gamma$ of Γ at a point $y \in \Gamma$, i.e. $P_\Gamma = 1 - \nu_\Gamma \otimes \nu_\Gamma$ with the outer unit normal field $\nu_\Gamma : \Gamma \longrightarrow \mathbb{R}^n$ of Ω. Analogously, Q_Γ denotes the projection onto the normal bundle $N\Gamma$ of Γ, i.e. $Q_\Gamma = 1 - P_\Gamma$.

Now, the boundary operators $\mathcal{B}^{\alpha,\beta}$ are defined via

$$P_\Gamma\mathcal{B}^{0,\beta}(u, \, p) := P_\Gamma[u]_\Gamma, \qquad P_\Gamma\mathcal{B}^{\pm 1,\beta}(u, \, p) := \mu P_\Gamma[\nabla u \pm \nabla u^\mathsf{T}]_\Gamma\, \nu_\Gamma$$

for $\beta \in \{-1, 0, +1\}$ and

$$Q_\Gamma \mathcal{B}^{\alpha,0}(u, p) \cdot \nu_\Gamma := [u]_\Gamma \cdot \nu_\Gamma,$$

$$Q_\Gamma \mathcal{B}^{\alpha,+1}(u, p) \cdot \nu_\Gamma := 2\mu\, \partial_\nu u \cdot \nu_\Gamma - [p]_\Gamma, \qquad Q_\Gamma \mathcal{B}^{\alpha,-1}(u, p) \cdot \nu_\Gamma := -[p]_\Gamma$$

for $\alpha \in \{-1, 0, +1\}$. Here and in the following chapters $[\cdot]_\Gamma$ denotes the trace of a function defined in Ω on the boundary Γ. We prefer this notation, which is motivated by the jump brackets $[\![\cdot]\!]_\Gamma$ as used in Section 2.1, since the L_p-setting requires traces to be obtained by a suitable trace operator. Accordingly, the normal derivative has to be understood as $\partial_\nu = [\nabla \cdot^\mathsf{T}]_\Gamma \nu_\Gamma$.

Note that the boundary operators $\mathcal{B}^{0,0}$ and $\mathcal{B}^{\pm 1,0}$ realise the three boundary conditions introduced in 2.19, which in their homogeneous version behave energy preserving for incompressible Newtonian flows as well as for Stokes flows. Furthermore, the boundary operators $\mathcal{B}^{0,\pm 1}$, $\mathcal{B}^{+1,+1}$ and $\mathcal{B}^{-1,-1}$ realise the four boundary conditions introduced in 2.22, which in their homogeneous version behave energy preserving for Stokes flows. Finally, the boundary operators $\mathcal{B}^{+1,-1}$ and $\mathcal{B}^{-1,+1}$ realise the two artificial boundary conditions introduced in 2.23.

Also note that the tangential part of the boundary operator $\mathcal{B}^{\alpha,\beta}$ neither depends on β nor on the pressure. Therefore, we will in slight abuse of notation make frequent use of the abbreviation

$$P_\Gamma \mathcal{B}^\alpha u = P_\Gamma \mathcal{B}^{\alpha,\beta}(u, p), \qquad \alpha, \beta \in \{-1, 0, +1\}.$$

Analogously, the normal part of the boundary operator $\mathcal{B}^{\alpha,\beta}$ does not depend on α, which motivates the abbreviation

$$Q_\Gamma \mathcal{B}^\beta(u, p) = Q_\Gamma \mathcal{B}^{\alpha,\beta}(u, p), \qquad \alpha, \beta \in \{-1, 0, +1\}.$$

This abbreviations allow the boundary condition to be split into two equations on the boundary, each of which is defined by one of the parameters, which will sometimes be convenient.

Our analysis of the Stokes equations $(\mathrm{S}\,|\,a,\,\Omega)$ is based on maximal L_p-regularity as presented in the next sections. In particular, we obtain strong solutions in an L_p-setting to the Stokes equations subject to one of the boundary conditions introduced in 2.19, 2.22 and 2.23. Moreover, the approach via maximal L_p-regularity allows the non-linear problem to be treated as a perturbation of the Stokes equations, which yields unique strong solutions on finite time intervals, provided the data satisfy a suitable smallness condition, as well as local-in-time unique strong solutions to the Navier-Stokes equations $(\mathrm{N}\,|\,a,\,\Omega)$, where no smallness condition on the data has to be imposed. Maximal L_p-regularity of the linear problem is obtained via a localisation procedure, which starts with an analysis of the Stokes equations in a halfspace as presented in Chapter 5. The halfspace results are then transferred to the more general geometry of a bent halfspace as presented in Chapter 6 by means of a perturbation argument. Finally, a suitable decomposition of the domain Ω allows to reduce the problem for a bounded smooth domain as considered here to finitely many problems in bent halfspaces. This final step is presented in Chapter 7.

Of course, some of the considered boundary conditions have already been treated in the literature. Concerning strong solutions to the Navier-Stokes equations subject to a homogeneous Dirichlet condition H. FUJITA and T. KATO as well as P. E. SOBOLEVSKII have established unique local strong solutions in an L_2-setting based on a semigroup approach already in 60's, cf. [FK62, FK64, Sob64]. Later on, Y. GIGA and T. MIYAKAWA as well as F. B. WEISSLER generalised these results to the L_p-setting, cf. [Gig85, GM85, Wei80]. The first approaches based on resolvent estimates in an L_p-setting are due to V. A. SOLONNIKOV, M. MCCRACKEN respectively S. UKAI, cf. [Sol77a, McC81, Uka87]. Finally, maximal L_p-regularity was established by V. A. SOLONNIKOV, W. BORCHERS and T. MIYAKAWA respectively W. DESCH, M. HIEBER and J. PRÜSS, cf. [Sol77a, BM88, DHP01]. A semigroup approach to the Stokes and Navier-Stokes equations, which in particular yields maximal L_p-L_q-regularity, was developed by T. KUBO and Y. SHI-BATA, cf. [KS04, KS05a, KS05b]. This approach was later generalised to Navier respectively Robin boundary conditions by Y. SHIBATA and R. SHIMADA, cf. [SS07a]. The extremal cases, i. e. the no-slip and the perfect slip condition, are also covered and the results apply to a halfspace, bent and perturbed halfspaces as well as to bounded and exterior domains. For a detailed presentation of the Fujita-Kato approach in the L_p-setting we refer to the monograph by H. SOHR, cf. [Soh01].

The Stokes and Navier-Stokes equations subject to Neumann boundary conditions have first been considered by V. A. SOLONNIKOV in a series of publications, cf. [Sol77b, Sol78, Sol88, Sol91]. Later on, Y. SHIBATA and S. SHIMIZU established maximal L_p-regularity, cf. [SS03, SS05, SS07b, SS08, SS11]. In all cases the motivation are either problems with free boundary or two-phase problems with an evolving phase-separating interface. As has already been mentioned in the introduction and in Section 2.3 the Stokes and Navier-Stokes equations subject to a Neumann boundary condition arise as a model problem in these situations.

A totally different approach to the Stokes and Navier-Stokes equations in an L_p-setting was developed in the 90's by G. GRUBB and V. A. SOLONNIKOV, cf. [GS90a, GS90b, GS91a, GS91b, Gru95a, Gru95b, Gru98]. In this series of publications the Stokes and Navier-Stokes equations are transformed into a system of pseudodifferential evolution equations and treated by an abstract pseudodifferential calculus. This way the authors are able to treat the Dirichlet, perfect slip and Neumann boundary conditions as well as the do-nothing type conditions (2.15) and (2.16) and establish maximal L_p-regularity in each case. As a remarkable fact this method is applicable for mixed order boundary conditions, which have not been treated in the literature before.

Another well-known approach to the Navier-Stokes equations dates back to the fundamental works of J. LERAY and E. HOPF, cf. [Ler34a, Ler34b, Hop51], who introduced the concept of weak solutions. This way one may construct global solutions without any smallness assumption on the initial data. However, the question of the uniqueness of these solutions still remains open in space dimension $n \geq 3$. Therefore, weak solutions to the Navier-Stokes equations are still an active field of research in mathematics and the concept is further developed nowadays. Nevertheless, the literature focuses on (homogeneous) Dirichlet conditions. Notable exceptions are the works of C. CONCA, F. MURAT, C. PARES, O. PIRONNEAU and M. THIRIET, cf. [CMP94, CPPT95], who consider boundary conditions prescribing the tangential vorticity or the pressure, the works of H. BELLOUT, J. NEUSTUPA and P. PENEL, cf. [BNP04, NP07, NP08, NP10], who consider boundary conditions of Navier type as generalised impermeability boundary conditions, and the article by F. BOYER and P. FABRIE, cf. [BF07], who consider Neumann type boundary conditions. For a more detailed overview of the theory of weak solutions and its development we refer to the monographs by O. A. LADYZHENSKAYA, R. TEMAM and G. P. GALDI, cf. [Lad69, Tem77, Gal94a, Gal94b].

3.1 Necessary Regularity Conditions

3.1 To establish a suitable functional analytic framework we consider the momentum equation of $(S \mid a, \Omega)^{\alpha, \beta}_{f,g,h,u_0}$ in the base space $L_p((0, a) \times \Omega, \mathbb{R}^n)$. Hence, we require

$$f \in \mathbb{Y}_f(a, \Omega) := L_p((0, a) \times \Omega, \mathbb{R}^n).$$

As our analysis will reveal, the Stokes equations form a degenerate parabolic system. Thus, a maximal regular solution should satisfy

$$u \in \mathbb{X}_u(a, \Omega) := H^1_p((0, a), L_p(\Omega, \mathbb{R}^n)) \cap L_p((0, a), H^2_p(\Omega, \mathbb{R}^n))$$

and

$$\nabla p \in L_p((0, a) \times \Omega, \mathbb{R}^n).$$

3.2 Here and throughout the following chapters $L_p((0, a), X)$, $L_p(\Omega, X)$ respectively $L_p((0, a) \times \Omega, X)$ denote the usual Lebesgue spaces with $1 \le p \le \infty$ on the time interval $(0, a)$, the domain Ω respectively the time-space cylinder $(0, a) \times \Omega$, the space X being an arbitrary Banach space. Moreover, we denote by $H^m_p((0, a), X)$, $H^m_p(\Omega, X)$ respectively $H^m_p((0, a) \times \Omega, X)$ the Bessel potential spaces of order $m \in \mathbb{N}$, where $1 < p < \infty$. Furthermore, we set

$$_0H^1_p((0, a), X) := \Big\{ \phi \in H^1_p((0, a), X) : \phi(0) = 0 \Big\}.$$

For a comprehensive introduction of the theory of Lebesgue and Bessel potential spaces we refer to the monographs by R. A. ADAMS and J. J. F. FOURNIER, H. AMANN and H. TRIEBEL, cf. [Tri98, Ama95, AF03, Ama09].

3.3 We restrict our considerations to bounded domains with boundary of class C^{3-}, i. e. we assume the boundary to be locally parametrised over $\mathbb{R}^{n-1}$ by C^{m-}-functions, which are the C^{m-1}-functions with Lipschitz continuous partial derivatives of order $m-1$, for $m = 3$. This implies the Bessel potential spaces $H^m_p(\Omega, X)$ to coincide with the Sobolev spaces

$$W^m_p(\Omega, X) = \Big\{ \phi \in L_p(\Omega, X) : \partial^\alpha \phi \in L_p(\Omega, X), \ |\alpha| \le m \Big\}$$

for $m = 1, 2, \ldots$, provided the underlying Banach space X is a UMD-space, cf. Remark 3.58. Note that the partial derivatives above have to be understood in the sense of distributions. The Sobolev spaces are canonically normed via

$$\|\phi\|_{W^m_p(\Omega, X)} = \left\{ \sum_{|\alpha| \le m} \|\partial^\alpha \phi\|^p_{L_p(\Omega, X)} \right\}^{1/p}, \qquad \phi \in W^m_p(\Omega, X).$$

3.4 The Sobolev spaces $W_p^m((0, a), X)$ are defined analogously for $1 < p < \infty$ and $m = 1, 2, \ldots$. Again, these spaces coincide with the corresponding Bessel potential spaces, provided the underlying Banach space X is a UMD-space, cf. Remark 3.58. For a comprehensive introduction of the theory of Sobolev spaces we again refer to the monographs by R. A. ADAMS and J. J. F. FOURNIER, H. AMANN and H. TRIEBEL, cf. [Tri98, Ama95, AF03, Ama09].

3.5 Since the momentum equation may only deliver regularity assertions for the pressure gradient as stated in 3.1, we require

$$p \in L_p((0, a), \dot{H}_p^1(\Omega)),$$

where

$$\dot{H}_p^1(\Omega) = \left\{ \phi \in L_{p,loc}(\Omega) : \nabla \phi \in L_p(\Omega, \mathbb{R}^n) \right\}$$

denotes the homogeneous Sobolev space of order one, which becomes a semi-normed space via

$$|\phi|_{\dot{H}_p^1(\Omega)} = \|\nabla \phi\|_{L_p(\Omega, \mathbb{R}^n)}, \qquad \phi \in \dot{H}_p^1(\Omega).$$

Note that the regularity assumptions on the boundary Γ of Ω imply the embedding

$$\dot{H}_p^1(\Omega) \hookrightarrow \left\{ \phi \in L_{p,loc}(\Omega) : \begin{array}{l} \phi \in H_p^1(\Omega'), \ \Omega' \subseteq \Omega \text{ open and bounded} \\ \text{with boundary of class } C^1 \end{array} \right\}$$

to be valid for all $1 < p < \infty$, cf. [Neč67, Chapitre 2, Théorème 7.6]. Thus, the functions in $\dot{H}_p^1(\Omega)$ exhibit a local H_p^1-regularity up to the boundary. Moreover, the domain Ω is assumed to be bounded and, therefore, the above embedding already implies the identity $\dot{H}_p^1(\Omega) = H_p^1(\Omega)$ to be valid algebraically.

3.6 Now, if we have to deal with a boundary condition prescribing the normal velocity, i. e. $\beta = 0$, the pressure is only defined up to a constant and we may fix a particular pressure by requiring

$$p \in \mathbb{X}_p^0(a, \Omega) := \left\{ q \in L_p((0, a), H_p^1(\Omega)) : (q)_\Omega = 0 \right\},$$

where

$$(\phi)_\Omega := \frac{1}{\mathcal{H}^n(\Omega)} \int_\Omega \phi \, d\mathcal{H}^n, \qquad \phi \in L_p(\Omega)$$

denotes the mean-value functional. Note that $\mathbb{X}_p^0(a, \Omega)$ with norm

$$\|q\|_{\mathbb{X}_p^0(a, \Omega)} := \|\nabla q\|_{L_p((0,a) \times \Omega, \mathbb{R}^n)}, \qquad q \in \mathbb{X}_p^0(a, \Omega)$$

forms a Banach space, since the Poincaré inequality is available for functions of zero mean, cf. [Eva98, Section 5.8, Theorem 1].

3.7 On the other hand, if the considered boundary condition involves the pressure, i. e. $\beta = \pm 1$, the pressure is uniquely defined in $\dot{H}_p^1(\Omega)$ and we require

$$p \in \mathbb{X}_p^{\pm 1}(a, \Omega) := L_p((0, a), \dot{H}_p^1(\Omega)).$$

Note that the spaces $\mathbb{X}_p^{\pm 1}(a,\,\Omega)$ are semi-normed spaces via

$$|q|_{\mathbb{X}_p^{\pm 1}(a,\,\Omega)} := \|\nabla q\|_{L_p((0,\,a)\times\Omega,\,\mathbb{R}^n)}, \qquad q \in \mathbb{X}_p^{\pm 1}(a,\,\Omega).$$

3.8 To derive the necessary regularity conditions for the right hand side g of the divergence equation it is necessary to preserve some of the time regularity of the velocity field that remains for its first spatial derivatives. This goal may be achieved by using the *Mixed Derivative Theorem*, which goes back to the work of P. E. SOBOLEVSKII, cf. Remark 3.60.

3.9 PROPOSITION. *Let $a > 0$ and let $\Omega \subseteq \mathbb{R}^n$ be a bounded domain with boundary $\Gamma = \partial\Omega$ of class C^{3-}. Let $1 < p < \infty$, $\tau \in (0,\,1]$ and $\sigma \in (0,\,2]$. Then the embeddings*

$$H_p^\tau((0,\,a),\,L_p(\Omega)) \cap L_p((0,\,a),\,H_p^\sigma(\Omega))$$

$$\hookrightarrow H_p^{(1-\theta)\tau}((0,\,a),\,H_p^{\theta\sigma}(\Omega)), \quad \theta \in [0,\,1]$$

are valid. Moreover, the embeddings

$$_0H_p^\tau((0,\,\bar{a}),\,L_p(\Omega)) \cap L_p((0,\,\bar{a}),\,H_p^\sigma(\Omega))$$

$$\hookrightarrow {}_0H_p^{(1-\theta)\tau}((0,\,\bar{a}),\,H_p^{\theta\sigma}(\Omega)), \quad \theta \in [0,\,1],\ \bar{a} \in (0,\,a]$$

are valid, where the embedding constants are independent of $\bar{a} \in (0,\,a]$. ☐

3.10 The function spaces in Proposition 3.9 denote the fractional order Bessel potential spaces, which are defined for $\sigma \in (0,\,\infty) \setminus \mathbb{N}$ by *complex interpolation* as

$$H_p^\sigma(\Omega) := [H_p^{[\sigma]}(\Omega),\,H_p^{[\sigma]+1}(\Omega)]_{\sigma-[\sigma]}.$$

Here $[\sigma]$ denotes the largest integer smaller than σ. For $\tau \in (0,\,1)$ the Bessel potential spaces $H_p^\tau((0,\,a),\,X)$ are defined analogously, the space X being an arbitrary Banach space. Moreover,

$$_0H_p^\tau((0,\,a),\,X) := [L_p((0,\,a),\,X),\,{}_0H_p^1((0,\,a),\,X)]_\tau, \qquad \tau \in (0,\,1).$$

For an introduction of interpolation theory and its applications to the theory of function spaces we refer to the monograph by H. TRIEBEL, cf. [Tri98].

3.11 The function spaces on the right hand side of the embeddings in Proposition 3.9 can be substituted by the fractional order Sobolev-Slobodeckij spaces, which are defined for $\sigma \in (0,\,\infty) \setminus \mathbb{N}$ as

$$W_p^\sigma(\Omega,\,X) := \left\{ \phi \in L_p(\Omega,\,X) : \|\phi\|_{W_p^\sigma(\Omega,\,X)} < \infty \right\},$$

where the norm is given as

$$\|\phi\|_{W_p^\sigma(\Omega,\,X)} = \left\{ \|\phi\|_{W_p^{[\sigma]}(\Omega,\,X)}^p + \sum_{|\alpha|=[\sigma]} \int_\Omega \int_\Omega \frac{\|\partial^\alpha\phi(x) - \partial^\alpha\phi(y)\|_X^p}{|x-y|^{n+(\sigma-[\sigma])p}}\,\mathrm{d}x\,\mathrm{d}y \right\}^{1/p}$$

for $\phi \in W_p^\sigma(\Omega,\,X)$, the space X being an arbitrary Banach space, cf. Remark 3.60. Note that these spaces may be obtained by *real interpolation* as

$$W_p^\sigma(\Omega,\,X) \cong \left(W_p^{[\sigma]}(\Omega,\,X),\,W_p^{[\sigma]+1}(\Omega,\,X)\right)_{\sigma-[\sigma],p}, \qquad \sigma \in (0,\,\infty) \setminus \mathbb{N}.$$

The Sobolev-Slobodeckij spaces $W_p^\tau((0, a), X)$ are defined analogously for $\tau \in (0, \infty) \backslash \mathbb{N}$. Moreover, we set

$$_0W_p^\tau((0, a), X) := \left\{ \phi \in W_p^\tau((0, a), X) : \phi(0) = 0 \right\}, \qquad \tau \in (1/p, 1).$$

See also Remark 3.61. For an introduction of the theory of Sobolev-Slobodeckij spaces we refer to the monographs by R. A. ADAMS and J. J. F. FOURNIER, H. AMANN and H. TRIEBEL, cf. [Tri98, Ama95, AF03, Ama09].

3.12 Applying Proposition 3.9 to the space $\mathbb{X}_u(a, \Omega)$ and using the regularity of the domain Ω, we obtain

$$\mathbb{X}_u(a, \Omega) \hookrightarrow H_p^{1/2}((0, a), H_p^1(\Omega, \mathbb{R}^n)),$$

which implies

$$g \in \mathbb{Y}_g(a, \Omega) := H_p^{1/2}((0, a), L_p(\Omega)) \cap L_p((0, a), H_p^1(\Omega)).$$

3.13 The necessary regularity of the right hand side

$$h \in \mathbb{Y}_h(a, \Gamma) := L_p((0, a), L_{p,loc}(\Gamma, \mathbb{R}^n))$$

of the boundary condition depends on the parameters $\alpha, \beta \in \{-1, 0, +1\}$. Again, we want to preserve as much time regularity as possible, if derivatives of the velocity field are involved on the left hand side. Moreover, a particular loss of spatial regularity occurs due to the application of the trace operator. Therefore, we are in need of its precise mapping properties in the L_p-setting. The results provided by the next proposition go back to the earliest works on Sobolev and Sobolev-Slobodeckij spaces and variants may be found in many classical sources, e. g. in the monographs by J. NEČAS and H. TRIEBEL, cf. [Neč67, Tri98]. A complete discussion of this topic as well as a detailed introduction to Sobolev-Slobodeckij spaces on the boundary of a domain are provided by J. MARSCHALL, cf. [Mar87]. See also Remarks 3.62 and 3.63.

3.14 PROPOSITION. *Let* $\Omega \subseteq \mathbb{R}^n$ *be a bounded domain with boundary* $\Gamma = \partial\Omega$ *of class* C^{3-}. *Let* $1 < p < \infty$. *Then the following assertions are valid.*

(i) Let $\sigma \in (1/p, 1 + 1/p) \cup (1 + 1/p, 2]$. *Then the trace operator*

$$[\,\cdot\,]_\Gamma : W_p^\sigma(\Omega) \longrightarrow W_p^{\sigma-1/p}(\Gamma)$$

is bounded and surjective and there exists a bounded linear operator

$$\mathcal{E}_\Omega : W_p^{\sigma-1/p}(\Gamma) \longrightarrow W_p^\sigma(\Omega),$$

which acts as a right inverse for the trace operator.

(ii) The trace operator

$$[\,\cdot\,]_\Gamma : \dot{H}_p^1(\Omega) \longrightarrow \dot{W}_p^{1-1/p}(\Gamma)$$

is bounded and surjective and there exists a bounded linear operator

$$\dot{\mathcal{E}}_\Omega : \dot{W}_p^{1-1/p}(\Gamma) \longrightarrow \dot{H}_p^1(\Omega),$$

which acts as a right inverse for the trace operator. $\qquad\square$

3.15 The homogeneous Sobolev-Slobodeckij space $\dot{W}_p^{1-1/p}(\Gamma)$ and the Sobolev-Slobodeckij spaces $W_p^\sigma(\Gamma)$ are defined by mapping the corresponding spaces on $\mathbb{R}^{n-1}$ to the boundary Γ via local coordinates. For the details, we refer to [Mar87]. Note, that the seminorm for the homogeneous space is given by

$$
|\phi|_{\dot{W}_p^{1-1/p}(\Gamma)} = \left\{ \int_\Gamma \int_\Gamma \frac{|\phi(x) - \phi(y)|^p}{|x-y|^{p-1}} \, d\sigma(x) \, d\sigma(y) \right\}^{1/p}, \qquad \phi \in \dot{W}_p^{1-1/p}(\Gamma).
$$

3.16 Unfortunately the mapping properties of the trace operator as stated in Proposition 3.14 are not appropriate to preserve the time regularity of the velocity field and its first derivatives. However, there are suitable trace theorems available for parabolic problems, which may be found e.g. in the monograph by R. DENK, M. HIEBER and J. PRÜSS, cf. [DHP03]. The two results that are needed for the analysis of the Stokes equations are provided by the following proposition, cf. Remark 3.64.

3.17 PROPOSITION. Let $\Omega \subseteq \mathbb{R}^n$ be a bounded domain with boundary $\Gamma = \partial\Omega$ of class C^{3-}. Let $1 < p < \infty$. Then the following assertions are valid.

(i) The trace operator

$$
[\,\cdot\,]_\Gamma : H_p^1((0,a), L_p(\Omega)) \cap L_p((0,a), H_p^2(\Omega))
$$
$$
\longrightarrow W_p^{1-1/2p}((0,a), L_p(\Gamma)) \cap L_p((0,a), W_p^{2-1/p}(\Gamma))
$$

is bounded.

(ii) The trace operator

$$
[\,\cdot\,]_\Gamma : H_p^{1/2}((0,a), L_p(\Omega)) \cap L_p((0,a), H_p^1(\Omega))
$$
$$
\longrightarrow W_p^{1/2-1/2p}((0,a), L_p(\Gamma)) \cap L_p((0,a), W_p^{1-1/p}(\Gamma))
$$

is bounded.   □

3.18 Combining Propositions 3.9 and 3.17 we obtain

$$
(3.1) \quad
\begin{aligned}
&[v]_\Gamma \in \; W_p^{1-1/2p}((0,a), L_p(\Gamma, \mathbb{R}^n)) \cap L_p((0,a), W_p^{2-1/p}(\Gamma, \mathbb{R}^n)), \\
&\partial_\nu v, \; [\partial_k v]_\Gamma \in W_p^{1/2-1/2p}((0,a), L_p(\Gamma, \mathbb{R}^n)) \cap L_p((0,a), W_p^{1-1/p}(\Gamma, \mathbb{R}^n))
\end{aligned}
$$

for all $v \in \mathbb{X}_u(a, \Omega)$ as well as

$$
[q]_\Gamma \in L_p((0,a), \dot{W}_p^{1-1/p}(\Gamma)), \qquad q \in \mathbb{X}_p^\beta(a, \Omega).
$$

Thus, if $u \in \mathbb{X}_u(a, \Omega)$ and $p \in \mathbb{X}_p^\beta(a, \Omega)$, we have

$$
P_\Gamma \mathcal{B}^0 u \; \in \; \mathbb{T}_h^0(a, \Gamma) \; \text{with}
$$
$$
\mathbb{T}_h^0(a, \Gamma) := W_p^{1-1/2p}((0,a), L_p(\Gamma, T\Gamma)) \cap L_p((0,a), W_p^{2-1/p}(\Gamma, T\Gamma)),
$$

$$
P_\Gamma \mathcal{B}^{\pm 1} u \; \in \; \mathbb{T}_h^{\pm 1}(a, \Gamma) \; \text{with}
$$
$$
\mathbb{T}_h^{\pm 1}(a, \Gamma) := W_p^{1/2-1/2p}((0,a), L_p(\Gamma, T\Gamma)) \cap L_p((0,a), W_p^{1-1/p}(\Gamma, T\Gamma)),
$$

$$
Q_\Gamma \mathcal{B}^0(u, p) \; \in \; \mathbb{N}_h^0(a, \Gamma) \; \text{with}
$$
$$
\mathbb{N}_h^0(a, \Gamma) := W_p^{1-1/2p}((0,a), L_p(\Gamma, N\Gamma)) \cap L_p((0,a), W_p^{2-1/p}(\Gamma, N\Gamma))
$$

as well as

$$Q_\Gamma \mathcal{B}^{\pm 1}(u, p) \in \mathbb{N}_h^{\pm 1}(a, \Gamma) := L_p((0, a), \dot{W}_p^{1-1/p}(\Gamma, N\Gamma)).$$

Here $T\Gamma$ denotes the tangent bundle of Γ and $N\Gamma$ denotes the normal bundle of Γ. Therefore, the regularity class for the boundary data h is given as

$$h \in \mathbb{Y}_h^{\alpha, \beta}(a, \Gamma) := \left\{ \eta \in \mathbb{Y}_h(a, \Gamma) : P_\Gamma \eta \in \mathbb{T}_h^\alpha(a, \Gamma), \; Q_\Gamma \eta \in \mathbb{N}_h^\beta(a, \Gamma) \right\}$$

for $\alpha, \beta \in \{-1, 0, +1\}$.

3.19 Last, but not least, we have to derive the necessary regularity condition for the initial datum u_0. Proposition 3.9 and Sobolev's embedding theorem suggest the embedding

$$\mathbb{X}_u(a, \Omega) \hookrightarrow BUC((0, a), W_p^{2-2/p}(\Omega, \mathbb{R}^n))$$

to be valid, where $BUC((0, a), X)$ denotes the space of bounded, uniformly continuous functions on the time interval $(0, a)$, the space X being an arbitrary Banach space. Indeed, this result is well-known even in the setting of abstract linear evolution equations, see e. g. the monograph by H. AMANN [Ama95] and Remark 3.65. Thus, we obtain

$$u_0 \in \mathbb{Y}_0(\Omega) := W_p^{2-2/p}(\Omega, \mathbb{R}^n).$$

3.2 Additional Regularity of the Pressure Trace

3.20 Since the Stokes equations subject to one of the energy preserving respectively artificial boundary conditions may also prove useful as model problems, e. g. for problems with a free surface or an evolving phase-separating interface, it is sometimes convenient to increase the regularity of the boundary data for $\beta = \pm 1$ to obtain an increased regularity of the pressure trace $[p]_\Gamma$.

3.21 Indeed, we have

$$Q_\Gamma \mathcal{B}^{\alpha, +1}(u, p) \cdot \nu_\Gamma = 2\mu \, \partial_\nu u \cdot \nu_\Gamma - [p]_\Gamma$$

for $\alpha \in \{-1, 0, +1\}$ and, thus, the regularity class of the data may also be chosen according to the regularity of $\partial_\nu u$, which is significantly higher than the regularity of $[p]_\Gamma$. Due to (3.1) we have

$$\partial_\nu u \in W_p^\gamma((0, a), L_p(\Gamma, \mathbb{R}^n)) \cap L_p((0, a), W_p^{1-1/p}(\Gamma, \mathbb{R}^n))$$

for all $\gamma \in [0, 1/2 - 1/2p]$. Our analysis will reveal that for $\beta = +1$ the pressure p belongs to the regularity class

$$p \in \mathbb{X}_{p,\gamma}^{+1}(a, \Omega) := \left\{ q \in \mathbb{X}_p^{+1}(a, \Omega) : \begin{array}{l} [q]_\Gamma \in W_p^\gamma((0, a), L_p(\Gamma)) \\ \cap L_p((0, a), W_p^{1-1/p}(\Gamma)) \end{array} \right\}$$

with $\gamma \in [0, 1/2 - 1/2p]$, if and only if the boundary data satisfies

$$Q_\Gamma h \in \mathbb{N}_{h,\gamma}^{+1}(a, \Gamma) := W_p^\gamma((0, a), L_p(\Gamma, N\Gamma)) \cap L_p((0, a), W_p^{1-1/p}(\Gamma, N\Gamma)).$$

3.22 If $\beta = -1$, the normal part of the boundary data equals the trace of the pressure. Hence, the pressure p belongs to the regularity class

$$p \in \mathbb{X}_{p,\gamma}^{-1}(a,\,\Omega) := \left\{ q \in \mathbb{X}_p^{-1}(a,\,\Omega) : \begin{array}{l} [q]_\Gamma \in W_p^\gamma((0,\,a),\,L_p(\Omega)) \\ \cap\, L_p((0,\,a),\,W_p^{1-1/p}(\Gamma)) \end{array} \right\}$$

with $\gamma \in [0,\,\infty)$, if and only if the boundary data satisfies

$$Q_\Gamma h \in \mathbb{N}_{h,\gamma}^{-1}(a,\,\Gamma) := W_p^\gamma((0,\,a),\,L_p(\Gamma,\,N\Gamma)) \cap L_p((0,\,a),\,W_p^{1-1/p}(\Gamma,\,N\Gamma)).$$

Thus, in this case the regularity of the pressure trace may be arbitrarily increased depending on the regularity of the data.

3.23 Now, an increased regularity of the pressure trace is optional. Moreover, for $\beta = 0$ the pressure does not appear in the boundary condition and the regularity of its trace may, thus, not be increased. Therefore, we simplify our notation by introducing

$$\mathbb{X}_{p,-\infty}^\beta(a,\,\Omega) := \mathbb{X}_p^\beta(a,\,\Omega), \qquad \mathbb{N}_{h,-\infty}^\beta(a,\,\Gamma) := \mathbb{N}_h^\beta(a,\,\Gamma)$$

for $\beta \in \{-1,\,0,\,+1\}$. This way, we obtain a maximal regular pressure $p \in \mathbb{X}_{p,\gamma}^\beta(a,\,\Omega)$, if and only if the boundary data satisfies

$$h \in \mathbb{Y}_{h,\gamma}^{\alpha,\beta}(a,\,\Gamma) := \left\{ \eta \in \mathbb{Y}_h(a,\,\Gamma) : P_\Gamma \eta \in \mathbb{T}_h^\alpha(a,\,\Gamma),\, Q_\Gamma \eta \in \mathbb{N}_{h,\gamma}^\beta(a,\,\Gamma) \right\}$$

with $\gamma = -\infty$ for $\beta = 0$, with $\gamma \in \{-\infty\} \cup [0,\,1/2 - 1/2p]$ for $\beta = 1$ respectively with $\gamma \in \{-\infty\} \cup [0,\,\infty)$ for $\beta = -1$.

3.24 Note, that $\mathbb{X}_{p,-\infty}^0(a,\,\Omega)$ constitutes a Banach space, whereas $\mathbb{X}_{p,-\infty}^{\pm 1}(a,\,\Omega)$ is seminormed via

$$|q|_{\mathbb{X}_{p,-\infty}^{\pm 1}(a,\,\Omega)} = |q|_{L_p((0,a),\,\dot{H}_p^1(\Omega))}, \qquad q \in \mathbb{X}_{p,-\infty}^{\pm 1}(a,\,\Omega).$$

However, for $\gamma \geq 0$, the spaces $\mathbb{X}_{p,\gamma}^{\pm 1}(a,\,\Omega)$ equipped with their natural norm

$$\|q\|_{\mathbb{X}_{p,\gamma}^{\pm 1}(a,\,\Omega)} = \max\left\{ |q|_{L_p((0,a),\,\dot{H}_p^1(\Omega))},\, \|[q]_\Gamma\|_{\mathbb{N}_{h,\gamma}^{\pm 1}(a,\,\Gamma)} \right\}, \qquad q \in \mathbb{X}_{p,\gamma}^{\pm 1}(a,\,\Gamma)$$

constitute Banach spaces, too. Analogously, the boundary data spaces $\mathbb{Y}_{h,-\infty}^{\alpha,\beta}(a,\,\Gamma)$ with $\alpha \in \{-1,\,0,\,+1\}$ and $\beta \in \{-1,\,+1\}$ are seminormed via

$$|\eta|_{\mathbb{Y}_{h,-\infty}^{\alpha,\beta}(a,\,\Gamma)} = \max\left\{ \|P_\Gamma \eta\|_{\mathbb{T}_h^\alpha(a,\,\Gamma)},\, |Q_\Gamma \eta|_{\mathbb{N}_{h,-\infty}^\beta(a,\,\Gamma)} \right\}, \qquad \eta \in \mathbb{Y}_{h,\gamma}^{\alpha,\beta}(a,\,\Gamma),$$

whereas the spaces $\mathbb{Y}_{h,-\infty}^{\alpha,0}(a,\,\Gamma)$ and $\mathbb{Y}_{h,\gamma}^{\alpha,\beta}(a,\,\Gamma)$ with $\alpha \in \{-1,\,0,\,+1\}$, $\beta \in \{-1,\,+1\}$ and $\gamma \geq 0$ constitute Banach spaces with their natural norm

$$\|\eta\|_{\mathbb{Y}_{h,\gamma}^{\alpha,\beta}(a,\,\Gamma)} = \max\left\{ \|P_\Gamma \eta\|_{\mathbb{T}_h^\alpha(a,\,\Gamma)},\, \|Q_\Gamma \eta\|_{\mathbb{N}_{h,\gamma}^\beta(a,\,\Gamma)} \right\}, \qquad \eta \in \mathbb{Y}_{h,\gamma}^{\alpha,\beta}(a,\,\Gamma).$$

Hence, continuous dependence of the solution on the data has in some cases to be understood w. r. t. seminorms, regardless of its uniqueness, which will always be guaranteed.

3.3 Necessary Compatibility Conditions

3.25 In addition to the regularity conditions derived in Sections 3.1 and 3.2 there are several compatibility conditions, which have to be satisfied by the data. First of all, the compatibility condition

$(C_1)_{g,u_0}$ $\operatorname{div} u_0 = g(0)$

is necessary, i. e. the right hand side of the divergence equation has to be compatible with the initial data, regardless of the particular boundary condition.

3.26 Moreover, the boundary condition implies the compatibility conditions

$$(C_2)_{h,u_0}^{\alpha} \qquad \begin{aligned} P_\Gamma[u_0]_\Gamma &= P_\Gamma h(0), \text{ if } \alpha = 0 \quad \text{and } p > \tfrac{3}{2}, \\ \mu P_\Gamma [\nabla u_0 \pm \nabla u_0^\mathsf{T}]_\Gamma\, \nu_\Gamma &= P_\Gamma h(0), \text{ if } \alpha = \pm 1 \text{ and } p > 3 \end{aligned}$$

to be necessary for $(S \,|\, a,\, \Omega)_{f,g,h,u_0}^{\alpha,\beta}$ to admit a maximal regular solution.

3.27 Last, but not least, there is a somewhat hidden compatibility condition, which stems from the divergence equation and the normal boundary condition. To reveal it we set $_0H_p^{-1}(\Omega) := H_{p'}^1(\Omega)'$ with $1/p + 1/p' = 1$ and define the linear functional

$$(\cdot, \cdot) : \mathbb{Y}_g(a,\, \Omega) \times \mathbb{N}_{h,-\infty}^0(a,\, \Gamma) \longrightarrow L_p((0,\, a),\, {}_0H_p^{-1}(\Omega))$$

for $\psi \in \mathbb{Y}_g(a,\, \Omega)$ and $\eta \in \mathbb{N}_{h,-\infty}^0(a,\, \Gamma)$ via

$$\langle \phi \,|\, (\psi,\, \eta) \rangle := \int_\Gamma [\phi]_\Gamma\, (\eta \cdot \nu_\Gamma)\, \mathrm{d}\mathcal{H}^{n-1} - \int_\Omega \phi\psi\, \mathrm{d}\mathcal{H}^n, \quad \phi \in H_{p'}^1(\Omega).$$

An integration by parts yields

$$\langle \phi \,|\, (\operatorname{div} u,\, Q_\Gamma[u]_\Gamma) \rangle = \int_\Omega \nabla\phi \cdot u\, \mathrm{d}\mathcal{H}^n, \quad \phi \in H_{p'}^1(\Omega)$$

and we infer

$$|\langle \phi \,|\, (\operatorname{div} u,\, Q_\Gamma[u]_\Gamma) \rangle| \le \|u\|_{\mathbb{X}_u(a,\,\Omega)} |\phi|_{\dot{H}_{p'}^1(\Omega)}, \quad \phi \in H_{p'}^1(\Omega)$$

as well as

$$|\langle \phi \,|\, \partial_t(\operatorname{div} u,\, Q_\Gamma[u]_\Gamma) \rangle| \le \|u\|_{\mathbb{X}_u(a,\,\Omega)} |\phi|_{\dot{H}_{p'}^1(\Omega)}, \quad \phi \in H_{p'}^1(\Omega).$$

Now, $H_{p'}^1(\Omega)$ and $\dot{H}_{p'}^1(\Omega)$ coincide algebraically. Hence, the definition $_0\dot{H}_p^{-1}(\Omega) := \dot{H}_{p'}^1(\Omega)'$ implies the compatibility condition

$$(C_3)_{g,h,u_0}^{\beta} \qquad \begin{aligned} &\left. \begin{aligned} Q_\Gamma[u_0]_\Gamma &= Q_\Gamma h, \quad \text{if } p > \tfrac{3}{2}, \\ \text{and } \ (g,\, Q_\Gamma h) &\in H_p^1((0,\, a),\, {}_0\dot{H}_p^{-1}(\Omega)), \end{aligned} \right\} && \text{if } \beta = 0, \\[1em] &\left. \begin{aligned} \text{there exists } \eta &\in \mathbb{N}_{h,-\infty}^0(a,\, \Gamma) \text{ such that} \\ Q_\Gamma[u_0]_\Gamma &= \eta, \quad \text{if } p > \tfrac{3}{2}, \\ \text{and } \ (g,\, \eta) &\in H_p^1((0,\, a),\, {}_0\dot{H}_p^{-1}(\Omega)), \end{aligned} \right\} && \text{if } \beta = \pm 1 \end{aligned}$$

to be necessary for $(S \mid a, \Omega)^{\alpha,\beta}_{f,g,h,u_0}$ to admit a maximal regular solution.

3.28 As our analysis will reveal, the regularity and compatibility conditions derived in Sections 3.1 and 3.2 and above are also sufficient to construct a unique maximal regular solution

$$(u, p) \in \mathbb{X}^{\beta}_{\gamma}(a, \Omega) := \mathbb{X}_u(a, \Omega) \times \mathbb{X}^{\beta}_{p,\gamma}(a, \Omega)$$

to $(S \mid a, \Omega)^{\alpha,\beta}_{f,g,h,u_0}$ for all $(f, g, h, u_0) \in \mathbb{Y}^{\alpha,\beta}_{\gamma}(a, \Omega)$, where the data space $\mathbb{Y}^{\alpha,\beta}_{\gamma}(a, \Omega)$ is defined to incorporate all necessary regularity and compatibility, i. e. to consist of all

$$(f, g, h, u_0) \in \mathbb{Y}_f(a, \Omega) \times \mathbb{Y}_g(a, \Omega) \times \mathbb{Y}^{\alpha,\beta}_{h,\gamma}(a, \Gamma) \times \mathbb{Y}_0(\Omega)$$

that satisfy the compatibility conditions $(C_1)_{g,u_0}$, $(C_2)^{\alpha}_{h,u_0}$ and $(C_3)^{\beta}_{g,u_0}$. The detailed results are stated as Theorem 3.30 below.

3.4 Maximal L_p-Regularity of the Stokes Equations

3.29 With the above preparations at hand, we are able to formulate our first main theorem, which postulates the maximal L_p-regularity of the Stokes equations $(S \mid a, \Omega)$ for a bounded, smooth domain.

3.30 THEOREM. *Let $a > 0$, let $\Omega \subseteq \mathbb{R}^n$ be a bounded domain with boundary $\Gamma = \partial\Omega$ of class C^{3-} and let $1 < p < \infty$ with $p \neq \frac{3}{2}$, 3. Let $\rho, \mu > 0$ and let $\alpha, \beta \in \{-1, 0, +1\}$. Moreover,*

- *let $\gamma = -\infty$, if $\beta = 0$;*
- *let $\gamma \in \{-\infty\} \cup [0, 1/2 - 1/2p]$, if $\beta = +1$;*
- *let $\gamma \in \{-\infty\} \cup [0, \infty)$, if $\beta = -1$.*

Then there exists a unique maximal regular solution

$$(u, p) \in \mathbb{X}^{\beta}_{\gamma}(a, \Omega)$$

to the Stokes equations $(S \mid a, \Omega)^{\alpha,\beta}_{f,g,h,u_0}$, if and only if the data satisfies

$$(f, g, h, u_0) \in \mathbb{Y}^{\alpha,\beta}_{\gamma}(a, \Omega).$$

Furthermore, the solutions depend continuously on the data.

3.31 The proof of Theorem 3.30 will be carried out in the next chapters. At this point we collect some interesting corollaries, which directly arise. First of all, we may recast the Stokes equations $(S \mid a, \Omega)$ as an operator equation as follows: The left hand side without the initial condition defines a linear operator

$$L^{\alpha,\beta}_{a,\gamma} : \mathbb{X}^{\beta}_{\gamma}(a, \Omega) \longrightarrow \mathbb{Y}_f(a, \Omega) \times \mathbb{Y}_g(a, \Omega) \times \mathbb{Y}^{\alpha,\beta}_{h,\gamma}(a, \Gamma)$$

between the solution space and the data space. As has been shown in Sections 3.1 and 3.2, this operator is bounded. Note that the Stokes equations $(S \mid a, \Omega)_{f,g,h,u_0}^{\alpha,\beta}$ are equivalent to

$$L_{a,\gamma}^{\alpha,\beta}(u, p) = (f, g, h), \qquad u(0) = u_0 \text{ in } \Omega.$$

Now, the parabolic equation

$$
\begin{aligned}
\rho \partial_t u - \mu \Delta u &= 0 && \text{in } (0, a) \times \Omega, \\
[u]_\Gamma &= e^{t \Delta_\Gamma} [u_0]_\Gamma && \text{on } (0, a) \times \partial\Omega, \\
u(0) &= u_0 && \text{in } \Omega
\end{aligned}
$$

(3.2)

defines a bounded linear solution operator $u^* : W_p^{2-2/p}(\Omega) \longrightarrow \mathbb{X}_u(a, \Omega)$, which follows by parabolic regularity theory, cf. [DHP03]. Here Δ_Γ denotes the Laplace-Beltrami operator on Γ and $e^{t\Delta_\Gamma}$ denotes the corresponding semigroup. Thus, every maximal regular solution $(u, p) \in \mathbb{X}_\gamma^\beta(a, \Omega)$ to the Stokes equations may be split as $(u, p) = (\bar{u}, \bar{p}) + (u^*(u_0), 0)$, where

$$\bar{u} \in {}_0\mathbb{X}_u(a, \Omega) := \Big\{ v \in \mathbb{X}_u(a, \Omega) : v(0) = 0 \Big\},$$

and the Stokes equations $(S \mid a, \Omega)_{f,g,h,u_0}^{\alpha,\beta}$ are equivalent to

$${}_0L_{a,\gamma}^{\alpha,\beta}(\bar{u}, \bar{p}) = (f, g, h) - L_{a,\gamma}^{\alpha,\beta}(u^*(u_0), 0),$$

where

$${}_0L_{a,\gamma}^{\alpha,\beta} : {}_0\mathbb{X}_\gamma^\beta(a, \Omega) \longrightarrow {}_0\mathbb{Y}_\gamma^{\alpha,\beta}(a, \Omega)$$

denotes the restriction of $L_{a,\gamma}^{\alpha,\beta}$ to

$${}_0\mathbb{X}_\gamma^\beta(a, \Omega) := {}_0\mathbb{X}_u(a, \Omega) \times \mathbb{X}_{p,\gamma}^\beta(a, \Omega)$$

and the data space ${}_0\mathbb{Y}_\gamma^{\alpha,\beta}(a, \Omega)$ is defined to consist of all

$$(f, g, h) \in \mathbb{Y}_f(a, \Omega) \times \mathbb{Y}_g(a, \Omega) \times \mathbb{Y}_{h,\gamma}^{\alpha,\beta}(a, \Gamma)$$

that satisfy the compatibility conditions $(C_1)_{g,0}$, $(C_2)_{h,0}^\alpha$ and $(C_3)_{g,h,0}^\beta$. Thus, we have the following equivalent formulation of Theorem 3.30.

3.32 COROLLARY. *Under the assumptions of Theorem 3.30 the bounded linear operator*

$${}_0L_{a,\gamma}^{\alpha,\beta} : {}_0\mathbb{X}_\gamma^\beta(a, \Omega) \longrightarrow {}_0\mathbb{Y}_\gamma^{\alpha,\beta}(a, \Omega)$$

constitutes an isomorphism between the solution space and the data space. □

3.33 Note, that the linear operator ${}_0L_{a,\gamma}^{\alpha,\beta}$ is bounded and, thus, due to the open mapping principle it constitutes an isomorphism, if and only if it is bijective. Analogously, it is sufficient to prove the existence and uniqueness of maximal regular solutions in order to prove Theorem 3.30. The additional assertion of continuous dependence is then a consequence of the open mapping principle. Also note, that the boundedness of the domain Ω implies the embedding

$$L_p((0, a) \times \Omega, X) \hookrightarrow L_2((0, a) \times \Omega, X)$$

to be valid, whenever $p \geq 2$. Hence, the computations in 2.15, 2.17 and 2.20, which lead to the energy balances for the energy preserving boundary conditions introduced in 2.19 and 2.22, may be carried out for the solutions delivered by Theorem 3.30. Thus, we obtain the following corollary.

3.34 COROLLARY. *Under the assumptions of Theorem 3.30 suppose in addition that $p \geq 2$ and that one of the energy preserving boundary conditions for Stokes flows as introduced in 2.19 and 2.22 is imposed. Then for the unique solution*

$$(u, p) \in \mathbb{X}_\gamma^\beta(a, \Omega)$$

to the Stokes equations $(S \mid a, \Omega)_{f,0,0,u_0}^{\alpha,\beta}$ the energy inequality

$$\frac{d}{dt} \int_\Omega \tfrac{1}{2}\rho\,|u|^2 \, d\mathcal{H}^3 \leq \int_\Omega \rho u \cdot f \, d\mathcal{H}^3.$$

is valid, provided $f \in \mathbb{Y}_f(a, \Omega)$ and $u_0 \in \mathbb{Y}_0(\Omega)$ satisfies $(C)_{0,0,u_0}^{\alpha,\beta}$. □

3.35 Note that this in particular implies the maximal regular solutions to be unique. However, the two artificial boundary conditions introduced in 2.23 are not covered by Corollary 3.34 and the restriction $p \geq 2$ has to be imposed. Therefore, we will give a proof of the uniqueness of maximal regular solutions in Chapter 7, which applies to all boundary conditions under consideration and all $1 < p < \infty$ with $p \neq \frac{3}{2}, 3$. Nevertheless, the uniqueness of the solutions implies the following corollary to be valid.

3.36 COROLLARY. *Under the assumptions of Theorem 3.30 the solution map $u_0 \mapsto u$ for the homogeneous Stokes equations $(S \mid a, \Omega)_{0,0,0,u_0}^{\alpha,\beta}$ generates a semiflow in*

$$\mathbb{Z}^{\alpha,\beta}(\Omega) := \left\{ v \in W_p^{2-2/p}(\Omega) \; : \; v \text{ satisfies } (C)_{0,0,v}^{\alpha,\beta} \right\},$$

the natural phase space of $(S \mid a, \Omega)_{0,0,0,u_0}^{\alpha,\beta}$ in the L_p-setting. □

3.5 Associated Stokes Operators

3.37 Sometimes it is convenient to have a more abstract view on the Stokes equations $(S \mid a, \Omega)_{f,0,0}^{\alpha,\beta}$ and to recast them as an evolution equation. This is possible by introducing the associated *Stokes operator*, which, of course, depends on the boundary condition. For $\alpha, \beta \in \{-1, 0, +1\}$ we employ the base spaces

$$X_p^0(\Omega) := L_{p,\sigma}(\Omega) := \mathrm{cls}\left(C_{0,\sigma}^\infty(\Omega), \, L_p(\Omega, \mathbb{R}^n), \, \|\cdot\|_{L_p(\Omega,\mathbb{R}^n)}\right),$$

i. e. $L_{p,\sigma}(\Omega)$ denotes the closure of the space $C_{0,\sigma}^\infty(\Omega)$ of solenoidal, compactly supported, smooth vector fields in $L_p(\Omega, \mathbb{R}^n)$, and

$$X_p^{\pm 1}(\Omega) := L_{p,s}(\Omega) := \left\{ v \in L_p(\Omega, \mathbb{R}^n) : \mathrm{div}\, v = 0 \right\},$$

where the divergence has to be understood in the distributional sense. Note that the inclusion $L_{p,\sigma}(\Omega) \subseteq L_{p,s}(\Omega)$ is always valid. These two spaces will be analysed in more detail in Section 4.1. For the following argumentation it is sufficient to observe that

$$H_p^2(\Omega, \mathbb{R}^n) \cap L_{p,\sigma}(\Omega) = \left\{ v \in H_p^2(\Omega, \mathbb{R}^n) \cap L_{p,s}(\Omega) \ : \ [v]_\Gamma \cdot \nu_\Gamma = 0 \right\},$$

i. e. the difference between the two spaces of solenoidal L_p-vector fields may be expressed in terms of a boundary condition.

3.38 Now, we set

$$D(A^{\alpha,\beta}) := \left\{ v \in H_p^2(\Omega, \mathbb{R}^n) \cap X_p^\beta(\Omega) \ : \ \begin{array}{c} \exists \, q \in \dot{H}_p^1(\Omega) : \\ -\mu\Delta v + \nabla q \in X_p^\beta(\Omega), \\ P_\Gamma \mathcal{B}^\alpha v = 0, \ Q_\Gamma \mathcal{B}^{\alpha,\beta}(v, q) = 0, \\ (q)_\Omega = 0, \ \text{if } \beta = 0 \end{array} \right\}$$

and define the Stokes operator $A^{\alpha,\beta} : D(A^{\alpha,\beta}) \subseteq X_p^\beta(\Omega) \longrightarrow X_p^\beta(\Omega)$ as

$$A^{\alpha,\beta}v := -\mu\Delta v + \nabla q, \qquad v \in D(A^{\alpha,\beta}),$$

where $q \in \dot{H}_p^1(\Omega)$ is chosen according to the definition of $D(A^{\alpha,\beta})$.

3.39 Employing Theorem 3.30 we may immediately derive the basic mapping properties of the Stokes operators. For $\beta = 0$ let $\gamma = -\infty$ and for $\beta = \pm 1$ let $\gamma = 1/2 - 1/2p$. Assume $(v_k)_k \subseteq D(A^{\alpha,\beta})$, set $\phi_k := A^{\alpha,\beta}v_k$ for $k = 1, 2, \ldots$ and suppose

$$v_k \to v \ \text{ in } X_p^\beta(\Omega) \qquad \text{and} \qquad \phi_k \to \phi \ \text{ in } X_p^\beta(\Omega) \qquad \text{as } k \to \infty.$$

Choose an $a > 0$, an $\varepsilon > 0$ and define $\chi \in C^\infty([0, a])$ as

$$\chi(t) = 1 - e^{-\varepsilon t}, \qquad 0 \le t \le a.$$

Now, choose $(q_k)_k \subseteq \dot{H}_p^1(\Omega)$ according to the definition of $D(A^{\alpha,\beta})$, such that

$$A^{\alpha,\beta}v_k = -\mu\Delta v_k + \nabla q_k, \qquad k = 1, 2, \ldots$$

and set

$$u_k := \chi v_k \in \mathbb{X}_u(a, \Omega), \quad p_k := \chi q_k \in \mathbb{X}_{p,\gamma}^\beta(a, \Omega),$$
$$\rho f_k := \chi \phi_k + \rho\varepsilon(1 - \chi)v_k \in \mathbb{Y}_f(a, \Omega), \qquad k = 1, 2, \ldots.$$

Then $(u_k, p_k) \in \mathbb{X}_\gamma^\beta(a, \Omega)$ uniquely solves the Stokes equations $(S \,|\, a, \Omega)_{f_k, 0, 0, 0}^{\alpha,\beta}$, where $k = 1, 2, \ldots$, and

$$\rho f_k \to \chi\phi + \rho\varepsilon(1 - \chi)v =: \rho f \ \text{ in } L_p((0, a) \times \Omega, \mathbb{R}^n) \qquad \text{as } k \to \infty$$

implies

$$u_k \to u \ \text{ in } \mathbb{X}_u(a, \Omega) \qquad \text{and} \qquad p_k \to p \ \text{ in } \mathbb{X}_{p,\gamma}^\beta(a, \Omega) \qquad \text{as } k \to \infty,$$

where $(u, p) \in \mathbb{X}_\gamma^\beta(a, \Omega)$ is the unique solution to the Stokes equations $(S \,|\, a, \Omega)_{f,0,0,0}^{\alpha,\beta}$. On the other hand, we have

$$u_k \to \chi v \quad \text{in } L_p((0, a) \times \Omega, \mathbb{R}^n) \qquad \text{as } k \to \infty$$

and, hence, $u = \chi v$. In particular

$$v \in D(A^{\alpha,\beta}) \qquad \text{and} \qquad A^{\alpha,\beta}v = -\mu\Delta v + \nabla q = \phi$$

with $\chi\nabla q = \nabla p$. Thus, we obtain a first result on the Stokes operators.

3.40 PROPOSITION. *Let $\Omega \subseteq \mathbb{R}^n$ be a bounded domain with boundary $\Gamma = \partial\Omega$ of class C^{3-} and let $1 < p < \infty$ with $p \neq \frac{3}{2}, 3$. Let $\rho, \mu > 0$ and let $\alpha, \beta \in \{-1, 0, +1\}$. Then the Stokes operator*

$$A^{\alpha,\beta} : D(A^{\alpha,\beta}) \subseteq X_p^\beta(\Omega) \longrightarrow X_p^\beta(\Omega)$$

as defined in 3.38 is closed.   □

3.41 As a consequence, the space $D_p^{\alpha,\beta}(\Omega)$ that denotes the domain $D(A^{\alpha,\beta})$ of $A^{\alpha,\beta}$ equipped with the graph norm of $A^{\alpha,\beta}$ constitutes a Banach space. Now, given any $v \in D(A^{\alpha,\beta})$ the associated pressure $q \in \dot{H}_p^1(\Omega)$ is uniquely defined via

$$(\nabla q \,|\, \nabla\phi)_\Omega = \mu(\Delta v \,|\, \nabla\phi)_\Omega, \quad \phi \in \dot{H}_{p'}^1(\Omega), \qquad (q)_\Omega = 0$$

for $\beta = 0$ respectively

$$(\nabla q \,|\, \nabla\phi)_\Omega = \mu(\Delta v \,|\, \nabla\phi)_\Omega, \quad \phi \in {}_0\dot{H}_{p'}^1(\Omega), \qquad [q]_\Gamma = 2\mu\,\partial_\nu v \cdot \nu_\Gamma$$

for $\beta = +1$ respectively

$$(\nabla q \,|\, \nabla\phi)_\Omega = \mu(\Delta v \,|\, \nabla\phi)_\Omega, \quad \phi \in {}_0\dot{H}_{p'}^1(\Omega), \qquad [q]_\Gamma = 0$$

for $\beta = -1$. Here $(\cdot \,|\, \cdot)_\Omega$ denotes the L_2-inner product over the domain Ω. Furthermore, $1/p + 1/p' = 1$ and

$$_0\dot{H}_{p'}^1(\Omega) := \text{cls}\left(C_0^\infty(\Omega), \dot{H}_{p'}^1(\Omega), |\cdot|_{\dot{H}_{p'}^1(\Omega)}\right).$$

This space as well as the above variational equations will be examined in detail in Section 4.1. At this point it is sufficient to note that there exists a unique solution $q \in \dot{H}_p^1(\Omega)$ for all $v \in D(A^{\alpha,\beta})$ such that

$$\|\nabla q\|_{L_p(\Omega, \mathbb{R}^n)} \leq M\|v\|_{H_p^2(\Omega, \mathbb{R}^n)}, \qquad v \in D(A^{\alpha,\beta})$$

with some constant $M > 0$, cf. Propositions 4.4 and 4.10. On one hand, this implies

$$D(A^{\alpha,\beta}) = \left\{ v \in H_p^2(\Omega, \mathbb{R}^n) \cap X_p^\beta(\Omega) \,:\, P_\Gamma \mathcal{B}^\alpha v = 0 \right\}.$$

On the other hand, we obtain

$$\begin{aligned}
\|v\|_{D_p^{\alpha,\beta}(\Omega)} &= \|v\|_{L_p(\Omega, \mathbb{R}^n)} + \|A^{\alpha,\beta}v\|_{L_p(\Omega, \mathbb{R}^n)} \\
&\leq \|v\|_{L_p(\Omega, \mathbb{R}^n)} + \mu\|\Delta v\|_{L_p(\Omega, \mathbb{R}^n)} + \|\nabla q\|_{L_p(\Omega, \mathbb{R}^n)} \\
&\leq \bar{M}\|v\|_{H_p^2(\Omega, \mathbb{R}^n)}
\end{aligned}$$

for all $v \in D(A^{\alpha,\beta})$ with some constant $\bar{M} > 0$. Hence, the identity

$$1 : \left(D(A^{\alpha,\beta}), \| \cdot \|_{H^2_p(\Omega, \mathbb{R}^n)} \right) \longrightarrow D^{\alpha,\beta}_p(\Omega)$$

is continuous and since $D(A^{\alpha,\beta})$ is a closed subspace of $H^2_p(\Omega, \mathbb{R}^n)$, the open mapping principle implies

$$D^{\alpha,\beta}_p(\Omega) \cong \left(D(A^{\alpha,\beta}), \| \cdot \|_{H^2_p(\Omega, \mathbb{R}^n)} \right).$$

Delaying the remaining argumentation to Remark 3.66, we have thus obtained our next result on the Stokes operators.

3.42 PROPOSITION. *Let $\Omega \subseteq \mathbb{R}^n$ be a bounded domain with boundary $\Gamma = \partial\Omega$ of class C^{3-} and let $1 < p < \infty$ with $p \neq \frac{3}{2}$, 3. Let $\rho, \mu > 0$ and let $\alpha, \beta \in \{-1, 0, +1\}$. Then the domain of the Stokes operator*

$$A^{\alpha,\beta} : D(A^{\alpha,\beta}) \subseteq X^\beta_p(\Omega) \longrightarrow X^\beta_p(\Omega)$$

as defined in 3.38 is characterised as

$$D^{\alpha,\beta}_p(\Omega) = \left(\left\{ v \in H^2_p(\Omega, \mathbb{R}^n) \cap X^\beta_p(\Omega) : P_\Gamma \mathcal{B}^\alpha v = 0 \right\}, \| \cdot \|_{H^2_p(\Omega, \mathbb{R}^n)} \right)$$

up to equivalence of norms. In particular, $A^{\alpha,\beta}$ is densely defined.   □

3.43 Now, note that due to the definition of the Stokes operators $A^{\alpha,\beta}$ the Stokes equations $(S \,|\, a, \Omega)^{\alpha,\beta}_{f,0,0,0}$ are equivalent to the abstract evolution equation

(3.3)
$$\rho\partial_t u + A^{\alpha,\beta} u = \rho f \quad \text{in } (0, a) \times \Omega,$$
$$u(0) = 0 \quad \text{in } \Omega,$$

whenever $f \in L_p((0, a), X^\beta_p(\Omega))$. This equivalence has to be understood in the following sense: For $\beta = 0$ let $\gamma = -\infty$ and for $\beta = \pm 1$ let $\gamma = 1/2 - 1/2p$. Now, if we denote by

$$(u, p) \in \mathbb{X}^\beta_\gamma(a, \Omega)$$

the unique maximal regular solution to the Stokes equations $(S \,|\, a, \Omega)^{\alpha,\beta}_{f,0,0,0}$, then

(3.4)
$$u \in H^1_p((0, a), X^\beta_p(\Omega)) \cap L_p((0, a), D^{\alpha,\beta}_p(\Omega))$$

and (3.3) is satisfied. Conversely, if u is a solution to (3.3) satisfying (3.4), then we may define $p \in \mathbb{X}^\beta_{p,\gamma}(a, \Omega)$ via

$$\nabla p := A^{\alpha,\beta} u + \mu \Delta u \quad \text{and} \quad (p)_\Omega = 0, \quad \text{if } \beta = 0$$

and $(u, p) \in \mathbb{X}^\beta_\gamma(a, \Omega)$ is a solution to the Stokes equations $(S \,|\, a, \Omega)^{\alpha,\beta}_{f,0,0,0}$. As a result, Theorem 3.30 and the L_p-theory of abstract parabolic evolution equations, cf. [Prü03, Proposition 1.2], implies the following properties of the Stokes operator to hold.

3.44 COROLLARY. Let $\Omega \subseteq \mathbb{R}^n$ be a bounded domain with boundary $\Gamma = \partial\Omega$ of class C^{3-} and let $1 < p < \infty$ with $p \neq \frac{3}{2}$, 3. Let ρ, $\mu > 0$ and let α, $\beta \in \{-1, 0, +1\}$. Then the Stokes operator

$$A^{\alpha,\beta} : D(A^{\alpha,\beta}) \subseteq X_p^\beta(\Omega) \longrightarrow X_p^\beta(\Omega)$$

as defined in 3.38 has the property of maximal L_p-regularity on finite time intervals. In particular, there exist $M \geq 1$ and $\omega > 0$, such that

$$\left\{ z \in \mathbb{C} : \operatorname{Re} z \geq \omega \right\} \subseteq \rho(-A^{\alpha,\beta})$$

and the estimate

$$\|z(z + A^{\alpha,\beta})^{-1}\|_{\mathcal{B}(X_p^\beta(\Omega))} \leq M, \qquad \operatorname{Re} z \geq \omega$$

is valid. Thus, $\omega + A^{\alpha,\beta}$ is sectorial with spectral angle $\phi_{\omega+A^{\alpha,\beta}} < \pi/2$ and $-A^{\alpha,\beta}$ generates an analytic semigroup in $X_p^\beta(\Omega)$. $\square$

3.6 Well-Posedness of the Navier-Stokes Equations

3.45 To treat the non-linear Navier-Stokes equations $(\mathrm{N}\,|\,a,\,\Omega)_{f,g,h,u_0}^{\alpha,\beta}$ it is convenient to recast them into an operator equation. In analogy to 3.31 the Navier-Stokes equations are equivalent to

$$L_{a,\gamma}^{\alpha,\beta}(u,\,p) = N_{a,\gamma}^{\alpha,\beta}(u,0) + (f,\,g,\,h), \qquad u(0) = u_0 \text{ in } \Omega,$$

where the non-linear operator

$$N_{a,\gamma}^{\alpha,\beta} : \mathbb{X}_\gamma^\beta(a,\,\Omega) \longrightarrow \mathbb{Y}_f(a,\,\Omega) \times \mathbb{Y}_g(a,\,\Omega) \times \mathbb{Y}_{h,\gamma}^{\alpha,\beta}(a,\,\Gamma)$$

is given as

$$N_{a,\gamma}^{\alpha,\beta}(v,\,q) = (-(v \cdot \nabla)v,\, 0,\, 0), \qquad (v,\,q) \in \mathbb{X}_\gamma^\beta(a,\,\Omega).$$

Now, Theorem 3.30 ensures the existence of a bounded linear solution operator

$$(u^*,\,p^*) : \mathbb{Y}_\gamma^{\alpha,\beta}(a,\,\Omega) \longrightarrow \mathbb{X}_\gamma^\beta(a,\,\Omega)$$

to the Stokes equations. Thus, thanks to Corollary 3.32 the Navier-Stokes equations $(\mathrm{N}\,|\,a,\,\Omega)_{f,g,h,u_0}^{\alpha,\beta}$ are equivalent to

$$(u,\,p) = (\bar{u},\,\bar{p}) + (u^*(f,\,g,\,h,\,u_0),\, p^*(f,\,g,\,h,\,u_0)),$$

$$(\bar{u},\,\bar{p}) = {}_0S_{a,\gamma}^{\alpha,\beta} N_{a,\gamma}^{\alpha,\beta}(\bar{u} + u^*(f,\,g,\,h,\,u_0),0) =: K_{a,\gamma}^{\alpha,\beta}(\bar{u},\,\bar{p}),$$

where

$$_0S_{a,\gamma}^{\alpha,\beta} : {}_0\mathbb{Y}_\gamma^{\alpha,\beta}(a,\,\Omega) \longrightarrow {}_0\mathbb{X}_\gamma^\beta(a,\,\Omega)$$

denotes the bounded linear inverse of $_0L_{a,\gamma}^{\alpha,\beta}$. This way we may employ the contraction mapping principle to obtain our next main result.

3.46 THEOREM. *Let $a > 0$ and let $\Omega \subseteq \mathbb{R}^n$ be a bounded domain with boundary $\Gamma = \partial\Omega$ of class C^{3-} and let $n + 2 < p < \infty$. Let $\rho, \mu > 0$ and let $\alpha, \beta \in \{-1, 0, +1\}$. Moreover,*

- *let $\gamma = -\infty$, if $\beta = 0$;*
- *let $\gamma \in \{-\infty\} \cup [0, 1/2 - 1/2p]$, if $\beta = +1$;*
- *let $\gamma \in \{-\infty\} \cup [0, \infty)$, if $\beta = -1$.*

Then there is $\varepsilon = \varepsilon(a) > 0$, such that the Navier-Stokes equations $(N \,|\, a, \Omega)^{\alpha,\beta}_{f,g,h,u_0}$ admit a unique maximal regular solution

$$(u, p) \in \mathbb{X}^{\beta}_{\gamma}(a, \Omega)$$

whenever the data $(f, g, h, u_0) \in \mathbb{Y}^{\alpha,\beta}_{\gamma}(a, \Omega)$ satisfies the smallness condition

$$\|(f, g, h, u_0)\|_{\mathbb{Y}^{\alpha,\beta}_{\gamma}(a, \Omega)} \leq \varepsilon.$$

Furthermore, the solutions depend continuously on the data.

3.47 Thus, the Navier-Stokes equations may be solved on finite time intervals, provided the data satisfies a suitable smallness condition. Theorem 3.46, which will be proved in the following paragraphs, is complemented by our next main result.

3.48 THEOREM. *Let $a > 0$ and let $\Omega \subseteq \mathbb{R}^n$ be a bounded domain with boundary $\Gamma = \partial\Omega$ of class C^{3-} and let $n + 2 < p < \infty$. Let $\rho, \mu > 0$ and let $\alpha, \beta \in \{-1, 0, +1\}$. Moreover,*

- *let $\gamma = -\infty$, if $\beta = 0$;*
- *let $\gamma \in \{-\infty\} \cup [0, 1/2 - 1/2p]$, if $\beta = +1$;*
- *let $\gamma \in \{-\infty\} \cup [0, \infty)$, if $\beta = -1$.*

Then for

$$(f, g, h, u_0) \in \mathbb{Y}^{\alpha,\beta}_{\gamma}(a, \Omega)$$

there exists a unique local-in-time strong solution (u, p) to the Navier-Stokes equations $(N \,|\, a^*, \Omega)^{\alpha,\beta}_{f,g,h,u_0}$ on a maximal time interval $(0, a^*)$ with

$$a^* = a^*(f, g, h, u_0) \in (0, a).$$

The solution satisfies

$$(u, p) \in \mathbb{X}^{\beta}_{\gamma}(\bar{a}, \Omega)$$

for all $\bar{a} \in (0, a^*)$. Furthermore, the solutions depend continuously on the data.

3.49 We prove Theorems 3.46 and 3.48 simultaneously employing the fixed point formulation of the Navier-Stokes equations introduced in 3.45. We start by fixing $a > 0$, $\bar{a} \in (0, a]$ and $(f, g, h, u_0) \in \mathbb{Y}^{\alpha,\beta}_{\gamma}(a, \Omega)$. Moreover, we denote by

$$(u^*, p^*) = (u^*, p^*)(f, g, h, u_0)$$

the unique maximal regular solution to the Stokes equations $(S \,|\, a,\, \Omega)_{f,g,h,u_0}^{\alpha,\beta}$, which satisfies

$$\|(u^*,\, p^*)\|_{\mathbb{X}_\gamma^\beta(a,\Omega)} \le M \|(f, g, h, u_0)\|_{\mathbb{Y}_\gamma^{\alpha,\beta}(a,\Omega)}$$

by Theorem 3.30 with some constant $M > 0$. According to 3.45 it remains to uniquely solve

$$(\bar{u},\, \bar{p}) = K_{\bar{a},\gamma}^{\alpha,\beta}(\bar{u},\, \bar{p}), \qquad (\bar{u},\, \bar{p}) \in {}_0\mathbb{X}_\gamma^\beta(\bar{a},\, \Omega)$$

in order to obtain the unique maximal regular solution to the Navier-Stokes equations $(N \,|\, \bar{a},\, \Omega)_{f,g,h,u_0}^{\alpha,\beta}$. Note that from this point on, we may assume $\gamma = 1/2 - 1/2p$, if $\beta = \pm 1$. Therefore, all function spaces, which occur in the next paragraphs, constitute Banach spaces.

3.50 First observe that

$$N_{\bar{a},\gamma}^{\alpha,\beta}(\,\cdot\, + u^*,\, 0) : {}_0\mathbb{X}_u(\bar{a},\, \Omega) \longrightarrow {}_0\mathbb{Y}_\gamma^{\alpha,\beta}(\bar{a},\, \Omega)$$

is Fréchet differentiable with

$$DN_{\bar{a},\gamma}^{\alpha,\beta}(\bar{u} + u^*,\, 0)\bar{v} = (-((\bar{u} + u^*) \cdot \nabla)\bar{v} - (\bar{v} \cdot \nabla)(\bar{u} + u^*),\, 0,\, 0), \quad \bar{u},\, \bar{v} \in {}_0\mathbb{X}_u(\bar{a},\, \Omega).$$

Now,

$$\|((\bar{u} + u^*) \cdot \nabla)\bar{v} + (\bar{v} \cdot \nabla)(\bar{u} + u^*)\|_{L_p((0,\bar{a})\times\Omega,\mathbb{R}^n)}$$

$$\le \|\bar{u} + u^*\|_{BUC((0,\bar{a}),\, BUC(\Omega,\mathbb{R}^n))} \cdot \|\bar{v}\|_{L_p((0,\bar{a}),\, H_p^1(\Omega,\mathbb{R}^n))}$$

$$+ \|\bar{u} + u^*\|_{BUC((0,\bar{a}),\, BUC^1(\Omega,\mathbb{R}^n))} \cdot \|\bar{v}\|_{L_p((0,\bar{a})\times\Omega,\mathbb{R}^n)}$$

$$\le c\bar{a}^{1/p}\|\bar{u} + u^*\|_{BUC((0,\bar{a}),\, BUC(\Omega,\mathbb{R}^n))} \cdot \|\bar{v}\|_{{}_0W_p^{1/2}((0,\bar{a}),\, H_p^1(\Omega,\mathbb{R}^n))}$$

$$+ c\bar{a}^{1/p}\|\bar{u} + u^*\|_{BUC((0,\bar{a}),\, BUC^1(\Omega,\mathbb{R}^n))} \cdot \|\bar{v}\|_{{}_0H_p^1((0,\bar{a}),\, L_p(\Omega,\mathbb{R}^n))},$$

where we used the embedding chain

$${}_0W_p^{1/2}((0,\, \bar{a}),\, H_p^1(\Omega,\, \mathbb{R}^n)) \hookrightarrow {}_0BUC((0,\, \bar{a}),\, H_p^1(\Omega,\, \mathbb{R}^n))$$

$$\hookrightarrow L_p((0,\, \bar{a}),\, H_p^1(\Omega,\, \mathbb{R}^n))$$

together with the estimates

$$\|\bar{v}\|_{L_p((0,\bar{a}),\, H_p^1(\Omega,\mathbb{R}^n))} \le \bar{a}^{1/p}\|\bar{v}\|_{{}_0BUC((0,\bar{a}),\, H_p^1(\Omega,\mathbb{R}^n))}$$

$$\le c\bar{a}^{1/p}\|\bar{v}\|_{{}_0W_p^{1/2}((0,\bar{a}),\, H_p^1(\Omega,\mathbb{R}^n))},$$

where the constant $c > 0$ is independent of $\bar{a} \in (0, a]$ thanks to the homogeneous initial condition, cf. [PSS07, Proposition 6.2 (a)] and Remark 4.79. As usual, we denote by the left subscript zero the subspaces with homogeneous initial value. Analogously, the embedding chain

$${}_0H_p^1((0,\, \bar{a}),\, L_p(\Omega,\, \mathbb{R}^n)) \hookrightarrow {}_0BUC((0,\, \bar{a}),\, L_p(\Omega,\, \mathbb{R}^n))$$

$$\hookrightarrow L_p((0,\, \bar{a}) \times \Omega,\, \mathbb{R}^n)$$

is valid together with the estimates

$$\|\bar{v}\|_{L_p((0,\bar{a})\times\Omega,\mathbb{R}^n)} \le \bar{a}^{1/p}\|\bar{v}\|_{0BUC((0,\bar{a}),L_p(\Omega,\mathbb{R}^n))}$$
$$\le c\bar{a}^{1/p}\|\bar{v}\|_{0H_p^1((0,\bar{a}),L_p(\Omega,\mathbb{R}^n))},$$

where the constant $c > 0$ is again independent of $\bar{a} \in (0, a]$. Employing Proposition 3.9 in its version for the Sobolev-Slobodeckij spaces as mentioned in 3.11 we obtain

$$\|DN_{\bar{a},\gamma}^{\alpha,\beta}(\bar{u} + u^*, 0)\|_{\mathcal{B}(_0\mathbb{X}_u(\bar{a},\Omega), _0\mathbb{Y}_\gamma^{\alpha,\beta}(\bar{a},\Omega))} \le c\bar{a}^{1/p}\|\bar{u} + u^*\|_{BUC((0,\bar{a}),BUC^1(\Omega,\mathbb{R}^n))}$$

for all $\bar{u} \in {}_0\mathbb{X}_u(\bar{a}, \Omega)$ with $c > 0$ still being independent of $\bar{a} \in (0, a]$.

3.51 By Proposition 3.9 and Sobolev's embedding theorem, we have

$$\|u^*\|_{BUC((0,\bar{a}),BUC^1(\Omega,\mathbb{R}^n))} \le \bar{M}\|(u^*, p^*)\|_{\mathbb{X}_\gamma^\beta(a,\Omega)} \le M\bar{M}\|(f,g,h,u_0)\|_{\mathbb{Y}_\gamma^{\alpha,\beta}(a,\Omega)}.$$

Note that the constants $M, \bar{M} > 0$ are independent of $\bar{a} \in (0, a]$, since (u^*, p^*) is defined on the whole interval $(0, a)$. Analogously,

$$\|\bar{u}\|_{0BUC((0,\bar{a}),BUC^1(\Omega,\mathbb{R}^n))} \le M_0\|\bar{u}\|_{0\mathbb{X}_u(\bar{a},\Omega)}, \qquad \bar{u} \in {}_0\mathbb{X}_u(\bar{a}, \Omega),$$

where $M_0 > 0$ is independent of $\bar{a} \in (0, a]$ again due to the homogeneous initial condition, cf. [PSS07, Proposition 6.2 (b)]. Hence, assuming

$$\|(f, g, h, u_0)\|_{\mathbb{Y}_\gamma^{\alpha,\beta}(a,\Omega)} \le \varepsilon$$

we obtain

$$\|DN_{\bar{a},\gamma}^{\alpha,\beta}(\bar{u} + u^*, 0)\|_{\mathcal{B}(_0\mathbb{X}_u(\bar{a},\Omega), _0\mathbb{Y}_\gamma^{\alpha,\beta}(\bar{a},\Omega))} \le c\bar{a}^{1/p}\left(M_0\|\bar{u}\|_{0\mathbb{X}_u(\bar{a},\Omega)} + \varepsilon M\bar{M}\right)$$

for all $\bar{u} \in {}_0\mathbb{X}_u(\bar{a}, \Omega)$, which implies

$$\|K_{\bar{a},\gamma}^{\alpha,\beta}(\bar{u}, \bar{p}) - K_{\bar{a},\gamma}^{\alpha,\beta}(\bar{v}, \bar{q})\|_{0\mathbb{X}_\gamma^\beta(\bar{a},\Omega)} \le c\bar{a}^{1/p}C\left(\delta M_0 + \varepsilon M\bar{M}\right)\|\bar{u} - \bar{v}\|_{0\mathbb{X}_\gamma^\beta(\bar{a},\Omega)}$$

for all $(\bar{u}, \bar{p}), (\bar{v}, \bar{q}) \in {}_0\mathbb{X}_\gamma^\beta(\bar{a}, \Omega)$ with $\|\bar{u}\|_{0\mathbb{X}_u(\bar{a},\Omega)}, \|\bar{v}\|_{0\mathbb{X}_u(\bar{a},\Omega)} \le \delta$. Here,

$$C := \sup\left\{\|_0S_{\bar{a},\gamma}^{\alpha,\beta}\|_{\mathcal{B}(_0\mathbb{Y}_\gamma^{\alpha,\beta}(\bar{a},\Omega), _0\mathbb{X}_\gamma^\beta(\bar{a},\Omega))} : \bar{a} \in (0, a]\right\}.$$

Finally,

$$\|K_{\bar{a},\gamma}^{\alpha,\beta}(0, 0)\|_{0\mathbb{X}_\gamma^\beta(\bar{a},\Omega)} \le C\|N_{a,\gamma}^{\alpha,\beta}(u^*, 0)\|_{0\mathbb{Y}_\gamma^{\alpha,\beta}(a,\Omega)} \le \varepsilon^2 CM^2\bar{M},$$

which implies

$$\|K_{\bar{a},\gamma}^{\alpha,\beta}(\bar{u}, \bar{p})\|_{0\mathbb{X}_\gamma^\beta(\bar{a},\Omega)} \le c\bar{a}^{1/p}\delta C\left(\delta M_0 + \varepsilon M\bar{M}\right) + \varepsilon^2 CM^2\bar{M}$$

for all $(\bar{u}, \bar{p}) \in {}_0\mathbb{X}_\gamma^\beta(\bar{a}, \Omega)$ with $\|\bar{u}\|_{0\mathbb{X}_\gamma^\beta(\bar{a},\Omega)} \le \delta$.

3.52 Hence, Theorem 3.46 is proved by the contraction mapping principle as follows: We set $\bar{a} = a$, choose $\delta > 0$, such that

$$ca^{1/p}\delta C\left(\tfrac{1}{2} + M_0\right) \le \frac{1}{2}$$

and then $\varepsilon > 0$, such that

$$\varepsilon M \bar{M}, \ \varepsilon^2 C M^2 \bar{M} \leq \frac{\delta}{2}.$$

This way $K_{a,\gamma}^{\alpha,\beta}$ constitutes a contraction in the closed ball with radius δ. Hence, the proof of Theorem 3.46 is complete. □

3.53 Analogously, Theorem 3.48 is proved by the contraction mapping principle as follows: We have $\varepsilon > 0$ given by the data, we set $\delta/2 := \varepsilon^2 C M^2 \bar{M}$ and then choose an $\bar{a} \in (0, a]$, such that

$$c\bar{a}^{1/p} C \left(\delta M_0 + \varepsilon M \bar{M} \right) \leq \frac{1}{2}.$$

This way $K_{\bar{a},\gamma}^{\alpha,\beta}$ constitutes a contraction in the closed ball with radius δ. Thus, the proof of Theorem 3.48 is complete. □

Remarks

3.54 The Fujita-Kato approach to the Navier-Stokes equations subject to a Dirichlet boundary condition has recently been generalised by M. MITREA and S. MONNIAUX to Lipschitz domains, cf. [MM08]. In a subsequent publication together with M. WRIGHT this approach is applied to the Navier-Stokes equations subject to a Neumann boundary condition in Lipschitz domains, cf. [MMW11].

3.55 As has been mentioned in Remark 2.27 there exists a complete L_p-theory including an $\mathcal{H}^\infty$-calculus for the Stokes equations with Robin boundary condition, which is due to J. SAAL, cf. [Saa06, Saa07], the thesis [Saa03] and Remark 2.27.

3.56 A complete L_p-theory for a class of non-Newtonian fluids was recently developed by D. BOTHE and J. PRÜSS, cf. [BP07]. This theory covers Dirichlet type boundary conditions as well as Neumann and slip type boundary conditions, which are modelled based on the stress tensor for non-Newtonian fluids. In fact, these boundary conditions are a generalisation of our energy preserving boundary conditions $(B \,|\, a, \, \Omega)^{+1,0}$ and $(B \,|\, a, \, \Omega)^{+1,+1}$ to the non-Newtonian case and behave energy preserving, too.

3.57 Classical sources concerning the theory of the Lebesgue spaces L_p, the Bessel potential spaces H_p^s, as well as the Sobolev respectively Sobolev-Slobodeckij spaces W_p^s include the works of J. NEČAS and H. TRIEBEL, cf. [Neč67, Tri98]. A more recent introduction to these scales of function spaces is given by R. A. ADAMS and J. J. F. FOURNIER, cf. [AF03]. However, it should be noted that all these sources focus on real respectively complex valued function spaces. The Banach space valued case is dealt with in the works of H. AMANN, cf. [Ama95, Ama09], where a complete theory of Banach space valued (tempered) distributions and function spaces is developed including the anisotropic case.

3.58 For $1 < p < \infty$ and $m = 1, 2, \ldots$ the Bessel potential spaces $H_p^m(\Omega, X)$ on a domain $\Omega \subseteq \mathbb{R}^n$ are defined by restricting the functions in $H_p^m(\mathbb{R}^n, X)$ to Ω. As usual, X denotes an arbitrary Banach space. Now, $H_p^m(\mathbb{R}^n, X) = W_p^m(\mathbb{R}^n, X)$, provided the underlying space X is a UMD-space, and, hence, $H_p^m(\Omega, X) = W_p^m(\Omega, X)$, provided there exists a bounded linear extension operator

$$\mathcal{E} : W_p^m(\Omega, X) \longrightarrow W_p^m(\mathbb{R}^n, X).$$

By [AF03, Theorem 4.32], which carries over to the Banach space valued case, this is true, whenever Ω satisfies a uniform cone property. In particular, a bounded interval $(0, a)$, the halfspace $\mathbb{R}_+^n$ and a bounded domain $\Omega \subseteq \mathbb{R}^n$ with boundary of class C^1 have this property. A similar argumentation may be employed to deal with a bent halfspace of class BUC^{m-}, cf. Remark 3.59. It should be noted that the UMD-property of the underlying space enters the discussion, because the identity $H_p^m(\mathbb{R}^n, X) = W_p^m(\mathbb{R}^n, X)$ is based on operator valued Fourier multiplier theorems. For a detailed introduction to this theory we refer to the monograph by R. DENK, M. HIEBER and J. PRÜSS, cf. [DHP03]. For a short introduction of this and related topics see Section 4.4.

3.59 The proof of Theorem 3.30 employs a localisation procedure, which starts with an analysis of the Stokes equations in a halfspace

$$\mathbb{R}_+^n := \mathbb{R}^{n-1} \times (0, \infty)$$

and transfers the results via bent halfspaces

$$\mathbb{R}_\omega^n := \left\{ (x, y) \in \mathbb{R}^{n-1} \times \mathbb{R} : y > \omega(x) \right\}$$

to a bounded domain. Of course, we have to assume the boundary of such a bent halfspace to be sufficiently regular by requiring $\omega \in BUC^{3-}(\mathbb{R}^{n-1})$. The Bessel potential, (homogeneous) Sobolev and Sobolev-Slobodeckij spaces on a (bent) halfspace are defined as on a bounded domain, see 3.3, 3.5 and 3.11. This way Propositions 3.9, 3.14 and 3.17 are also valid for (bent) halfspaces, cf. Remarks 3.60, 3.62, 3.63 and 3.64. Thus, we obtain the same necessary regularity conditions as for a bounded domain. However, since a (bent) halfspace is an unbounded domain, we have to modify the solution space $\mathbb{X}_{p,-\infty}^0(a, \Omega)$, if $\Omega = \mathbb{R}_+^n$ or $\Omega = \mathbb{R}_\omega^n$. In these cases the embedding

$$\dot{H}_p^1(\Omega) \hookrightarrow \left\{ \phi \in L_{p,loc}(\Omega) : \begin{array}{c} \phi \in H_p^1(\Omega'), \ \Omega' \subseteq \Omega \text{ open and bounded} \\ \text{with boundary of class } C^1 \end{array} \right\}$$

is valid, cf. 3.5, but $H_p^1(\Omega) \subsetneq \dot{H}_p^1(\Omega)$. Moreover, Poincaré's inequality is not available. To ensure the uniqueness of the pressure in these cases, we have to employ the space $\hat{H}_p^1(\Omega) := \dot{H}_p^1(\Omega)/\mathbb{R}$ and set

$$\mathbb{X}_{p,-\infty}^0(a, \Omega) := L_p((0, a), \hat{H}_p^1(\Omega)).$$

The solution spaces

$$\mathbb{X}_u(a, \Omega), \ {}_0\mathbb{X}_u(a, \Omega), \ \mathbb{X}_{p,\gamma}^{\pm 1}(a, \Omega)$$

and the data spaces

$$\mathbb{Y}_\gamma^{\alpha,\beta}(a, \Omega), \ {}_0\mathbb{Y}_\gamma^{\alpha,\beta}(a, \Omega), \ \mathbb{Y}_f(a, \Omega), \ \mathbb{Y}_g(a, \Omega), \ \mathbb{Y}_{h,\gamma}^{\alpha,\beta}(a, \Gamma), \ \mathbb{Y}_0(\Omega)$$

are defined as for a bounded domain. Moreover, we set

$$\mathbb{X}^\beta(a,\,\Omega) := \mathbb{X}_u(a,\,\Omega) \times \mathbb{X}^\beta_{p,\gamma}(a,\,\Omega) \quad \text{and} \quad {}_0\mathbb{X}^\beta(a,\,\Omega) := {}_0\mathbb{X}_u(a,\,\Omega) \times \mathbb{X}^\beta_{p,\gamma}(a,\,\Omega)$$

for $\alpha,\,\beta \in \{-1,\,0,\,+1\}$, $\gamma = -\infty$ for $\beta = 0$, $\gamma \in \{-\infty\} \cup [0,\,1/2 - 1/2p]$ for $\beta = +1$, $\gamma \in \{-\infty\} \cup [0,\,\infty)$ for $\beta = -1$ and, of course, with the modified pressure space for $\beta = 0$. Finally, the necessary compatibility conditions derived in Section 3.3 remain valid for (bent) halfspaces. Concerning (C$_3$) note that $H^1_p(\Omega) \subsetneq \dot{H}^1_p(\Omega)$ is at least a dense subspace in the (bent) halfspace case.

3.60 The mixed derivative theorem, Proposition 3.9, goes back to the work of P. E. So-BOLEVSKII. In fact, in [Sob75] it is proved that for two sectorial operators A and B in a Banach space X with spectral angles ϕ_A and ϕ_B, which commute in the resolvent sense and satisfy the parabolicity condition $\phi_A + \phi_B < \pi$, the coercivity estimate

$$\|Ax\|_X + t\|Bx\|_X \leq M\|Ax + tBx\|_X, \qquad x \in D(A) \cap D(B),\ t > 0$$

implies the estimate

$$\|A^{(1-\theta)}B^\theta x\|_X \leq C\|Ax + Bx\|_X, \qquad x \in D(A) \cap D(B),\ \theta \in [0,\,1],$$

provided the operators $A + tB$ are closed with their natural domain $D(A) \cap D(B)$ for all $t > 0$. Now, for $1 < p < \infty$ an application of this result to the operators $A = 1 + \partial_t$ with domain $D(A) = H^1_p(\mathbb{R},\,L_p(\mathbb{R}^n))$ and $B = 1 - \Delta$ with domain $D(B) = L_p(\mathbb{R},\,H^2_p(\mathbb{R}^n))$, which are sectorial in $X = L_p(\mathbb{R} \times \mathbb{R}^n)$ and admit a bounded $\mathcal{H}^\infty$-calculus with angles $\phi^\infty_A = \frac{\pi}{2}$ and $\phi^\infty_B = 0$, yields

$$H^1_p(\mathbb{R},\,L_p(\mathbb{R}^n)) \cap L_p(\mathbb{R},\,H^2_p(\mathbb{R}^n)) \hookrightarrow H^{1-\theta}_p(\mathbb{R},\,H^{2\theta}_p(\mathbb{R}^n)), \qquad \theta \in [0,\,1],$$

since the functional

$$\phi \mapsto \|A^{1-\theta}B^\theta \phi\|_{L_p(\mathbb{R} \times \mathbb{R}^n)} : H^{1-\theta}_p(\mathbb{R},\,H^{2\theta}_p(\mathbb{R}^n)) \longrightarrow [0,\,\infty)$$

defines an equivalent norm in $H^{1-\theta}_p(\mathbb{R},\,H^{2\theta}_p(\mathbb{R}^n))$. The corresponding embedding for the time-space cylinder $(0,\,a) \times \Omega$ then follows, whenever $a > 0$ and $\Omega \subseteq \mathbb{R}^n$ is a domain, which admits a bounded, linear extension operator

$$\mathcal{E} : H^s_p(\Omega) \longrightarrow H^s_p(\mathbb{R}^n), \qquad s \in [0,\,2],$$

which, however, is true for every domain $\Omega \subseteq \mathbb{R}^n$ by definition. Hence, the first embedding asserted in Proposition 3.9 is valid for $a > 0$, $\tau = 1$, $\sigma = 2$ and all domains $\Omega \subseteq \mathbb{R}^n$. An alternative proof may be given based on interpolation theory. Indeed, for any interpolation couple $\{X_0,\,X_1\}$ of Banach spaces we have

$$X_0 \cap X_1 \hookrightarrow [X_0,\,X_1]_\theta, \qquad X_0 \cap X_1 \hookrightarrow (X_0,\,X_1)_{\theta,p}, \qquad \theta \in (0,\,1),\ 1 < p < \infty$$

by definition. Thus, if we set $X_0 := H^\tau_p(\mathbb{R},\,L_p(\mathbb{R}^n))$ and $X_1 := L_p(\mathbb{R},\,H^{2\sigma}_p(\mathbb{R}^n))$, the first embedding asserted in Proposition 3.9 immediately follows for $a > 0$, $\tau \in (0,\,1]$, $\sigma \in (0,\,2]$ and all domains $\Omega \subseteq \mathbb{R}^n$ by interpolation theory, cf. [Ama09, Theorem 3.7.1], and the above extension argument. The analogous embeddings, where either the Bessel potential spaces on the right hand side or on both sides are substituted by the corresponding Sobolev-Slobodeckij spaces, may be obtained by the very same arguments. However, in this case a restriction on the domain $\Omega \subseteq \mathbb{R}^n$ is necessary, since a linear extension operator

$$\mathcal{E} : W_p^s(\Omega) \longrightarrow W_p^s(\mathbb{R}^n), \quad s \in [0, 2]$$

is required. Such an operator exists for a halfspace, a bent halfspace of type BUC^2 and a bounded domain with boundary of class C^2 as follows by [AF03, Theorem 4.26] and real interpolation. Hence, the first embedding asserted in Proposition 3.9 in the modified version as mentioned in 3.11 is valid for all these domains. The second embedding asserted in Proposition 3.9 together with the assertion on the uniformity of the embedding constants follows by an application of Corollary 4.67. Note that an analogous embedding is again also valid, if the Bessel potential spaces either on the right hand side or on both sides are substituted by the corresponding Sobolev-Slobodeckij spaces, provided $\tau \in (1/p, 1]$ in the second case, see Remark 4.79. A detailed proof of the properties of the operators $A = 1 + \partial_t$ and $B = 1 - \Delta$ and the embeddings asserted in Proposition 3.9 for the case $\Omega = \mathbb{R}^n$ was given by R. DENK, J. SAAL and J. SEILER, cf. [DSS08]. A similar result was proved by J. PRÜSS, J. SAAL and G. SIMONETT, cf. [PSS07, Lemma 6.3]. For a detailed introduction of the theory of sectorial operators we refer to the monograph by R. DENK, M. HIEBER and J. PRÜSS, cf. [DHP03]. For a short introduction of this theory and related topics see Section 4.4.

3.61 If $1 < p < \infty$ and $0 < \tau < 1/p$, then the Sobolev-Slobodeckij space $W_p^\tau((0, a), X)$ does not possess a trace at time $t = 0$. Here, X denotes an arbitrary Banach space. We therefore set

$$_0W_p^\tau((0, a), X) := W_p^\tau((0, a), X), \quad \tau \in (0, 1/p).$$

Also note, that the space of compactly supported, smooth functions $C_0^\infty((0, a), X)$ is a dense subspace of $W_p^\tau((0, a), X)$ for $\tau \in (0, 1/p)$.

3.62 Part (i) of the trace theorem, Proposition 3.14, is a consequence of [Mar87, Theorem 2]. Another consequence is, that the trace operator

$$[\,\cdot\,]_\Gamma : W_p^{1+1/p}(\Omega, X) \longrightarrow B_{p,p}^1(\Gamma, X),$$

$B_{p,p}^1(\Gamma, X)$ denoting the Besov space of order one, is bounded and surjective. However, $B_{p,p}^1(\Gamma, X) \neq W_p^1(\Gamma, X)$ and the existence of a bounded linear right inverse is not provided by [Mar87, Theorem 2]. This is the reason for excluding this case in Proposition 3.14.

3.63 Part (ii) of the trace theorem, Proposition 3.14, may not be found in the literature in this form. For a bounded domain $\Omega \subseteq \mathbb{R}^n$ with boundary of class C^1 we have $\dot{H}_p^1(\Omega, X) = H_p^1(\Omega, X)$ algebraically, cf. 3.5. Analogously,

$$\dot{W}_p^{1-1/p}(\Gamma, X) = W_p^{1-1/p}(\Gamma, X)$$

algebraically and the trace operator is bounded and surjective due to Poincaré's inequality. Based on these properties, Propositions 4.4 and 4.6 provide the existence of a bounded linear right inverse, since the proofs only employ the boundedness and surjectivity of the trace operator. For a halfspace, boundedness and surjectivity of the trace operator as

well as the existence of a bounded linear right inverse have been proved by L. D. KU-
DRJAVCEV, cf. [Kud68a, Kud68b, Theorems 2.4' and Theorem 2.7, Corollary 1]. These
properties may then be transferred to a bent halfspace of type BUC^1 by the techniques
introduced in Chapter 6. Moreover, Propositions 4.4 and 4.6 also apply in these cases
and provide the existence of a bounded linear right inverse for the trace operator based
on its boundedness and surjectivity.

3.64 A proof of the anisotropic trace theorem, Proposition 3.17, may be obtained by real
interpolation. Indeed, if $\{X_0, X_1\}$ is an interpolation couple of Banach spaces, $m \in \mathbb{N}$,
$\theta \in (0, 1)$ and $1 < p < \infty$, then the space

$$\mathbb{E}_p^m(X_0, X_1, \theta) := \left\{ \phi \in L_{p,loc}(\mathbb{R}_+, X_0 + X_1) : \begin{array}{l} y \mapsto y^{1-\theta-1/p}\phi^{(m)}(y) \in L_p(\mathbb{R}_+, X_0) \\ y \mapsto y^{1-\theta-1/p}\phi(y) \in L_p(\mathbb{R}_+, X_1) \end{array} \right\}$$

constitutes a Banach space, provided it is canonically normed via

$$\|\phi\|_{\mathbb{E}_p^m(X_0, X_1, \theta)} := \left(\|y \mapsto y^{1-\theta-1/p}\phi^{(m)}(y)\|_{L_p(\mathbb{R}_+, X_0)}^p + \|y \mapsto y^{1-\theta-1/p}\phi(y)\|_{L_p(\mathbb{R}_+, X_1)}^p \right)^{1/p}$$

for $\phi \in \mathbb{E}_p^m(X_0, X_1, \theta)$, cf. [Tri98, Lemma 1.8.1 (a)]. Moreover, the embedding

$$\mathbb{E}_p^m(X_0, X_1, \theta) \hookrightarrow BC^{m-1}([0, 1], X_0 + X_1)$$

is valid, cf. [Tri98, Lemma 1.8.1 (b)], and the trace spaces

$$\mathbb{T}_p^m(X_0, X_1, k, \theta) := \left\{ \phi^{(k)}(0) : \phi \in \mathbb{E}_p^m(X_0, X_1, \theta) \right\}, \qquad k = 0, 1, \ldots, m-1$$

are well-defined and constitute Banach spaces, provided they are canonically normed via

$$\|\psi\|_{\mathbb{T}_p^m(X_0, X_1, k, \theta)} := \inf \left\{ \|\phi\|_{\mathbb{E}_p^m(X_0, X_1, \theta)} : \phi \in \mathbb{E}_p^m(X_0, X_1, \theta),\ \psi = \phi^{(k)}(0) \right\}$$

for $\psi \in \mathbb{T}_p^m(X_0, X_1, k, \theta)$. This construction is also known as the *trace method for real
interpolation*, since

$$\mathbb{T}_p^m(X_0, X_1, k, \theta) \cong (X_0, X_1)_{\frac{(m-1)-k+\theta}{m}, p}, \qquad k = 0, 1, \ldots, m-1,$$

i. e. this is one of many possibilities to construct the real interpolation functor, cf.[Tri98,
Theorem 1.8.2]. Now, we set

$$X_0 = L_p(\mathbb{R}_+ \times \mathbb{R}^{n-1}), \qquad X_1 = H_p^1(\mathbb{R}_+, L_p(\mathbb{R}^{n-1})) \cap L_p(\mathbb{R}_+, H_p^2(\mathbb{R}^{n-1}))$$

and observe that

$$H_p^1(\mathbb{R}_+, L_p(\mathbb{R}_+^n)) \cap L_p(\mathbb{R}_+, H_p^2(\mathbb{R}_+^n)) \cong \mathbb{E}_p^m(X_0, X_1, 1-1/p)$$

for all $1 < p < \infty$ with $m = 2$. Thus, the trace spaces on the halfspace boundary $\partial\mathbb{R}_+^n$
are determined as

$$\mathbb{T}_p^m(X_0, X_1, k, 1-1/p) \cong (X_0, X_1)_{1-\frac{1}{2p}, p}$$
$$\cong W_p^{1-1/2p}(\mathbb{R}_+, L_p(\partial\mathbb{R}_+^n)) \cap L_p(\mathbb{R}_+, W_p^{2-1/p}(\partial\mathbb{R}_+^n))$$

with $k = 0$. Part (i) of Proposition 3.17 may now be obtained by a localisation argument. Part (ii) of Proposition 3.17 may be proved along the same lines. We first set

$$X_0 = L_p(\mathbb{R}_+ \times \mathbb{R}^{n-1}), \qquad X_1 = H_p^{1/2}(\mathbb{R}_+, L_p(\mathbb{R}^{n-1})) \cap L_p(\mathbb{R}_+, H_p^1(\mathbb{R}^{n-1})),$$

$m = 1$ and $k = 0$ to obtain the asserted embedding for a halfspace and then apply a localisation argument.

3.65 A short and elegant proof of the embedding employed in 3.19 is due to H. AMANN, cf. [Ama95, Section III.1.4]. First we employ the trace method of real interpolation as presented in Remark 3.64 to infer that the trace operators

$$[\,\cdot\,]_t : H_p^1(\mathbb{R}, L_p(\Omega)) \cap L_p(\mathbb{R}, H_p^2(\Omega)) \longrightarrow W_p^{2-2/p}(\Omega),$$

$$[\phi]_t := \phi(t), \qquad \phi \in H_p^1(\mathbb{R}, L_p(\Omega)) \cap L_p(\mathbb{R}, H_p^2(\Omega)), \qquad t \in \mathbb{R}$$

are bounded. Moreover, the translation group $\{\,\tau_s\,:\,s \in \mathbb{R}\,\}$, which is given as

$$(\tau_s \phi)(t) := \phi(t + s), \qquad \phi \in H_p^1(\mathbb{R}, L_p(\Omega)) \cap L_p(\mathbb{R}, H_p^2(\Omega)), \qquad s, t \in \mathbb{R},$$

is continuous. Thus, the asserted embedding for the time interval $\mathbb{R}$ follows by an application of [Ama95, Proposition 1.4.2] and the corresponding embedding for a time interval $(0, a)$ follows by an extension argument.

3.66 Concerning the domain $D(A^{\alpha,\beta})$ of the Stokes operator as defined in 3.38, we have $C_{0,\sigma}^\infty(\Omega) \subseteq D(A^{\alpha,0})$ for all $\alpha \in \{-1, 0, +1\}$. Hence, $D(A^{\alpha,0})$ is a dense subspace of in $X_p^0(\Omega)$. For $\beta = \pm 1$, we have to observe, that the operator

$$B^\alpha : \{\, v \in H_p^2(\Omega, \mathbb{R}^n) \,:\, P_\Gamma B^\alpha v = 0, \ [\operatorname{div} v]_\Gamma = 0 \,\} \subseteq L_p(\Omega, \mathbb{R}^n) \longrightarrow L_p(\Omega, \mathbb{R}^n),$$

$B^\alpha = -\mu\Delta$, has the property of maximal L_p-regularity on finite time intervals for all $1 < p < \infty$ with $p \neq \frac{3}{2}, 3$, cf. Proposition 4.21. Invoking the theory of abstract parabolic evolution equations again, we infer, that there exists some $\omega > 0$, such that the operator $\omega + B^\alpha$ is sectorial in $L_p(\Omega, \mathbb{R}^n)$, cf. [Prü03, Proposition 1.2]. Thus, given an $f \in X_p^{\pm 1}(\Omega) = L_{p,s}(\Omega)$, we obtain for every $k = 1, 2, \ldots$ a maximal regular solution $v_k \in H_p^2(\Omega)$ to the elliptic problem

$$kv_k + \omega v_k - \mu\Delta v_k = kf \qquad \text{in } \Omega,$$

$$P_\Gamma B^\alpha v_k = 0 \qquad \text{on } \partial\Omega,$$

$$[\operatorname{div} v_k]_\Gamma = 0 \qquad \text{on } \partial\Omega.$$

By construction, we have

$$k \operatorname{div} v_k + \omega \operatorname{div} v_k - \mu\Delta \operatorname{div} v_k = 0 \qquad \text{in } \Omega,$$

$$[\operatorname{div} v_k]_\Gamma = 0 \qquad \text{on } \partial\Omega,$$

which implies $\operatorname{div} v_k = 0$ for $k = 1, 2, \ldots$ by uniqueness of weak solutions to the Laplace equation with Dirichlet boundary condition. Therefore

$$D(A^{\alpha,\pm 1}) \ni v_k = k(k + \omega + B^\alpha)^{-1} f \to f \qquad \text{in } L_p(\Omega, \mathbb{R}^n) \text{ as } \quad k \to \infty,$$

cf. [DHP03, Proposition 1.2]. In particular, $D(A^{\alpha,\pm 1})$ is a dense subspace of $X_p^{\pm 1}(\Omega)$ for all $\alpha \in \{-1, 0, +1\}$. For a detailed introduction of the theory of sectorial operators we refer to the monograph by R. DENK, M. HIEBER and J. PRÜSS, cf. [DHP03]. For a short introduction of this theory and related topics see Section 4.4.

References

[AF03]   R. A. ADAMS and J. J. F. FOURNIER: Sobolev Spaces, *Pure and Applied Mathematics*, vol. 140. Academic Press, 2nd ed., 2003.

[Ama95]   H. AMANN: Linear and Quasilinear Parabolic Problems. Volume I. Abstract Linear Theory, *Monographs in Mathematics*, vol. 89. Birkhäuser, 1995.

[Ama09]   H. AMANN: Anisotropic Function Spaces and Maximal Regularity for Parabolic Problems. Part 1: Function Spaces, *Jindřich Nečas Center for Mathematical Modeling Lecture Notes*, vol. 6. MATFYZPRESS, 2009.

[BF07]   F. BOYER and P. FABRIE: *Outflow Boundary Conditions for the Incompressible Non-Homogeneous Navier-Stokes Equations*. Discrete Contin. Dyn. Syst. Ser. B, 7 (2), 219–250, 2007.

[BM88]   W. BORCHERS and T. MIYAKAWA: L^2 *Decay for the Navier-Stokes Flow in Halfspaces*. Math. Ann., 282, 139–155, 1988.

[BNP04]   H. BELLOUT, J. NEUSTUPA, and P. PENEL: *On the Navier-Stokes Equation with Boundary Conditions Based on Vorticity*. Math. Nachr., 269-270, 59–72, 2004.

[BP07]   D. BOTHE and J. PRÜSS: L_p-*Theory for a Class of Non-Newtonian Fluids*. SIAM J. Math. Anal., 39, 379–421, 2007.

[CMP94]   C. CONCA, F. MURAT, and O. PIRONNEAU: *The Stokes and Navier-Stokes Equations with Boundary Conditions Involing the Pressure*. Japan J. Math., 20 (2), 279–318, 1994.

[CPPT95] C. CONCA, C. PARÈS, O. PIRONNEAU, and M. THIRIET: *Navier-Stokes Equations with Imposed Pressure and Velocity Fluxes*. Int. J. Numer. Meth. Fluids, 20, 267–287, 1995.

[DHP01]   W. DESCH, M. HIEBER, and J. PRÜSS: L^p-*Theory of the Stokes Equation in a Half-Space*. J. Evol. Equ., 1, 115–142, 2001.

[DHP03]   R. DENK, M. HIEBER, and J. PRÜSS: $\mathcal{R}$-Boundedness, Fourier-Multipliers and Problems of Elliptic and Parabolic Type, *Mem. Amer. Math. Soc.*, vol. 166. American Mathematical Society, 2003.

[DSS08]   R. DENK, J. SAAL, and J. SEILER: *Inhomogeneous Symbols, the Newton Polygon, and Maximal L_p-Regularity*. Russian J. Math. Phys., 15 (2), 171–192, 2008.

[Eva98]   L. C. EVANS: Partial Differential Equations. American Mathematical Society, 1998.

[FK62]   H. FUJITA and T. KATO: *On the Nonstationary Navier-Stokes System*. Rend. Sem. Mat. Univ. Padova, 32, 234–260, 1962.

[FK64]   H. FUJITA and T. KATO: *On the Navier-Stokes Initial Value Problem. I*. Arch. Ration. Mech. Anal., 16 (4), 269–315, 1964.

[Gal94a] G. P. GALDI: An Introduction to the Mathematical Theory of the Navier-Stokes Equations, Volume 1: Linearized Steady Problems. Springer, 1994.

[Gal94b] G. P. GALDI: An Introduction to the Mathematical Theory of the Navier-Stokes Equations, Volume 2: Nonlinear Steady Problems. Springer, 1994.

[Gig85]   Y. GIGA: *Domains of Fractional Powers of the Stokes Operator in L_r Spaces*. Arch. Ration. Mech. Anal., 89, 251–265, 1985.

[GM85]   Y. GIGA and T. MIYAKAWA: *Solutions in L_r of the Navier-Stokes Initial Value Problem*. Arch. Ration. Mech. Anal., 89, 269–281, 1985.

[Gru95a] G. GRUBB: *Nonhomogeneous Time-Dependent Navier-Stokes Problems in L_p-Sobolev Spaces*. Differential and Integral Equations, 8 (5), 1013–1046, 1995.

[Gru95b] G. GRUBB: *Parameter-Elliptic and Parabolic Pseudodifferential Boundary Problems in Global L_p-Sobolev Spaces.* Math. Zeitschr., 218 (1), 43–90, 1995.

[Gru98] G. GRUBB: *Nonhomogeneous Navier-Stokes Problems in L_p Sobolev Spaces over Exterior and Interior Domains.* In: Theory of the Navier-Stokes Equations, Ser. Adv. Math. Appl. Sci., vol. 47 (J. G. HEYWOOD, K. MASUDA, R. RAUTMANN, and V. A. SOLONNIKOV, eds.), (46–63), World Scientific Publishing, 1998.

[GS90a] G. GRUBB and V. A. SOLONNIKOV: *Reduction of Basic Initial-Boundary Value Problems for the Stokes Equation to Initial-Boundary Value Problems for Systems of Pseudodifferential Equations.* J. Sov. Math., 49 (5), 1140–1147, 1990.

[GS90b] G. GRUBB and V. A. SOLONNIKOV: *Solution of Parabolic Pseudo-Differential Initial Boundary Value Problems.* J. Diff. Equ., 87 (2), 256–304, 1990.

[GS91a] G. GRUBB and V. A. SOLONNIKOV: *Boundary Value Problems for the Nonstationary Navier-Stokes Equations Treated by Pseudo-Differential Methods.* Math. Scand., 69 (2), 217–290, 1991.

[GS91b] G. GRUBB and V. A. SOLONNIKOV: *Reduction of Basic Initial-Boundary Value Problems for the Navier-Stokes Equations to Nonlinear Parabolic Systems of Pseudodifferential Equations.* J. Sov. Math., 56 (2), 2300–2308, 1991.

[Hop51] E. HOPF: *Über die Anfangswertaufgabe für die hydrodynamischen Grundgleichungen.* Math. Nachr., 4, 213–231, 1951.

[KS04] T. KUBO and Y. SHIBATA: *On some Properties of Solutions to the Stokes Equation in the Half-Space and Perturbed Half-Space.* Quad. Math., 15, 149–220, 2004.

[KS05a] T. KUBO and Y. SHIBATA: *On the Stokes and Navier-Stokes Equations in a Perturbed Half-Space.* Adv. Differential Equations, 10, 695–720, 2005.

[KS05b] T. KUBO and Y. SHIBATA: *On the Stokes and Navier-Stokes Flows in a Perturbed Half-Space.* Banach Center Publ., 70, 157–167, 2005.

[Kud68a] L. D. KUDRJAVCEV: *Imbedding Theorem for a Class of Functions Defined on the Entire Space or on a Halfspace. I.* Amer. Math. Soc. Transl., 74, 199–225, 1968.

[Kud68b] L. D. KUDRJAVCEV: *Imbedding Theorem for a Class of Functions Defined on the Entire Space or on a Halfspace. II.* Amer. Math. Soc. Transl., 74, 227–260, 1968.

[Lad69] O. A. LADYZHENSKAYA: The Mathematical Theory of Viscous Incompressible Flow. Gordon and Breach, 1969.

[Ler34a] J. LERAY: *Essai sur les Mouvements Plans d'un Liquide Visqueux que Limitent des Parois.* J. Math. Pures Appl., 13, 331–418, 1934.

[Ler34b] J. LERAY: *Sur le Mouvement d'un Liquide Visqueux Emplissant l'Espace.* Acta Math., 63, 193–248, 1934.

[Mar87] J. MARSCHALL: *The Trace of Sobolev-Slobodeckij Spaces on Lipschitz Domains.* Manuscripta Math., 58, 47–65, 1987.

[McC81] M. MCCRACKEN: *The Resolvent Problem for the Stokes Equation on Halfspaces in L_p.* SIAM J. Math. Anal., 12, 201–228, 1981.

[MM08] M. MITREA and S. MONNIAUX: *The Regularity of the Stokes Operator and the Fujita-Kato Approach to the Navier-Stokes Initial Value Problem in Lipschitz Domains.* J. Funct. Anal., 254, 1522–1574, 2008.

[MMW11] M. MITREA, S. MONNIAUX, and M. WRIGHT: *The Stokes Operator with Neumann Boundary Conditions in Lipschitz Domains.* J. Math. Sci. (N. Y.), 176 (3), 2011.

[Neč67] J. Nečas: Les Méthodes Directes en Théorie des Équations Elliptiques. Academia, 1967.

[NP07] J. Neustupa and P. Penel: *On Regularity of a Weak Solution to the Navier-Stokes Equation with Generalized Impermeability Boundary Conditions.* Nonlinear Anal., 66 (8), 1753–1769, 2007.

[NP08] J. Neustupa and P. Penel: *The Navier-Stokes Equation with Inhomogeneous Boundary Conditions Based on Vorticity.* Banach Center Publ., 81, 321–335, 2008.

[NP10] J. Neustupa and P. Penel: *Local in Time Strong Solvability of the Non-Steady Navier-Stokes Equations with Navier's Boundary Condition and the Question of the Inviscid Limit.* C. R., Math., Acad. Sci. Paris, 348 (19-20), 1093–1097, 2010.

[Prü03] J. Prüss: *Maximal Regularity for Evolution Equations in L_p-Spaces.* Conf. Sem. Mat. Univ. Bari, 285, 1–39, 2003.

[PSS07] J. Prüss, J. Saal, and G. Simonett: *Existence of Analytic Solutions for the Classical Stefan Problem.* Math. Ann., 338, 703–755, 2007.

[Saa03] J. Saal: Robin Boundary Conditions and Bounded $\mathcal{H}^\infty$-Calculus for the Stokes Operator. Logos, 2003.

[Saa06] J. Saal: *Stokes and Navier-stokes Equations with Robin Boundary Conditions in a Half-Space.* J. Math. Fluid Mech., 8, 211–241, 2006.

[Saa07] J. Saal: *The Stokes Operator with Robin Boundary Conditions in Solenoidal Subspaces of $L^1(\mathbb{R}_+^n)$ and $L^\infty(\mathbb{R}_+^n)$.* Commun. Partial Differ. Equations, 32 (3), 343–373, 2007.

[Sob64] P. E. Sobolevskii: *Study of Navier-Stokes Equations by Methods of the Theory of Parabolic Equations.* Soviet Math. Dokl., 5, 720–723, 1964.

[Sob75] P. E. Sobolevskii: *Fractional Powers of Coercively Positive Sums of Operators.* Soviet Math. Dokl., 16, 1638–1641, 1975.

[Soh01] H. Sohr: The Navier-Stokes Equations. Birkhäuser, 2001.

[Sol77a] V. A. Solonnikov: *Estimates for Solutions of Nonstationary Navier-Stokes Equations.* J. Sov. Math., 8, 467–529, 1977.

[Sol77b] V. A. Solonnikov: *Solvability of a Problem on the Motion of a Viscous Incompressible Fluid Bounded by a Free Surface.* Math. USSR Izv., 11, 1323–1358, 1977.

[Sol78] V. A. Solonnikov: *On the Solvability of the Second Initial-Boundary Value Problem for the Linear Nonstationary Navier-stokes System.* J. Sov. Math., 10, 141–193, 1978.

[Sol88] V. A. Solonnikov: *Unsteady Motion of a finite Mass of Fluid, Bounded by a Free Surface.* J. Sov. Math., 40, 672–686, 1988.

[Sol91] V. A. Solonnikov: *On an Initial-Boundary Value Problem for the Stokes System Arising in the Study of a Problem with a Free Surface.* Proc. Steklov Inst. Math., 3, 191–239, 1991.

[SS03] Y. Shibata and S. Shimizu: *On a Resolvent Estimate for the Stokes Sytem with Neumann Boundary Condition.* Differential Integral Equations, 16, 385–426, 2003.

[SS05] Y. Shibata and S. Shimizu: *L_p-L_q Maximal Regularity and Viscous Incompressible Flows with Free Surface.* Proc. Japan Acad. Ser. A Math. Sci., 81 (9), 151–155, 2005.

[SS07a] Y. Shibata and R. Shimada: *On the Stokes Equation with Robin Boundary Condition.* Adv. Stud. Pure Math., 47-1, 341–348, 2007.

[SS07b] Y. Shibata and S. Shimizu: *Decay Properties of the Stokes Semigroup in Exterior Domains with Neumann Boundary Condition.* J. Math. Soc. Japan, 59 (1), 1–34, 2007.

[SS08]   Y. SHIBATA and S. SHIMIZU: *On the L_p-L_q Maximal Regularity of the Neumann Problem for the Stokes Equations in a Bounded Domain.* J. reine angew. Math., 615, 157–209, 2008.

[SS11]   Y. SHIBATA and S. SHIMIZU: *Maximal L_p-L_q Regularity for the Stokes Equations; Model Problems.* Differential Equations, 251 (2), 373–419, 2011.

[Tem77]   R. TEMAM: Navier-Stokes Equations. Noth-Holland, 1977.

[Tri98]   H. TRIEBEL: Interpolation Theory. Function Spaces. Differential Operators. Wiley-VCH, 2nd ed., 1998.

[Uka87]   S. UKAI: *A Solution Formula for the Stokes Equation in $\mathbb{R}^n_+$.* Comm. Pure Appl. Math., 40 (5), 611–621, 1987.

[Wei80]   F. B. WEISSLER: *The Navier-Stokes Initial Value Problem in L^p.* Arch. Ration. Mech. Anal., 74, 219–230, 1980.

Chapter 4

Tools and Methods

The proof of Theorem 3.30, which will be carried out in the next chapters, requires a fair amount of tools and methods, which should be collected here. First of all, it will sometimes be convenient to construct auxiliary solutions for the pressure via weak elliptic problems as employed in 3.41 and for the velocity via parabolic systems as (3.2). Hence, an L_p-theory for these types of problems is needed. Of special interest will be parabolic systems with mixed order boundary conditions, which may not be treated directly based on the available methods as presented e. g. in [DHP03]. In fact, mixed order parabolic systems are still a field of active research, cf. [DS11].

Moreover, the analysis of the Stokes equations $(S \,|\, a, \Omega)$ in a bounded domain will be based on a localisation procedure, i. e. we will start with an analysis of these equations in the halfspace

$$\mathbb{R}^n_+ := \mathbb{R}^{n-1} \times (0, \infty)$$

and transfer the results via bent halfspaces

$$\mathbb{R}^n_\omega := \left\{ (x, y) \in \mathbb{R}^{n-1} \times \mathbb{R} : y > \omega(x) \right\}$$

to a bounded domain. Of course, we have to assume the boundary of such a bent halfspace to be sufficiently regular. In particular, we are interested in bent halfspaces of type BUC^{m-} for some $m \in \mathbb{N}$, which are defined by requiring

$$\omega \in BUC^{m-}(\mathbb{R}^{n-1}).$$

Now, the treatment of the Stokes equations in the halfspace will rely on harmonic analysis, in particular on operator valued functional calculi for the time derivative

$$\partial_t : {}_0H^1_p((0, a), X) \subseteq L_p((0, a), X) \longrightarrow L_p((0, a), X)$$

and the fractional powers of the Laplacian

$$(-\Delta)^{s/2} : \dot{H}^s_p(\mathbb{R}^{n-1}, X) \subseteq L_p(\mathbb{R}^{n-1}, X) \longrightarrow L_p(\mathbb{R}^{n-1}, X)$$

for $s \in \{1, 2\}$, the space X being an arbitrary Banach space. Thus, we will employ the theory of sectorial operators.

Finally, the transfer of the halfspace results to bent halfspaces and bounded domains will make frequent use of a class of Sobolev embeddings, which allow for the exploitation of available additional time regularity. These types of embeddings are inspired by [PSS07, Proposition 6.2 (a)] and have already been used in 3.50 to prove Theorems 3.46 and 3.48.

4.1 L_p-Theory for Elliptic Problems

4.1 Depending on the boundary condition under consideration we will construct auxiliary solutions $q \in \dot{H}_p^1(\Omega)$ for the pressure via the *Dirichlet problem* for the Laplacian

$$(\mathrm{L}_D \,|\, \Omega)_{f,h} \qquad \begin{aligned} -\Delta q &= \operatorname{div} f && \text{in } \Omega, \\ [q]_\Gamma &= h && \text{on } \partial\Omega \end{aligned}$$

with data $f \in L_p(\Omega, \mathbb{R}^n)$ and $h \in \dot{W}_p^{1-1/p}(\Gamma)$ respectively the *Neumann problem* for the Laplacian

$$(\mathrm{L}_N \,|\, \Omega)_{f,h} \qquad \begin{aligned} -\Delta q &= \operatorname{div} f && \text{in } \Omega, \\ \partial_\nu q &= h - [f]_\Gamma \cdot \nu && \text{on } \partial\Omega \end{aligned}$$

with data $f \in L_p(\Omega, \mathbb{R}^n)$ and $h \in \dot{W}_p^{-1/p}(\Gamma)$. We consider all types of domains that appear during the localisation procedure and assume $\Omega \subseteq \mathbb{R}^n$ to be the halfspace, a bent halfspace of type BUC^1 or a bounded domain with boundary of class C^1. As usual $\Gamma := \partial\Omega$. Moreover, $1 < p < \infty$ and

$$\dot{W}_p^{-1/p}(\Gamma) := \dot{W}_{p'}^{1-1/p'}(\Gamma)',$$

where $1/p + 1/p' = 1$. Note that the formulation of the Neumann problem requires a proper interpretation of the boundary condition, since ∇q, $f \in L_p(\Omega, \mathbb{R}^n)$ do not possess traces in the sense of Sobolev spaces. This will be clarified in 4.8.

4.2 Note that for all domains under consideration the trace space $\dot{W}_p^{1-1/p}(\Gamma)$ is well-defined and there exists a bounded linear trace operator

$$(4.1) \qquad\qquad [\,\cdot\,]_\Gamma : \dot{H}_p^1(\Omega) \longrightarrow \dot{W}_p^{1-1/p}(\Gamma),$$

which admits a bounded linear right inverse

$$(4.2) \qquad\qquad \dot{\mathcal{E}}_\Omega : \dot{W}_p^{1-1/p}(\Gamma) \longrightarrow \dot{H}_p^1(\Omega),$$

cf. Proposition 3.14 and Remark 3.63.

4.3 Due to the weak regularity assumptions on q and f problem $(\mathrm{L}_D \,|\, \Omega)_{f,h}$ has to be understood in a weak sense. To achieve the corresponding weak formulation we define

$$_0\dot{H}_p^1(\Omega) := \mathrm{cls}\left(C_0^\infty(\Omega),\ \dot{H}_p^1(\Omega),\ |\cdot|_{\dot{H}_p^1(\Omega)} \right)$$

as the closure of the space $C_0^\infty(\Omega)$ of compactly supported, smooth functions in the semi-normed space $\dot{H}_p^1(\Omega)$. This space has already been employed in 3.41 to construct the Stokes operators $A^{\alpha,\beta}$. Now, for all domains under consideration and all $1 < p < \infty$ the identity

$$_0\dot{H}_p^1(\Omega) = \left\{ q \in \dot{H}_p^1(\Omega) \ : \ [q]_\Gamma = 0 \right\}$$

is valid, cf. [SS96]. Hence, the Dirichlet problem $(L_D \,|\, \Omega)_{f,h}$ in the weak L_p-setting as formulated in 4.1 is equivalent to its weak formulation

$$(\mathcal{L}_D \,|\, \Omega)_{f,h} \qquad \begin{aligned} -(\nabla q \,|\, \nabla \phi)_\Omega &= (f \,|\, \nabla \phi)_\Omega, \quad \phi \in {}_0\dot{H}^1_{p'}(\Omega), \\ [q]_\Gamma &= h \qquad\qquad\quad \text{on } \partial\Omega. \end{aligned}$$

Following the argumentation of C. G. SIMADER and H. SOHR cf [SS96], this is exactly the right setting to obtain maximal regular solutions to $(L_D \,|\, \Omega)_{f,h}$ in a weak sense.

4.4 PROPOSITION. *Let $\Omega \subseteq \mathbb{R}^n$ be a halfspace, a bent halfspace of type BUC^1 or a bounded domain with boundary of class C^1 and let $\Gamma := \partial\Omega$. Moreover, let $1 < p < \infty$. Then the (weak) Dirichlet problem $(L_D \,|\, \Omega)_{f,h}$ respectively $(\mathcal{L}_D \,|\, \Omega)_{f,h}$ admits a unique maximal regular solution*

$$q \in \dot{H}^1_p(\Omega)$$

for all $f \in L_p(\Omega, \mathbb{R}^n)$ and $h \in \dot{W}^{1-1/p}_p(\Gamma)$. Moreover, the solution depends continuously on the data.

Proof. First note that the trace operator (4.1) is onto. Now, given

$$f \in L_p(\Omega, \mathbb{R}^n) \quad \text{and} \quad h \in \dot{W}^{1-1/p}_p(\Gamma),$$

we first choose $\bar{q} \in \dot{H}^1_p(\Omega)$ with $[\bar{q}]_\Gamma = h$ and then solve

$$-(\nabla q - \nabla \bar{q} \,|\, \nabla \phi)_\Omega = (f + \nabla \bar{q} \,|\, \nabla \phi)_\Omega, \qquad \phi \in {}_0\dot{H}^1_{p'}(\Omega)$$

to obtain $q - \bar{q} \in {}_0\dot{H}^1_p(\Omega)$. The possibility of obtaining such a solution follows from [SS96, Theorem II.1.1], which covers the case of a bounded domain, respectively [SS96, Lemma II.2.5], which covers the (bent) halfspace case. This proves the solvability of the (weak) Dirichlet problem in the desired maximal regularity class.

Now, if $q, \bar{q} \in \dot{H}^1_p(\Omega)$ are two solutions to $(\mathcal{L}_D \,|\, \Omega)_{f,h}$, then $q - \bar{q} \in {}_0\dot{H}^1_p(\Omega)$ solves

$$-(\nabla q - \nabla \bar{q} \,|\, \nabla \phi)_\Omega = 0, \qquad \phi \in {}_0\dot{H}^1_{p'}(\Omega),$$

which implies $q = \bar{q}$ again by [SS96, Theorem II.1.1] in case of a bounded domain respectively [SS96, Lemma II.2.5] in case of a (bent) halfspace. This proves uniqueness of maximal regular solutions. Finally, continuous dependence of the solutions on the data is a consequence of the open mapping principle. $\qquad\square$

4.5 As a main ingredient of the proof of the above result we exploited the fact that the trace operator (4.1) is onto. On the other hand, Proposition 4.4 implies the existence of a bounded linear extension operator (4.2), see also Remark 3.63.

4.6 PROPOSITION. *Let $\Omega \subseteq \mathbb{R}^n$ be a halfspace, a bent halfspace of type BUC^1 or a bounded domain with boundary of class C^1 and let $\Gamma := \partial\Omega$. Moreover, let $1 < p < \infty$. Then there exists a unique bounded linear extension operator*

$$\dot{\mathcal{E}}_\Omega : \dot{W}_p^{1-1/p}(\Gamma) \longrightarrow \dot{H}_p^1(\Omega)$$

that is characterised by

$$-\Delta\,\dot{\mathcal{E}}_\Omega\,h = 0 \text{ in } \Omega, \quad [\dot{\mathcal{E}}_\Omega\,h]_\Gamma = h \text{ on } \partial\Omega, \quad h \in \dot{W}_p^{1-1/p}(\Gamma).$$

Proof. Given $h \in \dot{W}_p^{1-1/p}(\Gamma)$, define $\dot{\mathcal{E}}_\Omega\,h \in \dot{H}_p^1(\Omega)$ to be the unique solution to the Dirichlet problem $(\mathrm{L}_D \,|\, \Omega)_{0,h}$. $\qquad\Box$

4.7 Another valuable consequence of Proposition 4.4 is the validity of the *Weyl decomposition*

$$L_p(\Omega, \mathbb{R}^n) = L_{p,s}(\Omega) \oplus \nabla_0\dot{H}_p^1(\Omega)$$

for all domains under consideration and all $1 < p < \infty$. Here

$$L_{p,s}(\Omega) := \Big\{\, v \in L_p(\Omega, \mathbb{R}^n) \,:\, \operatorname{div} v = 0 \,\Big\}$$

denotes the space of L_p-functions, which are divergence-free in the sense of distributions. This space has already been employed in 3.37 to construct the Stokes operators $A^{\alpha,\beta}$. Since both, $L_{p,s}(\Omega)$ and $\nabla_0\dot{H}_p^1(\Omega)$, are closed subspaces of $L_p(\Omega, \mathbb{R}^n)$, this direct decomposition is topological and the thereby induced bounded linear projection

$$\mathcal{W}_p : L_p(\Omega, \mathbb{R}^n) \longrightarrow L_p(\Omega, \mathbb{R}^n)$$

onto $L_{p,s}(\Omega)$ along $\nabla_0\dot{H}_p^1(\Omega)$ is called *Weyl projection*. Given $v \in L_p(\Omega, \mathbb{R}^n)$, we have

$$\mathcal{W}_p v = v - \nabla q \in L_{p,s}(\Omega),$$

where $q \in {}_0\dot{H}_p^1(\Omega)$ is obtained as the unique solution to the Dirichlet problem $(\mathrm{L}_D \,|\, \Omega)_{-v,0}$. Note that

$$(\mathcal{W}_p v \,|\, v')_\Omega = (\mathcal{W}_p v \,|\, \mathcal{W}_{p'} v')_\Omega + (\mathcal{W}_p v \,|\, (1 - \mathcal{W}_{p'})v')_\Omega = (\mathcal{W}_p v \,|\, \mathcal{W}_{p'} v')_\Omega$$
$$= (\mathcal{W}_p v \,|\, \mathcal{W}_{p'} v')_\Omega + ((1 - \mathcal{W}_p)v \,|\, \mathcal{W}_{p'} v')_\Omega = (v \,|\, \mathcal{W}_{p'} v')_\Omega$$

for all $v \in L_p(\Omega, \mathbb{R}^n)$ and $v' \in L_{p'}(\Omega, \mathbb{R}^n)$ with $1/p + 1/p' = 1$, since

$$(1 - \mathcal{W}_p)v \in \nabla_0\dot{H}_p^1(\Omega) \quad \text{and} \quad (1 - \mathcal{W}_{p'})v' \in \nabla_0\dot{H}_{p'}^1(\Omega).$$

Thus, the Weyl projections $\mathcal{W}_p$ and $\mathcal{W}_{p'}$ form a pair of dual operators.

4.8 Now, we consider the Neumann problem $(\mathrm{L}_N \,|\, \Omega)_{f,h}$ in the weak L_p-setting as formulated in 4.1. As has already been mentioned there, ∇q, $f \in L_p(\Omega, \mathbb{R}^n)$ do not possess traces in the sense of Sobolev spaces. However, we may employ a *generalised normal trace*

(4.3) $$[\,\cdot\,]_\nu : L_{p,s}(\Omega) \longrightarrow \dot{W}_p^{-1/p}(\Gamma)$$

that is defined via

$$\langle\,\psi\,|\,[v]_\nu\,\rangle := (\,\nabla\dot{\mathcal{E}}_\Omega\,\psi\,|\,v\,)_\Omega, \qquad \psi \in \dot{W}_{p'}^{1-1/p'}(\Gamma),\ v \in L_{p,s}(\Omega).$$

Note that for $\phi \in \dot{H}_{p'}^1(\Omega)$ we have $\dot{\mathcal{E}}_\Omega\,[\phi]_\Gamma - \phi \in {}_0\dot{H}_{p'}^1(\Omega)$ and, therefore,

$$\mathcal{W}_{p'}\{\,\nabla\dot{\mathcal{E}}_\Omega\,[\phi]_\Gamma - \nabla\phi\,\} = 0.$$

Hence,

$$(\,\nabla\dot{\mathcal{E}}_\Omega\,[\phi]_\Gamma\,|\,v\,)_\Omega = (\,\nabla\dot{\mathcal{E}}_\Omega\,[\phi]_\Gamma\,|\,\mathcal{W}_p v\,)_\Omega = (\,\mathcal{W}_{p'}\nabla\dot{\mathcal{E}}_\Omega\,[\phi]_\Gamma\,|\,v\,)_\Omega$$
$$= (\,\mathcal{W}_{p'}\nabla\phi\,|\,v\,)_\Omega = (\,\nabla\phi\,|\,\mathcal{W}_p v\,)_\Omega = (\,\nabla\phi\,|\,v\,)_\Omega$$

for all $\phi \in \dot{H}_{p'}^1(\Omega)$ and $v \in L_{p,s}(\Omega)$ and, thus, the *generalised principle of partial integration*

$$\langle\,[\phi]_\Gamma\,|\,[v]_\nu\,\rangle = (\,\nabla\phi\,|\,v\,)_\Omega, \qquad \phi \in \dot{H}_{p'}^1(\Omega),\ v \in L_{p,s}(\Omega)$$

is available. In particular, we have

$$\langle\,\psi\,|\,[v]_\nu\,\rangle = (\,\nabla\phi\,|\,v\,)_\Omega, \qquad \psi \in \dot{W}_{p'}^{1-1/p'}(\Gamma),\ \phi \in \dot{H}_{p'}^1(\Omega),\ v \in L_{p,s}(\Omega),\ [\phi]_\Gamma = \psi,$$

which implies the definition of the generalised normal trace to be independent of the particular choice of the extension operator. With this notation the Neumann problem $(\mathrm{L}_N\,|\,\Omega)_{f,h}$ has to be understood as

$$-\Delta q = \operatorname{div} f \qquad \text{in } \Omega,$$
$$[\nabla q + f]_\nu = h \qquad \text{on } \partial\Omega,$$

which has a well-defined boundary condition, since the Laplace equation especially implies $\nabla q + f \in L_{p,s}(\Omega)$. Note that Proposition 4.12 below reveals the existence of a bounded linear extension operator

$$(4.4) \qquad\qquad \hat{\mathcal{E}}_\Omega : \dot{W}_p^{-1/p}(\Gamma) \longrightarrow L_{p,s}(\Omega),$$

which in particular implies the generalised normal trace operator to be onto.

4.9 Another issue concerning the Neumann problem $(\mathrm{L}_N\,|\,\Omega)_{f,h}$, which has to be addressed, is the fact that uniqueness of solutions has to be understood as uniqueness up to a constant. Therefore, we set

$$\hat{H}_p^1(\Omega) := \dot{H}_p^1(\Omega)/\mathbb{R}$$

and note that the Neumann problem $(\mathrm{L}_N\,|\,\Omega)_{f,h}$ in the weak L_p-setting as formulated in 4.1 is equivalent to the problem of the construction of a unique maximal regular solution $q \in \hat{H}_p^1(\Omega)$ to the weak formulation

$$(\mathcal{L}_N\,|\,\Omega)_{f,h} \qquad -(\,\nabla q\,|\,\nabla\phi\,)_\Omega = (\,f\,|\,\nabla\phi\,)_\Omega - \langle\,[\phi]_\Gamma\,|\,h\,\rangle, \qquad \phi \in \hat{H}_{p'}^1(\Omega).$$

The next proposition, which is due to C. G. SIMADER and H. SOHR, cf. [SS92], shows that this is exactly the right setting to obtain maximal regular solutions to $(\mathrm{L}_N\,|\,\Omega)_{f,h}$ in a weak sense.

4.10 PROPOSITION. Let $\Omega \subseteq \mathbb{R}^n$ be a halfspace, a bent halfspace of type BUC^1 or a bounded domain with boundary of class C^1 and let $\Gamma := \partial\Omega$. Moreover, let $1 < p < \infty$. Then the (weak) Neumann problem $(\mathrm{L}_N \,|\, \Omega)_{f,h}$ respectively $(\mathcal{L}_N \,|\, \Omega)_{f,h}$ admits a unique maximal regular solution

$$q \in \hat{H}_p^1(\Omega)$$

for all $f \in L_p(\Omega, \mathbb{R}^n)$ and $h \in \dot{W}_p^{-1/p}(\Gamma)$. Moreover, the solution depends continuously on the data.   □

4.11 A first interesting consequence of Proposition 4.10 is the pendant of Proposition 4.6 for the generalised normal trace operator (4.3), i. e. the existence of a bounded linear extension operator (4.4).

4.12 PROPOSITION. Let $\Omega \subseteq \mathbb{R}^n$ be a halfspace, a bent halfspace of type BUC^1 or a bounded domain with boundary of class C^1 and let $\Gamma := \partial\Omega$. Moreover, let $1 < p < \infty$. Then there exist a unique bounded linear extension operator

$$\hat{\mathcal{E}}_\Omega : \dot{W}_p^{-1/p}(\Gamma) \longrightarrow L_{p,s}(\Omega)$$

that is characterised by

$$\hat{\mathcal{E}}_\Omega h \in \nabla\hat{H}_p^1(\Omega), \quad [\hat{\mathcal{E}}_\Omega h]_\nu = h \text{ on } \partial\Omega, \quad h \in \dot{W}_p^{-1/p}(\Gamma).$$

Proof. Given $h \in \dot{W}_p^{-1/p}(\Gamma)$, define $\hat{\mathcal{E}}_\Omega h \in \hat{H}_p^1(\Omega)$ to be the unique solution to the Neumann problem $(\mathrm{L}_N \,|\, \Omega)_{0,h}$.   □

4.13 Last, but not least, Proposition 4.10 implies the *Helmholtz decomposition*

$$L_p(\Omega, \mathbb{R}^n) = L_{p,\sigma}(\Omega) \oplus \nabla\hat{H}_p^1(\Omega)$$

to be valid for all domains under consideration and all $1 < p < \infty$. Here

$$L_{p,\sigma}(\Omega) := \mathrm{cls}\left(C_{0,\sigma}^\infty(\Omega), \, L_p(\Omega, \mathbb{R}^n), \, \|\cdot\|_{L_p(\Omega,\mathbb{R}^n)}\right)$$

denotes the closure of the space $C_{0,\sigma}^\infty(\Omega)$ of solenoidal, compactly supported, smooth vector fields in $L_p(\Omega, \mathbb{R}^n)$. This space has already been employed in 3.37 to construct the Stokes operators $A^{\alpha,\beta}$. Unfortunately, the validity of the Helmholtz decomposition based on Proposition 4.10 is not as obvious as the validity of the Weyl decomposition based on Proposition 4.4 as presented in 4.7. Indeed, one first has to observe the following annihilator identities.

4.14 PROPOSITION. Let $\Omega \subseteq \mathbb{R}^n$ be an open set. Moreover, let $1 < p, \, p' < \infty$ with $1/p + 1/p' = 1$. Then, the annihilator identities

$$L_{p,\sigma}(\Omega)^\perp = \nabla\hat{H}_{p'}^1(\Omega) \qquad \text{and} \qquad (\nabla\hat{H}_p^1(\Omega))^\perp = L_{p',\sigma}(\Omega)$$

are valid.

Proof. If $v \in L_{p,\sigma}(\Omega)^\perp \subseteq L_{p'}(\Omega, \mathbb{R}^n)$, we have

$$(\,\phi\,|\,v\,)_\Omega = 0, \qquad \phi \in C_{0,\sigma}^\infty(\Omega).$$

As has been proved by G. DE RAHM, cf. [Rha60, Théorème 17'], this implies $v = \nabla q$ with $q \in \mathcal{D}'(\Omega)$. Now, $q \in \mathcal{D}'(\Omega)$ and $\nabla q \in L_{p'}(\Omega, \mathbb{R}^n)$ imply $q \in \hat{H}_{p'}^1(\Omega)$, i.e. $v \in \nabla \hat{H}_{p'}^1(\Omega)$, cf. [Neč67, Chapitre 2, Remarque 7.3].

Conversely, if $v \in \nabla \hat{H}_{p'}^1(\Omega) \subseteq L_{p'}(\Omega, \mathbb{R}^n)$, then $v = \nabla q$ for some $q \in \hat{H}_{p'}^1(\Omega)$ and we infer

$$(\,\phi\,|\,v\,)_\Omega = (\,\phi\,|\,\nabla q\,)_\Omega = 0, \qquad \phi \in C_{0,\sigma}^\infty(\Omega).$$

Since $C_{0,\sigma}^\infty(\Omega)$ is a dense subspace of $L_{p,\sigma}(\Omega)$, we further infer $v \in L_{p,\sigma}(\Omega)^\perp$. This proves the first of the asserted annihilator identities.

Finally, reflexivity of the involved spaces implies

$$(\nabla \hat{H}_p^1(\Omega))^\perp = (L_{p',\sigma}(\Omega)^\perp)^\perp = L_{p',\sigma}(\Omega)$$

and the second annihilator identity is proved. □

4.15 Now, the continuity of the generalised normal trace (4.3), the principle of partial integration as presented in 4.8 together with Proposition 4.14 imply

$$L_{p,\sigma}(\Omega) \subseteq \left\{ v \in L_{p,s}(\Omega) \,:\, [v]_\nu = 0 \right\} \subseteq (\nabla \hat{H}_{p'}^1(\Omega))^\perp = L_{p,\sigma}(\Omega)$$

and, thus, the characterisation

$$L_{p,\sigma}(\Omega) = \left\{ v \in L_{p,s}(\Omega) \,:\, [v]_\nu = 0 \right\}$$

is valid for all domains under consideration and all $1 < p < \infty$. Therefore, the Helmholtz decomposition as stated in 4.13 is valid based on the generalised principle of partial integration and Proposition 4.10. Moreover, since both, $L_{p,\sigma}(\Omega)$ and $\nabla \hat{H}_p^1(\Omega)$, are closed subspaces of $L_p(\Omega, \mathbb{R}^n)$, this direct decomposition is topological and the thereby induced bounded linear projection

$$\mathcal{H}_p : L_p(\Omega, \mathbb{R}^n) \longrightarrow L_p(\Omega, \mathbb{R}^n)$$

onto $L_{p,\sigma}(\Omega)$ along $\nabla \hat{H}_p^1(\Omega)$ is called *Helmholtz projection*. Given $v \in L_p(\Omega, \mathbb{R}^n)$, we have

$$\mathcal{H}_p v = v - \nabla q \in L_{p,\sigma}(\Omega),$$

where $q \in \hat{H}_p^1(\Omega)$ is obtained as the unique solution to the Neumann problem $(L_N \,|\, \Omega)_{-v,0}$. Analogously to the corresponding Weyl projections, the Helmholtz projections $\mathcal{H}_p$ and $\mathcal{H}_{p'}$ form a pair of dual operators for $1/p + 1/p' = 1$.

4.2 L_p-Theory for Parabolic Problems

4.16 Auxiliary solutions for the velocity may conveniently be obtained via parabolic problems with boundary conditions involving the divergence

$$\rho \partial_t u - \mu \Delta u = \rho f \quad \text{in } (0, a) \times \Omega,$$

$(\mathrm{P} \mid a,\, \Omega)^\alpha_{f,h,u_0}$
$$\mathcal{B}^{\alpha,\mathrm{div}} u = h \quad \text{on } (0, a) \times \partial\Omega,$$

$$u(0) = u_0 \quad \text{in } \Omega.$$

with $\alpha \in \{-1,\, 0,\, +1\}$, where the boundary condition is given by the linear operator

$$P_\Gamma \mathcal{B}^{\alpha,\mathrm{div}} u = P_\Gamma \mathcal{B}^\alpha u \quad \text{and} \quad Q_\Gamma \mathcal{B}^{\alpha,\mathrm{div}} u \cdot \nu_\Gamma = [\mathrm{div}\, u]_\Gamma.$$

As in Chapter 3, we assume $a > 0$ and $1 < p < \infty$ with $p \neq \frac{3}{2}$, 3 as well as $\rho > 0$, $\mu > 0$. However, $\Omega \subseteq \mathbb{R}^n$ may be the halfspace, a bent halfspace of type BUC^{3-} or a bounded domain with boundary of class C^{3-}. As usual $\Gamma := \partial\Omega$ and $\nu_\Gamma : \Gamma \longrightarrow \mathbb{R}^n$ denotes the outer unit normal field of Ω. Moreover, P_Γ respectively Q_Γ denote the projections onto the tangent bundle respectively the normal bundle of Γ. Thus, given $\alpha,\, \beta \in \{-1,\, 0,\, +1\}$ the boundary operator $\mathcal{B}^{\alpha,\mathrm{div}}$ realises the same tangential boundary condition as the boundary operator $\mathcal{B}^{\alpha,\beta}$ introduced in Chapter 3. However, here we assume the set of boundary conditions to be completed by prescribing the divergence instead of a normal boundary condition.

4.17 Since the parabolic problems $(\mathrm{P} \mid a,\, \Omega)^\alpha_{f,h,u_0}$ will prove very useful in our analysis of the Stokes equations, we will establish an L_p-theory for these problems along the next paragraphs. The aim is to show, that the system $(\mathrm{P} \mid a,\, \Omega)^\alpha_{f,h,u_0}$ admits a unique maximal regular solution

$$u \in \mathbb{X}_u(a,\, \Omega) := H^1_p((0,\, a),\, L_p(\Omega,\, \mathbb{R}^n)) \cap L_p((0,\, a),\, H^2_p(\Omega,\, \mathbb{R}^n)),$$

whenever the data satisfies the necessary and sufficient regularity and compatibility conditions.

4.18 As a base space we employ

$$f \in \mathbb{Y}_f(a,\, \Omega) := L_p((0,\, a) \times \Omega).$$

The necessary regularity of the right hand side

$$h \in \mathbb{Y}_h(a,\, \Gamma) := L_p((0,\, a),\, L_{p,loc}(\Gamma,\, \mathbb{R}^n))$$

of the boundary condition depends on the parameter $\alpha \in \{-1,\, 0,\, +1\}$. As in 3.18 we may employ Propositions 3.9 and 3.17 to obtain

$$P_\Gamma \mathcal{B}^{0,\mathrm{div}} u \in \mathbb{T}^0_h(a,\, \Gamma) \text{ with}$$

$$\mathbb{T}^0_h(a,\, \Gamma) := W^{1-1/2p}_p((0,\, a),\, L_p(\Gamma,\, T\Gamma)) \cap L_p((0,\, a),\, W^{2-1/p}_p(\Gamma,\, T\Gamma)),$$

$$P_\Gamma \mathcal{B}^{\pm 1,\mathrm{div}} u \in \mathbb{T}^{\pm 1}_h(a,\, \Gamma) \text{ with}$$

$$\mathbb{T}^{\pm 1}_h(a,\, \Gamma) := W^{1/2-1/2p}_p((0,\, a),\, L_p(\Gamma,\, T\Gamma)) \cap L_p((0,\, a),\, W^{1-1/p}_p(\Gamma,\, T\Gamma)),$$

$$Q_\Gamma \mathcal{B}^{\alpha,\mathrm{div}} u \in \mathbb{N}^{\mathrm{div}}_h(a,\, \Gamma) \text{ with}$$

$$\mathbb{N}^{\mathrm{div}}_h(a,\, \Gamma) := W^{1/2-1/2p}_p((0,\, a),\, L_p(\Gamma,\, N\Gamma)) \cap L_p((0,\, a),\, W^{1-1/p}_p(\Gamma,\, N\Gamma))$$

as necessary regularity conditions, if $u \in \mathbb{X}_u(a, \Omega)$. Therefore, the regularity class for the boundary data h is given as

$$h \in \mathbb{Y}_h^{\alpha, \mathrm{div}}(a, \Gamma) := \left\{ \eta \in \mathbb{Y}_h(a, \Gamma) : P_\Gamma \eta \in \mathbb{T}_h^\alpha(a, \Gamma), \, Q_\Gamma \eta \in \mathbb{N}_h^{\mathrm{div}}(a, \Gamma) \right\}$$

for $\alpha \in \{ -1, 0, +1 \}$. Finally, we may argue as in 3.19 to obtain

$$u_0 \in \mathbb{Y}_0(\Omega) := W_p^{2-2/p}(\Omega, \mathbb{R}^n)$$

as the right regularity class for the initial data. Note that the above definitions of the function spaces coincide with those used in Chapter 3 with the exceptions that $\mathbb{N}_h^{\mathrm{div}}(a, \Gamma)$ and $\mathbb{Y}_h^{\alpha, \mathrm{div}}(a, \Gamma)$ do not appear there and that we generalise the definitions of Chapter 3 to a larger class of domains here, see also Remarks 3.60 and 3.62.

4.19 In addition to the above regularity conditions, the compatibility conditions

$$(\mathrm{C})_{h, u_0}^{\alpha, \mathrm{div}} \qquad \begin{aligned} P_\Gamma[u_0]_\Gamma &= P_\Gamma h(0), && \text{if } \alpha = 0 \quad \text{and } p > \tfrac{3}{2}, \\ \mu P_\Gamma[\nabla u_0 \pm \nabla u_0^\mathsf{T}]_\Gamma \, \nu_\Gamma &= P_\Gamma h(0), && \text{if } \alpha = \pm 1 \text{ and } p > 3, \\ [\mathrm{div}\, u_0]_\Gamma &= h(0) \cdot \nu_\Gamma, && \text{if } p > 3 \end{aligned}$$

are necessary for $(\mathrm{P} \,|\, a, \Omega)_{f, h, u_0}^\alpha$ to admit a maximal regular solution.

4.20 To formulate our result concerning the maximal L_p-regularity of the parabolic system $(\mathrm{P} \,|\, a, \Omega)_{f, h, u_0}^\alpha$ is is convenient to introduce the data space $\mathbb{Y}^{\alpha, \mathrm{div}}(a, \Omega)$, which is defined to consist of all

$$(f, h, u_0) \in \mathbb{Y}_f(a, \Omega) \times \mathbb{Y}_h^{\alpha, \mathrm{div}}(a, \Gamma) \times \mathbb{Y}_0(\Omega)$$

satisfy the compatibility condition $(\mathrm{C})_{h, u_0}^{\alpha, \mathrm{div}}$. With this notation we obtain the following proposition, which will be proved in the following paragraphs.

4.21 PROPOSITION. *Let $a > 0$ and let $\Omega \subseteq \mathbb{R}^n$ be a halfspace, a sufficiently flat bent halfspace of type BUC^{3-} or a bounded domain with boundary of class C^{3-}. Let $1 < p < \infty$ with $p \neq \tfrac{3}{2}, 3$, let $\rho, \mu > 0$ and let $\alpha \in \{ -1, 0, +1 \}$. Then there exists a unique maximal regular solution*

$$u \in \mathbb{X}_u(a, \Omega)$$

to the parabolic problem $(\mathrm{P} \,|\, a, \Omega)_{f, h, u_0}^\alpha$, if and only if the data satisfies

$$(f, h, u_0) \in \mathbb{Y}^{\alpha, \mathrm{div}}(a, \Omega).$$

Furthermore, the solutions depend continuously on the data.

4.22 Concerning the proof of Proposition 4.21 one should distinguish the two cases $\alpha = 0$ and $\alpha = \pm 1$. In the second case all boundary conditions are of order one. Therefore, the system fits into the context of the abstract parabolic problems for which a complete L_p-theory developed by R. DENK, M. HIEBER and J. PRÜSS is available, cf. [DHP03]. Indeed, the system satisfies for $\alpha = \pm 1$ all necessary and sufficient conditions, which reveal maximal L_p-regularity – especially the *Lopatinskii-Shapiro condition*. For the complete theory we refer to [DHP03].

4.23 For $\alpha = 0$ the boundary conditions are of mixed order. Unfortunately, there is no L_p-theory available that is able to cover such systems. However, the parabolic problem $(\mathrm{P} \,|\, a,\, \Omega)$ may be treated as the Stokes equations $(\mathrm{S} \,|\, a,\, \Omega)$ by a localisation procedure, which starts with an analysis of the halfspace case and successively transfers the results to bent halfspaces and bounded domains. The argumentation is much simpler as compared to the case of the Stokes equations, since the latter form a degenerate parabolic problem and additional efforts are necessary to take care of the divergence constraint and the pressure. Therefore, we restrict the proof of Proposition 4.21 for $\alpha = 0$ to a brief overview in the next paragraphs and refer to Chapters 5, 6 and 7, where the localisation procedure is carried out for the Stokes equations in all details.

4.24 If $\alpha = 0$ and $\Omega = \mathbb{R}^n_+$ is a halfspace, we may decompose the desired solution as $u = (v,\, w)$ into a tangential vector field $v : (0,\, a) \times \mathbb{R}^n_+ \longrightarrow \mathbb{R}^{n-1}$ and a normal component $w : (0,\, a) \times \mathbb{R}^n_+ \longrightarrow \mathbb{R}$. The system $(\mathrm{P} \,|\, a,\, \Omega)^\alpha_{f,h,u_0}$ then reads

$$
\begin{aligned}
\rho \partial_t v - \mu \Delta v &= \rho f_v && \text{in } (0,\, a) \times \mathbb{R}^n_+, \\
\rho \partial_t w - \mu \Delta w &= \rho f_w && \text{in } (0,\, a) \times \mathbb{R}^n_+, \\
[v]_y &= h_v && \text{on } (0,\, a) \times \partial\mathbb{R}^n_+, \\
\mathrm{div}_x\,[v]_y + [\partial_y w]_y &= h_w && \text{on } (0,\, a) \times \partial\mathbb{R}^n_+, \\
v(0) &= v_0 && \text{in } \mathbb{R}^n_+, \\
w(0) &= w_0 && \text{in } \mathbb{R}^n_+,
\end{aligned}
$$

where we decomposed $f = (f_v,\, f_w)$, $h = (h_v,\, h_w)$ and $u_0 = (v_0,\, w_0)$ into a tangential part and a normal part analogously to the solution u. Moreover, we decomposed the spatial variable into a tangential part $x \in \mathbb{R}^{n-1}$ and a normal part $y > 0$. Of course, div_x denotes the divergence w.r.t. x and ∂_y denotes the partial derivative w.r.t. y. Moreover, $[\,\cdot\,]_y$ denotes the trace operator for the halfspace, where

$$
\partial\mathbb{R}^n_+ = \Big\{ (x,\, y) \in \mathbb{R}^{n-1} \times \mathbb{R} \,:\, y = 0 \Big\}.
$$

Obviously, this system may first be solved for a unique maximal regular solution v, which may then be used as part of the data of the parabolic boundary value problem for w. Hence, w may in a second step be obtained as the unique maximal regular solution to this remaining problem.

4.25 If $\alpha = 0$ and $\Omega = \mathbb{R}^n_\omega$ is a bent halfspace of type BUC^{3-}, we may first solve the plain Dirichlet problem

$$
\begin{aligned}
\rho \partial_t u - \mu \Delta u &= \rho f && \text{in } (0,\, a) \times \Omega, \\
[u]_\Gamma &= P_\Gamma h + Q_\Gamma e^{t\Delta_\Gamma}[u_0]_\Gamma && \text{on } (0,\, a) \times \partial\Omega, \\
u(0) &= u_0 && \text{in } \Omega,
\end{aligned}
$$
(4.5)

which forms a system of n decoupled parabolic initial boundary value problems with Dirichlet condition. Here, Δ_Γ denotes the Laplace-Beltrami operator on Γ and $e^{t\Delta_\Gamma}$ denotes the corresponding semigroup. Hence, we may in the sequel assume $f = 0$, $P_\Gamma h = 0$ and $u_0 = 0$. Therefore, we may repeat exactly the same arguments and estimates as in

Chapter 6 to reduce the problem to the halfspace case. However, this requires the boundary of the bent halfspace to be sufficiently flat, which has to be understood in the sense that there exists some constant $\delta > 0$ such that the assertions of Proposition 4.21 are valid for all bent halfspaces $\mathbb{R}^n_\omega$ of type BUC^{3-} subject to the additional flatness condition

$$\|\nabla \omega\|_{L_\infty(\mathbb{R}^{n-1})} < \delta.$$

For the details in case of the Stokes equations we refer to Chapter 6.

4.26 Finally, if $\alpha = 0$ and $\Omega \subseteq \mathbb{R}^n$ is a bounded domain with boundary of class C^{3-}, we may first solve the plain Dirichlet problem (4.5) and again assume $f = 0$, $P_\Gamma h = 0$ and $u_0 = 0$. Thus, we may repeat the arguments and estimates used in Chapter 7 to reduce the problem to the whole space case respectively the bent halfspace case. However, the whole procedure is much simpler as in the case of the Stokes equations, since we do not need to solve auxiliary problems to ensure the divergence constraint to be satisfied and we do not need to estimate any pressure terms. This completes the proof of Proposition 4.21. □

4.3 Special Solutions in a Halfspace

4.27 The main reason for the introduction and analysis of the elliptic problems $(\mathrm{L}_D \mid \Omega)_{f,h}$ and $(\mathrm{L}_N \mid \Omega)_{f,h}$ and the parabolic problems $(\mathrm{P} \mid a, \Omega)^\alpha_{f,h,u_0}$ is their impact as a source for auxiliary solutions to the Stokes equations. However, in some cases the Stokes equations may be completely solved via a *splitting scheme*, which is a pair of an elliptic problem, which determines the pressure, and a parabolic problem, which determines the velocity. Of course, in order to completely reduce the Stokes equations to such a pair of problems the latter should be coupled in one direction only. An example for such a splitting scheme, which is applicable for the cases $\alpha \in \{-1, 0, +1\}$ and $\beta = -1$ and various types of domains, will be presented in Section 4.5 below. Another example, which is applicable for $\alpha \in \{-1, 0, +1\}$ and $\beta \in \{0, +1\}$ but restricted to the halfspace case, will be developed in Chapter 5. To simplify this development it will be convenient to have representations of special solutions to the elliptic problems $(\mathrm{L}_D \mid \Omega)_{f,h}$ and $(\mathrm{L}_N \mid \Omega)_{f,h}$ and the parabolic problems $(\mathrm{P} \mid a, \Omega)^\alpha_{f,h,u_0}$ in a halfspace at hand. Such representations will be derived along the next paragraphs.

4.28 Aiming at a splitting scheme for the Stokes equations in a halfspace we are interested in maximal regular solutions to the parabolic problem $(\mathrm{P} \mid \infty, \mathbb{R}^n_+)^\alpha_{-\nabla p,0,0}$. Now, this system lacks the maximal L_p-regularity property, since the considered time interval is unbounded. However, since the operator

$$B^\alpha : D(B^\alpha) \subseteq L_p(\mathbb{R}^n_+, \mathbb{R}^n) \longrightarrow L_p(\mathbb{R}^n_+, \mathbb{R}^n), \qquad B^\alpha v := -\mu \Delta v, \quad v \in D(B^\alpha)$$

with domain

$$D(B^\alpha) := \left\{ v \in H^2_p(\mathbb{R}^n_+) : \mathcal{B}^{\alpha,\mathrm{div}} v = 0 \right\} \subseteq L_p(\mathbb{R}^n_+, \mathbb{R}^n)$$

has the property of maximal L_p-regularity on finite time intervals due to Proposition 4.21, the same is true for the shifted operators $\varepsilon + B^\alpha$ on the unbounded time interval $(0, \infty)$, provided $\varepsilon > 0$ is chosen to be sufficiently large. This is a result of the L_p-theory for abstract parabolic evolution equations, cf. [Prü03, Section 1]. Therefore, we study the modified parabolic problem

$$(P \mid \infty, \mathbb{R}_+^n)_{-\nabla p,0,0}^{\alpha,\varepsilon} \qquad \begin{aligned} \rho\varepsilon u + \rho\partial_t u - \mu\Delta u &= -\nabla p &&\text{in } (0, \infty) \times \mathbb{R}_+^n, \\ \mathcal{B}^{\alpha,\mathrm{div}} u &= 0 &&\text{on } (0, \infty) \times \partial\mathbb{R}_+^n, \\ u(0) &= 0 &&\text{in } \mathbb{R}_+^n, \end{aligned}$$

where the shift $\varepsilon > 0$ is chosen to be sufficiently large to ensure the existence of a maximal regular solution $u \in \mathbb{X}_u(\infty, \mathbb{R}_+^n)$. The right hand side

$$\nabla p \in L_p((0, \infty) \times \mathbb{R}_+^n)$$

is assumed to be obtained as a solution to the Dirichlet problem

$$(L_D \mid \infty, \mathbb{R}_+^n)_{0,h} \qquad \begin{aligned} -\Delta p &= 0 &&\text{in } (0, \infty) \times \mathbb{R}_+^n, \\ [p]_\Gamma &= h &&\text{on } (0, \infty) \times \partial\mathbb{R}_+^n \end{aligned}$$

for a given boundary data $h \in L_p((0, \infty), \dot{W}_p^{1-1/p}(\Gamma))$ respectively the Neumann problem

$$(L_N \mid \infty, \mathbb{R}_+^n)_{0,h} \qquad \begin{aligned} -\Delta p &= 0 &&\text{in } (0, \infty) \times \mathbb{R}_+^n, \\ \partial_\nu p &= h &&\text{on } (0, \infty) \times \partial\mathbb{R}_+^n \end{aligned}$$

for a given boundary data $h \in L_p((0, \infty), \dot{W}_p^{-1/p}(\Gamma))$. Note that the elliptic problems $(L_D \mid \infty, \mathbb{R}_+^n)_{0,h}$ respectively $(L_N \mid \infty, \mathbb{R}_+^n)_{0,h}$ are equivalent to the elliptic problems $(L_D \mid \mathbb{R}_+^n)_{0,h}$ respectively $(L_N \mid \mathbb{R}_+^n)_{0,h}$, which have been considered in Section 4.1, lifted to the $L_p((0, \infty) \times \mathbb{R}_+^n)$-setting. Thus, Propositions 4.4 and 4.10 ensure the existence of a unique solution in the desired regularity class.

4.29 The motivation for this setup is the observation that the shifted parabolic problem $(P \mid \infty, \mathbb{R}_+^n)_{-\nabla p,0,0}^{\alpha,\varepsilon}$ in conjunction with the Dirichlet problem $(L_D \mid \infty, \mathbb{R}_+^n)_{0,h}$ delivers a unique maximal regular solution

$$(u, p) \in \mathbb{X}_u(\infty, \mathbb{R}_+^n) \times \mathbb{X}_{p,-\infty}^{+1}(\infty, \mathbb{R}_+^n),$$

where

$$\mathbb{X}_{p,-\infty}^{+1}(\infty, \mathbb{R}_+^n) := L_p((0, \infty), \dot{H}_p^1(\mathbb{R}_+^n))$$

denotes the halfspace version of the maximal regularity space $\mathbb{X}_{p,-\infty}^{+1}(a, \Omega)$, which was introduced as the regularity class for the pressure in Section 3.2 for a bounded, smooth domain $\Omega \subseteq \mathbb{R}^n$, see also Remark 3.59. In an analogous way the shifted parabolic problem $(P \mid \infty, \mathbb{R}_+^n)_{-\nabla p,0,0}^{\alpha,\varepsilon}$ in conjunction with the Neumann problem $(L_N \mid \infty, \mathbb{R}_+^n)_{0,h}$ delivers a unique maximal regular solution

$$(u, p) \in \mathbb{X}_u(\infty, \mathbb{R}_+^n) \times \mathbb{X}_{p,-\infty}^0(\infty, \mathbb{R}_+^n),$$

where

$$\mathbb{X}_{p,-\infty}^0(\infty, \mathbb{R}_+^n) := L_p((0, \infty), \hat{H}_p^1(\mathbb{R}_+^n))$$

denotes the halfspace version of the maximal regularity space $\mathbb{X}_{p,-\infty}^0(a, \Omega)$, which was introduced as the regularity class for the pressure in Section 3.2 for a bounded, smooth domain $\Omega \subseteq \mathbb{R}^n$, see also Remark 3.59. However, here we employ the space $\hat{H}_p^1(\mathbb{R}_+^n)$ to ensure the uniqueness of the pressure instead of the space

$$\left\{ q \in \dot{H}_p^1(\Omega) : (q)_\Omega = 0 \right\},$$

which was used for a bounded, smooth domain $\Omega \subseteq \mathbb{R}^n$ in Section 3.2. Now observe that this construction implies

$$\rho\varepsilon \operatorname{div} u + \rho\partial_t \operatorname{div} u - \mu\Delta \operatorname{div} u = 0 \quad \text{in } (0, \infty) \times \mathbb{R}_+^n,$$

$$[\operatorname{div} u]_\Gamma = 0 \quad \text{on } (0, \infty) \times \partial\mathbb{R}_+^n,$$

$$(\operatorname{div} u)(0) = 0 \quad \text{in } \mathbb{R}_+^n,$$

which ensures $\operatorname{div} u = 0$ by uniqueness of weak solutions to this parabolic initial boundary value problem with Dirichlet boundary condition. Thus, we obtain a maximal regular solution to the shifted variant of the Stokes equations $(\mathrm{S} \mid \infty, \mathbb{R}_+^n)_{0,0,\bar{h},0}^{\alpha,\beta}$, where the boundary data satisfies $P_\Gamma \bar{h} = 0$. In summary, we relinquished the normal boundary condition and imposed a boundary condition involving the divergence instead. This way we obtain a splitting scheme, which may be successively solved first for the pressure and then for the velocity, and the remaining *degree of freedom*, which is available due to the relinquishment of the normal boundary condition, is the boundary condition for the pressure, which has to be chosen appropriately. The complete splitting scheme for a halfspace and for all parameters $\alpha \in \{-1, 0, +1\}$ and $\beta \in \{0, +1\}$ will be developed in Chapter 5 based on a representation of the solutions to $(\mathrm{P} \mid \infty, \mathbb{R}_+^n)_{-\nabla p,0,0}^{\alpha,\varepsilon}$, which will be derived along the next paragraphs.

4.30 Now, assume

$$u \in \mathbb{X}_u((0, \infty), \mathbb{R}_+^n)$$

to be the unique maximal regular solution to $(\mathrm{P} \mid \infty, \mathbb{R}_+^n)_{-\nabla p,0,0}^{\alpha,\varepsilon}$. Splitting the spatial variable as

$$(x, y) \in \mathbb{R}^{n-1} \times (0, \infty)$$

into a tangential part $x \in \mathbb{R}^{n-1}$ and a normal part $y > 0$ we may employ a *Laplace transformation*

$$\hat{\phi}(\lambda) := \mathcal{L}\phi(\lambda) := \int_0^\infty e^{-\lambda t}\phi(t)\,\mathrm{d}t, \quad \operatorname{Re}\lambda \geq 0, \qquad \phi \in L_1((0, \infty), X)$$

in time and a *Fourier transformation*

$$\tilde{\phi}(\xi) := \mathcal{F}\phi(\xi) := \int_{\mathbb{R}^{n-1}} e^{-ix\cdot\xi}\phi(x)\,\mathrm{d}x, \quad \xi \in \mathbb{R}^{n-1}, \qquad \phi \in L_1(\mathbb{R}^{n-1}, X)$$

in the tangential variable, where X denotes an arbitrary Banach space. This way we formally obtain the ordinary differential equations

$$\omega^2\hat{v}(\lambda, \xi, y) - \mu\partial_y^2\hat{v}(\lambda, \xi, y) = -i\xi\hat{p}(\lambda, \xi, y), \quad \operatorname{Re}\lambda \geq 0, \xi \in \mathbb{R}^{n-1}, y > 0,$$

$$\omega^2\hat{w}(\lambda, \xi, y) - \mu\partial_y^2\hat{w}(\lambda, \xi, y) = -\partial_y\hat{p}(\lambda, \xi, y), \quad \operatorname{Re}\lambda \geq 0, \xi \in \mathbb{R}^{n-1}, y > 0,$$

where the solution is split analogously to the spatial variable as $u = (v, w)$ into a tangential part $v : [0, \infty) \times \mathbb{R}_+^n \longrightarrow \mathbb{R}^{n-1}$ and a normal part $w : [0, \infty) \times \mathbb{R}_+^n \longrightarrow \mathbb{R}$, and where we abbreviated

$$\omega := \sqrt{\rho \lambda_\varepsilon + \mu |\xi|^2}$$

and $\lambda_\varepsilon := \varepsilon + \lambda$. Moreover $\hat{v}$, $\hat{w}$ and $\hat{p}$ denote the Laplace-Fourier transforms of the unknown functions v, w and p.

4.31 If $\alpha = 0$, the boundary conditions read

$$[\hat{v}]_y(\lambda, \xi) = 0, \quad \operatorname{Re}\lambda \geq 0, \xi \in \mathbb{R}^{n-1},$$

$$i\xi^\mathsf{T}[\hat{v}]_y(\lambda, \xi) + [\partial_y \hat{w}]_y(\lambda, \xi) = 0, \quad \operatorname{Re}\lambda \geq 0, \xi \in \mathbb{R}^{n-1}$$

and we obtain

$$\hat{v}(\lambda, \xi, y) = -\int_0^\infty G_-(\lambda, \xi, y, \eta)\, i\xi\hat{p}(\lambda, \xi, \eta)\, \mathrm{d}\eta, \quad \operatorname{Re}\lambda \geq 0, \xi \in \mathbb{R}^{n-1}, y > 0,$$

$$\hat{w}(\lambda, \xi, y) = -\int_0^\infty G_+(\lambda, \xi, y, \eta)\, \partial_y \hat{p}(\lambda, \xi, \eta)\, \mathrm{d}\eta, \quad \operatorname{Re}\lambda \geq 0, \xi \in \mathbb{R}^{n-1}, y > 0$$

where

$$G(\lambda, \xi, y, \eta) = \frac{1}{2\sqrt{\mu}\omega} e^{-\frac{\omega}{\sqrt{\mu}}|y-\eta|}, \quad \operatorname{Re}\lambda \geq 0, \xi \in \mathbb{R}^{n-1}, y \in \mathbb{R}, \eta \in \mathbb{R}$$

denotes the fundamental solution to the ordinary differential equation

$$\omega^2 \hat{\phi}(\lambda, \xi, y) - \mu \partial_y^2 \hat{\phi}(\lambda, \xi, y) = \hat{f}(\lambda, \xi, y), \quad \operatorname{Re}\lambda \geq 0, \xi \in \mathbb{R}^{n-1}, y \in \mathbb{R}$$

and

$$G_\pm(\lambda, \xi, y, \eta) = G(\lambda, \xi, y, \eta) \pm G(\lambda, \xi, y, -\eta),$$
$$\operatorname{Re}\lambda \geq 0, \xi \in \mathbb{R}^{n-1}, y > 0, \eta > 0.$$

4.32 Hence, if p is obtained as a solution to the Neumann problem $(\mathrm{L}_N \mid \infty, \mathbb{R}_+^n)_{0,h}$, we have

$$(4.6) \qquad -\partial_y \hat{p}(\lambda, \xi, y) = |\xi| e^{-|\xi|y} \widehat{(-\Delta_\Gamma)^{-1/2}h}(\lambda, \xi), \quad \operatorname{Re}\lambda \geq 0, \xi \in \mathbb{R}^{n-1}, y > 0,$$

where Δ_Γ denotes the Laplacian on $\mathbb{R}^{n-1}$, which implies

$$(\mathrm{M})^{0,0} \qquad
\begin{aligned}
[\hat{w}]_y(\lambda, \xi) &= \int_0^\infty G_+(\lambda, \xi, 0, \eta)\, |\xi| e^{-|\xi|\eta} \widehat{(-\Delta_\Gamma)^{-1/2}h}(\lambda, \xi)\, \mathrm{d}\eta \\
&= \frac{1}{\omega(\omega + |\zeta|)} \hat{h}(\lambda, \xi), \quad \operatorname{Re}\lambda \geq 0, \xi \in \mathbb{R}^{n-1},
\end{aligned}$$

where we have set $\zeta := \sqrt{\mu}\xi$, i.e. the velocity satisfies the Dirichlet type boundary condition $\mathcal{M}^{0,0}[w]_y = h$, where

$$\mathcal{M}^{0,0} : D(\mathcal{M}^{0,0}) \subseteq L_p((0, \infty), \dot{W}_p^{-1/p}(\mathbb{R}^{n-1})) \longrightarrow L_p((0, \infty), \dot{W}_p^{-1/p}(\mathbb{R}^{n-1}))$$

is given as

$$D(\mathcal{M}^{0,0}) := {}_0H_p^1((0, \infty), \dot{W}_p^{-1/p}(\mathbb{R}^{n-1})) \cap L_p((0, \infty), W_p^{2-1/p}(\mathbb{R}^{n-1})),$$

$$\mathcal{M}^{0,0} := \rho\varepsilon + \rho\partial_t + (\rho\varepsilon + \rho\partial_t - \mu\Delta_\Gamma)^{1/2}(-\mu\Delta_\Gamma)^{1/2} - \mu\Delta_\Gamma.$$

Thus, $\mathcal{M}^{0,0}$ is defined via its Laplace-Fourier symbol

$$m^{0,0}(\lambda, \xi) = \omega(\omega + |\zeta|), \qquad \operatorname{Re}\lambda \geq 0,\ \xi \in \mathbb{R}^{n-1}.$$

4.33 If $\alpha = \pm 1$, the boundary conditions read

$$\mp\mu[\partial_y\hat{v}]_y(\lambda, \xi) - \mu\nabla_x[\hat{w}]_y(\lambda, \xi) = 0, \qquad \operatorname{Re}\lambda \geq 0,\ \xi \in \mathbb{R}^{n-1},$$

$$i\xi^\mathsf{T}[\hat{v}]_y(\lambda, \xi) + [\partial_y\hat{w}]_y(\lambda, \xi) = 0, \qquad \operatorname{Re}\lambda \geq 0,\ \xi \in \mathbb{R}^{n-1}$$

and we obtain

$$\hat{v}(\lambda, \xi, y) = -\int_0^\infty K_v^\pm(\lambda, \xi, y, \eta)\, i\xi\hat{p}(\lambda, \xi, \eta)\, \mathrm{d}\eta, \qquad \operatorname{Re}\lambda \geq 0,\ \xi \in \mathbb{R}^{n-1},\ y > 0,$$

$$\hat{w}(\lambda, \xi, y) = -\int_0^\infty K_w^\pm(\lambda, \xi, y, \eta)\, \partial_y\hat{p}(\lambda, \xi, \eta)\, \mathrm{d}\eta, \qquad \operatorname{Re}\lambda \geq 0,\ \xi \in \mathbb{R}^{n-1},\ y > 0,$$

where

$$K_v^\pm(\lambda, \xi, y, \eta) := \left(1 \pm \frac{(\omega + |\zeta|)|\zeta|}{(\omega + 2|\zeta|)^2 \pm |\zeta|^2}\right) G_+(\lambda, \xi, y, \eta)$$

$$\mp \frac{(\omega + |\zeta|)|\zeta|}{(\omega + 2|\zeta|)^2 \pm |\zeta|^2} G_-(\lambda, \xi, y, \eta)$$

and

$$K_w^\pm(\lambda, \xi, y, \eta) := \left(1 - \frac{(\omega + 2|\zeta|)|\zeta| \mp |\zeta|^2}{(\omega + 2|\zeta|)^2 \pm |\zeta|^2}\right) G_+(\lambda, \xi, y, \eta)$$

$$+ \frac{(\omega + 2|\zeta|)|\zeta| \mp |\zeta|^2}{(\omega + 2|\zeta|)^2 \pm |\zeta|^2} G_-(\lambda, \xi, y, \eta)$$

for $\operatorname{Re}\lambda \geq 0,\ \xi \in \mathbb{R}^{n-1},\ y > 0,\ \eta > 0$.

4.34 Hence, if p is obtained as a solution to the Neumann problem $(\mathrm{L}_N \mid \infty, \mathbb{R}_+^n)_{0,h}$, then (4.6) implies

$$(\mathrm{M})^{\pm 1,0} \qquad [\hat{w}]_y(\lambda, \xi) = \int_0^\infty K_w^\pm(\lambda, \xi, 0, \eta)\, |\xi| e^{-|\xi|\eta}\, \widehat{(-\Delta_\Gamma)^{-1/2}h}(\lambda, \xi)\, \mathrm{d}\eta$$

$$= \frac{1}{\omega^2 \pm |\zeta|^2}\hat{h}(\lambda, \xi), \qquad \operatorname{Re}\lambda \geq 0,\ \xi \in \mathbb{R}^{n-1},$$

i.e. the velocity satisfies the Dirichlet type boundary condition $\mathcal{M}^{\pm 1,0}[w]_y = h$, where

$$\mathcal{M}^{+1,0} : D(\mathcal{M}^{+0}) \subseteq L_p((0, \infty), \dot{W}_p^{-1/p}(\mathbb{R}^{n-1})) \longrightarrow L_p((0, \infty), \dot{W}_p^{-1/p}(\mathbb{R}^{n-1}))$$

is given as

$$D(\mathcal{M}^{+1,0}) := {}_0H^1_p((0,\,\infty),\,\dot{W}^{-1/p}_p(\mathbb{R}^{n-1})) \cap L_p((0,\,\infty),\,W^{2-1/p}_p(\mathbb{R}^{n-1})),$$
$$\mathcal{M}^{+1,0} := \rho\varepsilon + \rho\partial_t - 2\mu\Delta_\Gamma$$

and

$$\mathcal{M}^{-1,0} : D(\mathcal{M}^{-1,0}) \subseteq L_p((0,\,\infty),\,\dot{W}^{-1/p}_p(\mathbb{R}^{n-1})) \longrightarrow L_p((0,\,\infty),\,\dot{W}^{-1/p}_p(\mathbb{R}^{n-1}))$$

is given as

$$D(\mathcal{M}^{-1,0}) := {}_0H^1_p((0,\,\infty),\,\dot{W}^{-1/p}_p(\mathbb{R}^{n-1})) \cap L_p((0,\,\infty),\,W^{2-1/p}_p(\mathbb{R}^{n-1})),$$
$$\mathcal{M}^{-1,0} := \rho\varepsilon + \rho\partial_t.$$

Thus, $\mathcal{M}^{\pm1,0}$ is defined via its Laplace-Fourier symbol

$$m^{\pm1,0}(\lambda,\,\xi) = \omega^2 \pm |\zeta|^2, \qquad \operatorname{Re}\lambda \geq 0,\,\xi \in \mathbb{R}^{n-1}.$$

4.35 Similarly, if p is obtained as a solution to the Dirichlet problem $(\mathrm{L} \mid \infty,\, \mathbb{R}^n_+)^N_{0,h}$, we have

$$-\partial_y\hat{p}(\lambda,\,\xi,\,y) = |\xi|e^{-|\xi|y}\,\hat{h}(\lambda,\,\xi), \qquad \operatorname{Re}\lambda \geq 0,\,\xi \in \mathbb{R}^{n-1},\,y > 0,$$

which implies

$$-2\mu[\partial_y w]_y + [p]_y$$

$$(\mathrm{M})^{\pm1,+1} \qquad = -2\mu\int_0^\infty (\partial_y K^\pm_w)(\lambda,\,\xi,\,0,\,\eta)\,|\xi|e^{-|\xi|\eta}\,\hat{h}(\lambda,\,\xi)\,\mathrm{d}\eta + \hat{h}(\lambda,\,\xi)$$

$$= \frac{(\omega^2 \mp |\zeta|^2) + 2\frac{\omega}{\omega+|\zeta|}(|\zeta|^2 \pm |\zeta|^2)}{\omega^2 \pm |\zeta|^2}\hat{h}(\lambda,\,\xi), \qquad \operatorname{Re}\lambda \geq 0,\,\xi \in \mathbb{R}^{n-1},$$

i.e. the Neumann type boundary condition $\mathcal{M}^{\pm1,+1}(-2\mu[\partial_y w]_y + [p]_y) = h$ is satisfied, where

$$\mathcal{M}^{+1,+1} : L_p((0,\,\infty),\,\dot{W}^{1-1/p}_p(\mathbb{R}^{n-1})) \longrightarrow L_p((0,\,\infty),\,\dot{W}^{1-1/p}_p(\mathbb{R}^{n-1}))$$

is given as

$$\mathcal{M}^{+1,+1} := (\rho\varepsilon + \rho\partial_t - 2\mu\Delta_\Gamma)(\rho\varepsilon + \rho\partial_t - \mu\Delta^*_\Gamma)^{-1}$$

and

$$\mathcal{M}^{-1,+1} : L_p((0,\,\infty),\,\dot{W}^{1-1/p}_p(\mathbb{R}^{n-1})) \longrightarrow L_p((0,\,\infty),\,\dot{W}^{1-1/p}_p(\mathbb{R}^{n-1}))$$

is given as

$$\mathcal{M}^{-1,+1} := (\rho\varepsilon + \rho\partial_t)(\rho\varepsilon + \rho\partial_t - 2\mu\Delta_\Gamma)^{-1}.$$

Thus, $\mathcal{M}^{\pm1,+1}$ is defined via its Laplace-Fourier symbol

$$m^{\pm1,+1}(\lambda,\,\xi) = \frac{\omega^2 \pm |\zeta|^2}{(\omega^2 \mp |\zeta|^2) + 2\frac{\omega}{\omega+|\zeta|}(|\zeta|^2 \pm |\zeta|^2)}, \qquad \operatorname{Re}\lambda \geq 0,\,\xi \in \mathbb{R}^{n-1}.$$

Here the pseudodifferential operator

$$-\Delta_\Gamma^* : D(-\Delta_\Gamma^*) \subseteq L_p((0, \infty), \dot{W}_p^{1-1/p}(\mathbb{R}^{n-1})) \longrightarrow L_p((0, \infty), \dot{W}_p^{1-1/p}(\mathbb{R}^{n-1}))$$

with domain

$$D(-\Delta_\Gamma^*) := L_p((0, \infty), \dot{W}_p^{3-1/p}(\mathbb{R}^{n-1}))$$

is defined via its Laplace-Fourier symbol

$$s(\lambda, \xi) := 4\frac{\omega}{\omega + |\zeta|}|\xi|^2, \quad \mathrm{Re}\,\lambda \geq 0, \ \xi \in \mathbb{R}^{n-1}.$$

4.36 Up to now the above calculations are formally. To obtain rigorous results, we need a precise knowledge of the mapping properties of the operators $\mathcal{M}^{\alpha,\beta}$ for $\alpha \in \{-1, 0, +1\}$ and $\beta \in \{0, +1\}$ and $-\Delta_\Gamma^*$. Moreover, we have to clarify their precise relation to the derived Laplace-Fourier symbols $m^{\alpha,\beta}$ and s. Note that the definition of some of these operators involve *fractional powers* or the definition of an operator based on a given Laplace-Fourier symbol. Therefore, we need to introduce the theory of *sectorial operators* and the concept of a *functional calculus* for operators of this type. This will be done in the next section along with a derivation of the missing rigorous results for the above operators.

4.4 Sectorial Operators and Functional Calculus

4.37 We consider a linear operator $A : D(A) \subseteq X \longrightarrow X$ in a Banach space X with dense domain $D(A)$. To avoid trivialities, we assume A to be unbounded. Note that this is true for most of the operators introduced in Section 4.3. Also note that the space X_A, which is defined to be $D(A)$ equipped with the graph norm of A, is a Banach space and A is bounded as $A : X_A \longrightarrow X$, if and only if A is closed. These observations together with the comments 4.36 motivate to employ the theory of sectorial operators. Therefore, we will shortly summarise this theory as presented in the monograph by R. DENK, M. HIEBER and J. PRÜSS, cf. [DHP03]. However, we restrict the presentation to the collection of results, which are relevant to treat the operators introduced in Section 4.3.

4.38 DEFINITION. *Let X be a Banach space and let $A : D(A) \subseteq X \longrightarrow X$ be a closed, densely defined linear operator with dense range. Then A is called sectorial, if*

$$(0, \infty) \subseteq \rho(-A) \quad \text{and} \quad \|t(t + A)^{-1}\|_{\mathcal{B}(X)} \leq M, \quad t > 0$$

holds with some constant $M > 0$.

4.39 For more details we refer to [DHP03, Sections 1.1 – 1.3]. As a first important result for a sectorial operator A, [DHP03, Proposition 1.2] implies A to be one-to-one and $D(A^k) \cap R(A^k)$ to be a dense subspace of X. Moreover, employing the holomorphy of the resolvent on $\rho(A)$ we infer that

$$\Sigma_\phi \subseteq \rho(-A) \quad \text{and} \quad \|z(z + A)^{-1}\|_{\mathcal{B}(X)} \leq M_\phi, \quad z \in \Sigma_\phi$$

for some sector

$$\Sigma_\phi := \{\, z \in \mathbb{C} \setminus \{\, 0 \,\} \; : \; |\arg z| < \phi \,\}$$

with angle $\phi > 0$ and some constant $M_\phi > 0$. Therefore, we call

$$\phi_A := \inf \left\{ \phi \in [0, \pi) \; : \; \Sigma_{\pi - \phi} \subseteq \rho(-A), \; \sup_{z \in \Sigma_{\pi - \phi}} \|z(z + A)^{-1}\|_{\mathcal{B}(X)} < \infty \right\}$$

the *spectral angle* of A. Obviously, we have $\phi_A \in [0, \pi)$ and

$$\phi_A \geq \sup \{\, |\arg z| \; : \; z \in \sigma(A) \,\}.$$

4.40 A first prominent example of a sectorial operator is the time derivative

$$G_s : {}_0H_p^1((0, \infty), \dot{H}_p^s(\mathbb{R}^{n-1})) \subseteq L_p((0, \infty), \dot{H}_p^s(\mathbb{R}^{n-1}))$$
$$\longrightarrow L_p((0, \infty), \dot{H}_p^s(\mathbb{R}^{n-1})), \qquad G_s := \rho \partial_t,$$

which is well-known to be invertible and sectorial with spectral angle $\phi_{G_s} = \frac{\pi}{2}$ for every $s \in \mathbb{R}$, see [HP98, DSS08] and Remark 4.72.

4.41 Another prominent example of a sectorial operator is the Laplacian

$$D_s : L_p((0, \infty), \dot{H}_p^{s+2}(\mathbb{R}^{n-1})) \subseteq L_p((0, \infty), \dot{H}_p^s(\mathbb{R}^{n-1}))$$
$$\longrightarrow L_p((0, \infty), \dot{H}_p^s(\mathbb{R}^{n-1})), \qquad D_s := -\mu \Delta_\Gamma,$$

which is well-known to be sectorial with spectral angle $\phi_{D_s} = 0$ for every $s \in \mathbb{R}$, see [DHP01, DSS08] and Remark 4.73.

4.42 To simplify our argumentation concerning the operators $\mathcal{M}^{\alpha, \beta}$ as defined in 4.32, 4.34 and 4.35 we are interested in the mapping properties of the operators

$$M_s^{\alpha, 0} : {}_0H_p^1((0, \infty), \dot{H}_p^s(\mathbb{R}^{n-1})) \cap L_p((0, \infty), \dot{H}_p^{s+2}(\mathbb{R}^{n-1}))$$
$$\subseteq L_p((0, \infty), \dot{H}_p^s(\mathbb{R}^{n-1})) \longrightarrow L_p((0, \infty), \dot{H}_p^s(\mathbb{R}^{n-1}))$$

for $\alpha \in \{\, -1, \, 0, \, +1 \,\}$ with

$$M_s^{0,0} := \rho\varepsilon + G_s + (\rho\varepsilon + G_s + D_s)^{1/2} D_s^{1/2} + D_s,$$
$$M_s^{+1,0} := \rho\varepsilon + G_s + 2D_s, \qquad \text{and} \qquad M_s^{-1,0} := \rho\varepsilon + G_s$$

for $s \in \mathbb{R}$. Note that due to the embedding

$$W_p^{2-1/p}(\mathbb{R}^{n-1}) \hookrightarrow \dot{W}_p^{2-1/p}(\mathbb{R}^{n-1})$$

we obtain the closedness of $\mathcal{M}^{\alpha, 0}$ via the closedness of the $M_s^{\alpha, 0}$ for all $s \in \mathbb{R}$ and real interpolation. Moreover, we define the operators

$$M_s^{-1,+1} : L_p((0, \infty), \dot{H}_p^s(\mathbb{R}^{n-1})) \longrightarrow L_p((0, \infty), \dot{H}_p^s(\mathbb{R}^{n-1}))$$

for $s \in \mathbb{R}$ as $M_s^{-1,+1} := (\rho\varepsilon + G_s)(\rho\varepsilon + G_s + 2D_s)^{-1}$ and note that the boundedness of the $M_s^{-1,+1}$ for all $s \in \mathbb{R}$ and real interpolation imply the boundedness of $\mathcal{M}^{-1,+1}$. Of course, the discussion in the next paragraphs will also reveal that the fractional powers and the inverse operators involved in the definition of the $M_s^{\alpha,\beta}$ are well-defined. Moreover, we will clarify the relation between these operators and the symbols $m^{\alpha,\beta}$ as defined in 4.32, 4.34 and 4.35. The operator $\mathcal{M}^{+1,+1}$ has a more complicated structure and will be treated separately. On the other hand, the operators $M_s^{-1,0}$ inherit all properties of the operators G_s. In particular, $M_s^{-1,0}$ is closed for every $s \in \mathbb{R}$.

4.43 We may obtain first results for the operators $M_s^{\alpha,\beta}$ by studying the sums $G_s + D_s$ with domain $D(G_s) \cap D(D_s)$. Due to the famous result of G. DA PRATO and P. GRISVARD on the sum of commuting sectorial operators, see [PG75] and Remark 4.74, we infer that

$$G_s + D_s : D(G_s) \cap D(D_s) \subseteq L_p((0, \infty), \dot{H}_p^s(\mathbb{R}^{n-1}))$$
$$\longrightarrow L_p((0, \infty), \dot{H}_p^s(\mathbb{R}^{n-1}))$$

is closable, its closure being sectorial with spectral angle $\frac{\pi}{2}$. However, based on the sectoriality of the operators no statement may be made concerning the closedness of their sum. To obtain such results, the notion of a *functional calculus* is needed.

4.44 For a sectorial operator $A : D(A) \subseteq X \longrightarrow X$ with spectral angle $\phi_A \geq 0$ we may extend the *Dunford functional calculus* for bounded linear operators via

$$f \mapsto f(A) := \frac{1}{2\pi i} \int\limits_{\partial \Sigma_\psi} f(z)(z - A)^{-1} \, \mathrm{d}z, \qquad f \in \mathcal{H}_0(\Sigma_\phi)$$

to a functional calculus for A. Here $\phi \in (\phi_A, \pi]$, $\psi \in (\phi_A, \phi)$ and

$$\mathcal{H}_\omega(\Sigma_\phi) := \left\{ f : \Sigma_\phi \longrightarrow \mathbb{C} : \begin{array}{c} f \text{ is holomorphic and} \\ \sup\limits_{|z| \leq 1} |z^{-\kappa} f(z)| + \sup\limits_{|z| \geq 1} |z^\kappa f(z)| < \infty \\ \text{for some } \kappa > -\omega \end{array} \right\}$$

with $\omega \geq 0$ denotes the space of polynomial decaying, holomorphic functions on Σ_ϕ. For the details we refer to [DHP03, Section 1.4]. With this definition, we have $f(A) \in \mathcal{B}(X)$. However, the class of admitted functions is quite restrictive. To define e. g. fractional or imaginary powers of a sectorial operator we need to include the holomorphic functions

$$\pi_s \in \mathcal{H}(\Sigma_\pi), \qquad \pi_s(z) = z^s, \quad s \in \Sigma_\pi,$$

which realise the powers for $s \in \mathbb{C}$. These belong to one of the bigger function classes with an appropriate $\omega > 0$. Indeed, $\pi_s \in \mathcal{H}_\omega(\Sigma_\pi)$ for all $\omega > -\mathrm{Re}\,s$.

4.45 To construct such a functional calculus we consider the function

$$\psi \in \mathcal{H}(\Sigma_\pi), \qquad \psi(z) = \frac{z}{(1+z)^2}, \quad z \in \Sigma_\pi$$

and observe, that

$$\psi^k f \in \mathcal{H}_0(\Sigma_\phi), \qquad f \in \mathcal{H}_\omega(\Sigma_\phi), \ k > \omega.$$

Moreover, $\psi(A)$ is bounded and injective with range $D(A) \cap R(A)$ and, thus, admits the inverses

$$\psi(A)^{-k} : D(A^k) \cap R(A^k) \subseteq X \longrightarrow X, \qquad k = 1, 2, \ldots$$

Hence, a reasonable extension of the functional calculus for the class $\mathcal{H}_0(\Sigma_\phi)$ to the classes $\mathcal{H}_\omega(\Sigma_\phi)$ is given via

$$f(A) = \psi(A)^{-k}(\psi^k f)(A), \quad D(f(A)) = \left\{ x \in X : (\psi^k f)(A)x \in D(A^k) \cap R(A^k) \right\}$$

for $f \in \mathcal{H}_\omega(\Sigma_\phi)$ and $k > \omega$. For the details we refer to [DHP03, Section 2.1]. In particular, the above definition is independent of the choice of $k > \omega$ and $f(A)$ is a densely defined, closed linear operator in X with $D(A^k) \cap R(A^k) \subseteq D(f(A))$.

4.46 Based on the above considerations, the complex powers $A^s = \pi_s(A)$ are densely defined, closed linear operators in X. The real powers A^σ with $|\sigma| < \pi/\phi_A$ are even sectorial with a spectral angle of at most $|\sigma|\phi_A$, cf. [DHP03, Theorem 2.1]. Moreover, the complex powers allow for the definition of an important class of sectorial operators for which a more refined result on the sum of two such operators is available.

4.47 DEFINITION. *Let X be a Banach space and let $A : D(A) \subseteq X \longrightarrow X$ be a sectorial operator. Then A is said to admit bounded imaginary powers, if*

$$A^{i\sigma} \in \mathcal{B}(X), \quad \sigma \in \mathbb{R} \quad \text{and} \quad \|A^{i\sigma}\|_{\mathcal{B}(X)} \leq C, \quad \sigma \in [-1, 1]$$

holds with some constant $C > 0$.

4.48 Now, if $A : D(A) \subseteq X \longrightarrow X$ is a sectorial operator in a Banach space X that admits bounded imaginary powers, the set $\left\{ A^{i\sigma} : \sigma \in \mathbb{R} \right\}$ forms a strongly continuous group of bounded linear operators in X and the growth bound

$$\theta_A = \limsup_{|\sigma|\to\infty} |\sigma|^{-1} \log \|A^{i\sigma}\|_{\mathcal{B}(X)}$$

is called the *power angle* of A. For the details we refer to [DHP03, Section 2.3]. We only note that the operator G_s as defined in 4.40 admits bounded imaginary powers with power angle $\theta_{G_s} = \frac{\pi}{2}$ for every $s \in \mathbb{R}$ and that the operator D_s as defined in 4.41 admits bounded imaginary powers with power angle $\theta_{D_s} = 0$ for every $s \in \mathbb{R}$, see Remarks 4.72 and 4.73. Thus, the famous result of G. DORE and A. VENNI, see [DV87] and Remark 4.75, on the sum of commuting sectorial operators admitting bounded imaginary powers in the extended version given by J. PRÜSS and H. SOHR, see [PS90] and Remark 4.76, applies and we infer that

$$G_s + D_s : D(G_s) \cap D(D_s) \subseteq L_p((0, \infty), \dot{H}_p^s(\mathbb{R}^{n-1}))$$
$$\longrightarrow L_p((0, \infty), \dot{H}_p^s(\mathbb{R}^{n-1}))$$

is sectorial, invertible and admits bounded imaginary powers with power angle $\frac{\pi}{2}$ for every $s \in \mathbb{R}$. Therefore, the operators $M_s^{+1,0}$ have the same properties. In particular, $M_s^{+1,0}$ is closed for every $s \in \mathbb{R}$. Moreover, the operators $M^{-1,+1}$ are well-defined and bounded for all $s \in \mathbb{R}$.

4.49 Now, the fractional powers $D_s^{1/2}$ are sectorial and admit bounded imaginary powers with power angle 0 for every $s \in \mathbb{R}$. Based on [DHP03, Theorem 2.5] their domain equipped with the graph norm may be computed as the complex interpolation space

$$[L_p((0, \infty), \dot{H}_p^{s+2}(\mathbb{R}^{n-1})), L_p((0, \infty), \dot{H}_p^s(\mathbb{R}^{n-1}))]_{1/2}$$
$$= L_p((0, \infty), \dot{H}_p^{s+1}(\mathbb{R}^{n-1})).$$

On the other hand,

$$D_s^{1/2} : L_p((0, \infty), \dot{H}_p^{s+1}(\mathbb{R}^{n-1})) \longrightarrow L_p((0, \infty), \dot{H}_p^s(\mathbb{R}^{n-1}))$$

is an isomorphism for every $s \in \mathbb{R}$ by the definition of the homogeneous Bessel potential spaces. Furthermore, the fractional powers $(G_s + D_s)^{1/2}$ are sectorial, invertible and admit bounded imaginary powers with power angle $\frac{\pi}{4}$ for every $s \in \mathbb{R}$. Based on [DHP03, Theorem 2.5] their domain equipped with the graph norm may be computed as the complex interpolation space

$$[{}_0H_p^1((0, \infty), \dot{H}_p^s(\mathbb{R}^{n-1})) \cap L_p((0, \infty), \dot{H}_p^{s+2}(\mathbb{R}^{n-1})), L_p((0, \infty), \dot{H}_p^s(\mathbb{R}^{n-1}))]_{1/2}$$
$$= [{}_0H_p^1((0, \infty), \dot{H}_p^s(\mathbb{R}^{n-1})), L_p((0, \infty), \dot{H}_p^s(\mathbb{R}^{n-1}))]_{1/2}$$
$$\cap [L_p((0, \infty), \dot{H}_p^{s+2}(\mathbb{R}^{n-1})), L_p((0, \infty), \dot{H}_p^s(\mathbb{R}^{n-1}))]_{1/2}$$
$$= {}_0H_p^{1/2}((0, \infty), \dot{H}_p^s(\mathbb{R}^{n-1})) \cap L_p((0, \infty), \dot{H}_p^{s+1}(\mathbb{R}^{n-1})).$$

Therefore, the operators

$$(\rho\varepsilon + G_s + D_s)^{1/2} : {}_0H_p^{1/2}((0, \infty), \dot{H}_p^s(\mathbb{R}^{n-1})) \cap L_p((0, \infty), \dot{H}_p^{s+1}(\mathbb{R}^{n-1}))$$
$$\longrightarrow L_p((0, \infty), \dot{H}_p^s(\mathbb{R}^{n-1}))$$

are isomorphisms for all $s \in \mathbb{R}$ and the same is true for the operators

$$(\rho\varepsilon + G_s + D_s)^{1/2} D_s^{1/2} : {}_0H_p^{1/2}((0, \infty), \dot{H}_p^{s+1}(\mathbb{R}^{n-1})) \cap L_p((0, \infty), \dot{H}_p^{s+2}(\mathbb{R}^{n-1}))$$
$$\longrightarrow L_p((0, \infty), \dot{H}_p^s(\mathbb{R}^{n-1})).$$

In particular, the operators $M_s^{0,0}$ are closed for all $s \in \mathbb{R}$.

4.50 To construct the operator $-\Delta_\Gamma^*$, which was employed in 4.35 to define the operator $\mathcal{M}^{+1,+1}$, we first need a further extension of the functional calculus defined in 4.44 to the class

$$\mathcal{H}^\infty(\Sigma_\phi) := \Big\{ f : \Sigma_\phi \longrightarrow \mathbb{C} : f \text{ is holomorphic and bounded} \Big\}.$$

4.51 DEFINITION. *Let X be a Banach space and let $A : D(A) \subseteq X \longrightarrow X$ be a sectorial operator with spectral angle $\phi_A \geq 0$. Then A is said to admit a bounded $\mathcal{H}^\infty$-calculus, if the estimate*

$$\|f(A)\|_{\mathcal{B}(X)} \leq K_\phi \|f\|_\infty, \qquad f \in \mathcal{H}_0(\Sigma_\phi)$$

holds for some $\phi \in (\phi_A, \pi)$ and some constant $K_\phi > 0$.

4.52 The reason for the above notion is the fact that the uniform boundedness of the operators $f(A)$ for $f \in \mathcal{H}_0(\Sigma)$ implies the functional calculus to be uniquely extensible to the class $\mathcal{H}^\infty(\Sigma_\phi)$. The infimum of all angles $\phi \in (\phi_A, \pi)$, for which such a uniform estimate is valid, is called the $\mathcal{H}^\infty$-angle of A and denoted by ϕ_A^∞. For the details we refer to [DHP03, Section 2.4]. We only note that the operator G_s as defined in 4.40 admits a bounded $\mathcal{H}^\infty$-calculus with $\mathcal{H}^\infty$-angle $\phi_{G_s}^\infty = \frac{\pi}{2}$ for every $s \in \mathbb{R}$ and that the operator D_s as defined in 4.41 admits a bounded $\mathcal{H}^\infty$-calculus with $\mathcal{H}^\infty$-angle $\phi_{D_s}^\infty = 0$ for every $s \in \mathbb{R}$,

see Remarks 4.72 and 4.73. Furthermore, the celebrated theorem of N. J. KALTON and
L. WEIS, see [KW01] and Remark 4.77, applies and we infer that the operator $G_s + D_s$
as defined in 4.48 admits a bounded $\mathcal{H}^\infty$-calculus with $\mathcal{H}^\infty$-angle $\phi_{G_s + D_s}^\infty = \frac{\pi}{2}$.

4.53 Now, we have to relate the symbols obtained in Section 4.3 to the corresponding
operators. First observe that

$$\mathcal{L}G_0\varphi(\lambda) = \rho\lambda\mathcal{L}\varphi(\lambda), \quad \operatorname{Re}\lambda \geq 0, \quad \varphi \in C_0^\infty((0, \infty), L_p(\mathbb{R}^{n-1})).$$

Thus, the symbol of G_0 is $\rho\lambda$ and G_0 is uniquely defined by this relation, since the space
$C_0^\infty((0, \infty), L_p(\mathbb{R}^{n-1}))$ is a dense subspace of $_0H_p^1((0, \infty), L_p(\mathbb{R}^{n-1}))$. Furthermore, the
definition of the functional calculus as introduced in 4.44 implies

$$\mathcal{L}f(G_0)\varphi(\lambda) = \frac{1}{2\pi i} \int\limits_{\partial\Sigma_\psi} \frac{f(z)}{z - \rho\lambda} \mathcal{L}\varphi(\lambda)\,dz$$

$$= f(\rho\lambda)\mathcal{L}\varphi(\lambda), \quad \operatorname{Re}\lambda \geq 0, \qquad \begin{array}{l} \varphi \in C_0^\infty((0, \infty), L_p(\mathbb{R}^{n-1})), \\ f \in \mathcal{H}_0(\Sigma_\phi) \end{array}$$

with $\frac{\pi}{2} < \psi < \phi < \pi$. Hence, the symbol of $f(G_0)$ is $f(\rho\lambda)$ and this relation remains valid
for all $f \in \mathcal{H}^\infty(\Sigma_\phi)$ with $\phi > \frac{\pi}{2}$ by the $\mathcal{H}^\infty$-calculus of G_0. Analogously

$$\mathcal{F}D_0\varphi(\xi) = |\zeta|^2\mathcal{F}\varphi(\xi), \quad \xi \in \mathbb{R}^{n-1}, \quad \varphi \in L_p((0, \infty), C_0^\infty(\mathbb{R}^{n-1})).$$

Thus, the symbol of D_0 is $|\zeta|^2$ and D_0 is uniquely defined by this relation, since the space
$L_p((0, \infty), C_0^\infty(\mathbb{R}^{n-1}))$ is a dense subspace of $L_p((0, \infty), \hat{H}_p^2(\mathbb{R}^{n-1}))$. Moreover, we have

$$\mathcal{F}f(D_0)\varphi(\xi) = \frac{1}{2\pi i} \int\limits_{\partial\Sigma_\psi} \frac{f(z)}{z - |\zeta|^2} \mathcal{F}\varphi(\xi)\,dz$$

$$= f(|\zeta|^2)\mathcal{F}\varphi(\xi), \quad \xi \in \mathbb{R}^{n-1}, \qquad \begin{array}{l} \varphi \in L_p((0, \infty), C_0^\infty(\mathbb{R}^{n-1})), \\ f \in \mathcal{H}_0(\Sigma_\phi) \end{array}$$

with $0 < \psi < \phi < \pi$. Hence, the symbol of $f(D_0)$ is $f(|\zeta|^2)$ and this relation remains
valid for all $f \in \mathcal{H}^\infty(\Sigma_\phi)$ with $\phi > 0$ by the $\mathcal{H}^\infty$-calculus of D_0. Employing both, the
Laplace and the Fourier transformation, we infer that the symbol of $G_0 + D_0$ is ω^2 and
that the symbol of $G_0 + 2D_0$ is $\omega^2 + |\zeta|^2$.

4.54 To compute the symbols of fractional powers recall the definition of the power
functions π_s with $s \in \Sigma_\pi$ introduced in 4.44 and the definition of the function ψ introduced
in 4.45. In particular, we have $\psi\pi_{1/2} \in \mathcal{H}_0(\Sigma_\pi)$ and infer

$$\mathcal{F}D_0^{1/2}\varphi(\xi) = \mathcal{F}\pi_{1/2}(D_0)\varphi(\xi)$$

$$= \mathcal{F}\psi(D_0)^{-1}(\psi\pi_{1/2})(D_0)\varphi(\xi)$$

$$= \psi(|\zeta|^2)^{-1}(\psi\pi_{1/2})(|\zeta|^2)\mathcal{F}\varphi(\xi)$$

$$= |\zeta|\mathcal{F}\varphi(\xi), \quad \xi \in \mathbb{R}^{n-1}, \quad \varphi \in L_p((0, \infty), C_0^\infty(\mathbb{R}^{n-1})).$$

As the not surprising result the symbol of the operator $D_0^{1/2}$ is given by $|\zeta|$. Analogously, we have the operator-symbol relations

$$(G_0 + D_0)^{1/2} \sim \omega \quad \text{and} \quad (G_0 + D_0)^{1/2} D_0^{1/2} \sim \omega|\zeta|.$$

Therefore, the operators $\mathcal{M}^{\alpha,\beta}$ introduced in Section 4.3 indeed belong to the symbols $m^{\alpha,\beta}$ introduced there for $(\alpha,\ \beta) \in \{(-1,\ 0),\ (-1,\ +1),\ (0,\ 0),\ (0,\ +1),\ (+1,\ 0)\}$. Recall that all these operators are closed as has been proved in the previous paragraphs.

4.55 Finally, an elegant way to rigorously define the operator $-\Delta_\Gamma^*$ is to employ a *joint* $\mathcal{H}^\infty$-*calculus*. Assume A and B to be two sectorial operators in a Banach space X with domains $D(A)$ respectively $D(B)$ and spectral angels $\phi_A \geq 0$ and $\phi_B \geq 0$. If A and B commute, then the functional calculus introduced in 4.44 may be extended to a joint functional calculus

$$f(A,\ B) := \frac{1}{(2\pi i)^2} \int\limits_{\partial\Sigma_\psi} \int\limits_{\partial\Sigma_{\psi'}} f(z,\ w)(z - A)^{-1}(w - B)^{-1}\,dw\,dz, \quad f \in \mathcal{H}_0(\Sigma_\phi \times \Sigma_{\phi'}),$$

where $\phi_A < \psi < \phi$ and $\phi_B < \psi' < \phi'$. For the details we refer to the work of N. J. KALTON and L. WEIS, cf. [KW01]. Due to [KW01, Theorem 6.1] this calculus applied to the operators $A = \rho\varepsilon + G_s + D_s$ and $B = D_s$ even extends to a bounded joint $\mathcal{H}^\infty$-calculus. Thus, we set

$$f^*(z,\ w) = \frac{\sqrt{\rho\varepsilon + z}}{\sqrt{\rho\varepsilon + z} + \sqrt{w}}, \quad z \in \Sigma_{\pi-\psi},\ w \in \Sigma_{\psi/2}$$

with some $0 < \psi < \frac{\pi}{2}$ and infer that the operators

$$f^*(G_s + D_s,\ D_s) : L_p((0,\ \infty),\ \dot{H}_p^s(\mathbb{R}^{n-1})) \longrightarrow L_p((0,\ \infty),\ \dot{H}_p^s(\mathbb{R}^{n-1}))$$

are bounded for all $s \in \mathbb{R}$. Indeed, $\mathrm{Re}(\sqrt{\rho\varepsilon + z} + \sqrt{w}) \geq \mathrm{Re}(\sqrt{z} + \sqrt{w}) > 0$ for $z \in \Sigma_{\pi-\psi}$ and $w \in \Sigma_{\psi/2}$ and, hence,

$$\mathrm{Re}(\sqrt{z} + \sqrt{w}) \geq C_\psi, \quad z \in \bar{\Sigma}_{\pi-\psi},\ w \in \bar{\Sigma}_{\psi/2},\ |z| + |w| = 1$$

with some constant $C_\psi > 0$. Thus,

$$|\sqrt{z} + \sqrt{w}| \geq \mathrm{Re}(\sqrt{z} + \sqrt{w}) = \mathrm{Re}\left(\frac{\sqrt{z}}{\sqrt{|z| + |w|}} + \frac{\sqrt{w}}{\sqrt{|z| + |w|}}\right)\sqrt{|z| + |w|}$$

$$\geq C_\psi\sqrt{|z| + |w|}, \quad z \in \Sigma_{\pi-\psi},\ w \in \Sigma_{\psi/2}$$

and f^* is bounded by $1/C_\psi$. Furthermore, the Laplace and Fourier transformations reveal the symbol of the operator $f^*(G_0 + D_0,\ D_0)$ to be $\frac{\omega}{\omega + |\zeta|}$. Thus, we may define the operators D_s^* as

$$D_s^* : L_p((0,\ \infty),\ \dot{H}_p^{s+2}(\mathbb{R}^{n-1})) \subseteq L_p((0,\ \infty),\ \dot{H}_p^s(\mathbb{R}^{n-1}))$$

$$\longrightarrow L_p((0,\ \infty),\ \dot{H}_p^s(\mathbb{R}^{n-1})), \quad D_s^* := (4/\mu)f^*(G_s + D_s,\ D_s)D_s.$$

By the bounded joint $\mathcal{H}^\infty$-calculus of $G_s + D_s$ and D_s the operators D_s^* are closed for all $s \in \mathbb{R}$. Hence, we obtain $-\Delta_\Gamma^*$ as defined in Section 4.3 by real interpolation.

4.56 Analogously, we set

$$f(z, w) = \frac{\rho\varepsilon + z + 2w}{\rho\varepsilon + z + 4\frac{\sqrt{\rho\varepsilon + z + w}}{\sqrt{\rho\varepsilon + z + w} + \sqrt{w}}w}, \qquad z \in \Sigma_{\pi - \psi}, \ w \in \Sigma_{\psi/2}$$

with some $0 < \psi < \frac{\pi}{2}$ and infer that the operators

$$M_s^{+1,+1} : L_p((0, \infty), \dot{H}_p^s(\mathbb{R}^{n-1}))$$

$$\longrightarrow L_p((0, \infty), \dot{H}_p^s(\mathbb{R}^{n-1})), \quad M_s^{+1,+1} := f(G_s, D_s)$$

are bounded for every $s \in \mathbb{R}$, by the joint $\mathcal{H}^\infty$-calculus of G_s and D_s. Hence, we obtain the boundedness of the operator $\mathcal{M}^{+1,+1}$ as defined in 4.35 by real interpolation. Last, but not least, the symbol of $\mathcal{M}^{+1,+1}$ is given by $m^{+1,+1}$ as an application of the Laplace and Fourier transformations reveals.

4.5 A Splitting Scheme for the Stokes Equations

4.57 With the results of Sections 4.1 and 4.2 at hand we may develop a splitting scheme, which is applicable for a large class of domains and completely solves the Stokes equations in the case $\beta = -1$, i.e. if a plain pressure condition is imposed. However, we first have to deal with an inhomogeneous divergence condition in $(S \mid a, \Omega)$ by defining a suitable pressure, whose gradient forces the velocity field to develop the requested divergence. Recall that the linear functional

$$(\cdot \mid \cdot) : L_p((0, a), L_p(\Omega)) \times L_p((0, a), L_p(\Gamma, N\Gamma)) \longrightarrow L_p((0, a), {}_0H_p^{-1}(\Omega)),$$

$$\langle \phi \mid (\psi \mid \eta) \rangle := \int_\Gamma [\phi]_\Gamma (\eta \cdot \nu_\Gamma) \, d\mathcal{H}^{n-1} - \int_\Omega \phi\psi \, d\mathcal{H}^n, \qquad \phi \in H_{p'}^1(\Omega),$$

$$\psi \in L_p((0, a), L_p(\Omega)), \ \eta \in L_p((0, a), L_p(\Gamma, N\Gamma)),$$

where $1/p + 1/p' = 1$, has already been employed in 3.27.

4.58 PROPOSITION. *Let $a > 0$ and let $\Omega \subseteq \mathbb{R}^n$ be a halfspace, a bent halfspace of type BUC^{3-} or a bounded domain with boundary $\Gamma = \partial\Omega$ of class C^{3-}. Let $1 < p < \infty$ with $p \neq \frac{3}{2}$, 3, let $\rho, \mu > 0$ and let*

$$g \in H_p^{1/2}((0, a), L_p(\Omega)) \cap L_p((0, a), H_p^1(\Omega)),$$

such that there exists

$$\eta \in W_p^{1-1/2p}((0, a), L_p(\Gamma, N\Gamma)) \cap L_p((0, a), W_p^{2-1/p}(\Gamma, N\Gamma))$$

$$\text{with} \quad (g, \eta) \in H_p^1((0, a), {}_0\dot{H}_p^{-1}(\Omega)).$$

Then there exists

$$q \in L_p((0, a), {}_0\dot{H}_p^1(\Omega))$$

such that

$$-\mathrm{div}\nabla q = (\rho\partial_t - \mu\Delta)g$$

holds in the sense of distributions.

Proof. We define $\hat{q} \in H_p^1((0, a), \hat{H}_p^1(\Omega))$ be the unique solution to

$$(\nabla \phi \,|\, \nabla \hat{q})_\Omega = \langle \phi \,|\, (g, \eta) \rangle, \qquad \phi \in \hat{H}_{p'}^1(\Omega),$$

which is available thanks to Proposition 4.10, cf. also [SS92]. On one hand, we have $\partial_t \nabla \hat{q} \in L_p((0, a), L_p(\Omega, \mathbb{R}^n))$, and, on the other hand,

$$(\nabla \phi \,|\, \nabla \hat{q})_\Omega = \langle \phi \,|\, (g, \eta) \rangle = -\int_\Omega \phi g \, d\mathcal{H}^n, \qquad \phi \in C_0^\infty(\Omega)$$

implies $\Delta \hat{q} = \operatorname{div} \nabla \hat{q} = g$ and, hence, $\Delta \nabla \hat{q} = \nabla \Delta \hat{q} = \nabla g \in L_p((0, a), L_p(\Omega, \mathbb{R}^n))$. Finally, we define $q \in L_p((0, a), {}_0\dot{H}_p^1(\Omega))$ via $\nabla q = -(1 - \mathcal{W}_p)(\rho \partial_t - \mu \Delta) \nabla \hat{q}$ and infer that

$$-\operatorname{div} \nabla q = \operatorname{div}(1 - \mathcal{W}_p)(\rho \partial_t - \mu \Delta) \nabla \hat{q} = \operatorname{div}(\rho \partial_t - \mu \Delta) \nabla \hat{q}$$

$$= (\rho \partial_t - \mu \Delta) \operatorname{div} \nabla \hat{q} = (\rho \partial_t - \mu \Delta) g$$

holds in the sense of distributions. □

4.59 Using Proposition 4.58, we may formulate the promised splitting scheme which completely solves the Stokes equations in the case $\beta = -1$. Moreover, it may serve to eliminate almost all data for the Stokes equations $(S \,|\, a, \Omega)$ in the cases $\beta \in \{ 0, +1 \}$.

4.60 THEOREM. *Let $a > 0$ and let $\Omega \subseteq \mathbb{R}^n$ be a halfspace, a sufficiently flat bent halfspace of type BUC^{3-} or a bounded domain with boundary $\Gamma = \partial \Omega$ of class C^{3-}. Let $1 < p < \infty$ with $p \neq \frac{3}{2}, 3$, let $\rho, \mu > 0$ and let $\alpha, \beta \in \{ -1, 0, +1 \}$. Moreover,*

- *let $\gamma = -\infty$, if $\beta = 0$;*
- *let $\gamma \in \{ -\infty \} \cup [0, 1/2 - 1/2p]$, if $\beta = +1$;*
- *let $\gamma \in \{ -\infty \} \cup [0, \infty)$, if $\beta = -1$.*

Then the unique maximal regular solution

$$(u, p) \in \mathbb{X}_\gamma^\beta(a, \Omega)$$

to the Stokes equations $(S \,|\, a, \Omega)_{f,g,h,u_0}^{\alpha,\beta}$ with data

$$(f, g, h, u_0) \in \mathbb{Y}_\gamma^{\alpha,\beta}(a, \Omega)$$

may be obtained as follows:

1. *If $\beta = 0$, choose $\eta = Q_\Gamma h$, define*

$$\tilde{q} \in L_p((0, a), {}_0\dot{H}_p^1(\Omega))$$

according to Proposition 4.58 based on g and η and set

$$\bar{q} = \tilde{q} + \mathbb{R} \quad \text{for a (bent) halfspace,} \quad \bar{q} = \hat{q} - (\tilde{q})_\Omega \quad \text{for a bounded domain.}$$

If $\beta = \pm 1$, choose $\eta \in \mathbb{N}_{h,-\infty}^0(a, \Gamma)$ via compatibility condition $(C_3)_{g,h,u_0}^\beta$, define

$$\tilde{q} \in L_p((0, a), {}_0\dot{H}_p^1(\Omega))$$

according to Proposition 4.58 based on g and η and set

$$\bar{q} = \tilde{q} - \dot{\mathcal{E}}_\Omega(h \cdot \nu_\Gamma).$$

2. Choose $v \in \mathbb{X}_u(a, \Omega)$ to be the unique maximal regular solution to the parabolic problem

$$\rho \partial_t v - \mu \Delta v = \rho \mathcal{W}_p f - \nabla \bar{q} \quad \text{in } (0, a) \times \Omega,$$

$$P_\Gamma \mathcal{B}^\alpha v = P_\Gamma h \qquad\qquad \text{on } (0, a) \times \partial\Omega,$$

$$[div\, v]_\Gamma = [g]_\Gamma \qquad\qquad \text{on } (0, a) \times \partial\Omega,$$

$$v(0) = u_0 \qquad\qquad\quad \text{in } \Omega.$$

and define $q \in \mathbb{X}_{p,\gamma}^\beta(a, \Omega)$ by means of $\nabla q = \nabla \bar{q} + \rho(1 - \mathcal{W}_p)f$ and $(q)_\Omega = 0$ for a bounded domain and $\beta = 0$. If $\beta = -1$, then $(u, p) = (v, q)$ already constitutes the unique maximal regular solution to the Stokes equations $(S \,|\, a, \Omega)_{f,g,h,u_0}^{\alpha,\beta}$.

3. If $\beta \in \{0, +1\}$, the solution to the Stokes equations $(S \,|\, a, \Omega)_{f,g,h,u_0}^{\alpha,\beta}$ splits as $u = v + \bar{u}$ and $p = q + \bar{p}$, where

$$(\bar{u}, \bar{p}) \in {}_0\mathbb{X}_\gamma^\beta(a, \Omega)$$

denotes the unique maximal regular solution to the Stokes equations $(S \,|\, a, \Omega)_{0,0,\bar{h},0}^{\alpha,\beta}$ with

$$\bar{h} = Q_\Gamma(h - [v]_\Gamma), \text{ if } \beta = 0 \quad \text{respectively} \quad \bar{h} = -2\mu\, Q_\Gamma \partial_\nu v, \text{ if } \beta = +1.$$

We have $P_\Gamma \bar{h} = 0$ and

$$Q_\Gamma \bar{h} \in {}_0 H_p^1((0, a), \dot{W}_p^{-1/p}(\Gamma, N\Gamma)) \cap \mathbb{N}_{h,-\infty}^0(a, \Gamma), \quad \text{if } \beta = 0,$$

respectively

$$Q_\Gamma \bar{h} \in \mathbb{N}_{h,1/2-1/2p}^{+1}(a, \Gamma), \quad \text{if } \beta = +1.$$

Proof. First note that by construction we always have

$$\tilde{q} \in L_p((0, a), {}_0\dot{H}_p^1(\Omega)), \qquad -\text{div}\,\nabla\tilde{q} = (\rho\partial_t - \mu\Delta)g \quad \text{in } \mathcal{D}'(\Omega)$$

and, therefore,

$$\bar{q} \in \mathbb{X}_{p,\gamma}^\beta(a, \Omega), \qquad -\text{div}\,\nabla\bar{q} = (\rho\partial_t - \mu\Delta)g \quad \text{in } \mathcal{D}'(\Omega)$$

and $\bar{q}$ satisfies the boundary condition

$$-[\bar{q}]_\Gamma = h \cdot \nu_\Gamma, \quad \text{if } \beta \in \{-1, +1\}.$$

Hence, $q \in \mathbb{X}_{p,\gamma}^\beta(a, \Omega)$ enjoys the same property on the boundary. Moreover, $v \in \mathbb{X}_u(a, \Omega)$ by the maximal regularity property of the parabolic system, cf. Proposition 4.21. Furthermore,

$$\rho\partial_t v - \mu\Delta v + \nabla q = \rho\partial_t v - \mu\Delta v + \nabla\bar{q} + \rho(1 - \mathcal{W}_p)f = \rho f \quad \text{in } (0, a) \times \Omega,$$

which shows that the momentum equation is valid for v and q. Applying the divergence in the sense of distributions to the partial differential equation used to obtain v and using the properties of $\bar{q}$ as well as compatibility condition $(C_1)_{g,u_0}$ we derive

$$\rho\partial_t\,(\text{div}\,v - g) - \mu\Delta\,(\text{div}\,v - g) = 0 \quad \text{in } (0, \infty) \times \Omega,$$

$$[\text{div}\,v - g]_\Gamma = 0 \quad \text{on } (0, \infty) \times \partial\Omega,$$

$$(\text{div}\,v - g)(0) = 0 \quad \text{in } \Omega,$$

which ensures $\operatorname{div} v = g$ by uniqueness of weak solutions to this parabolic initial boundary value problem with Dirichlet boundary condition. Last, but not least, v satisfies the tangential part of the desired boundary condition and the initial condition $v(0) = u_0$.

Now, if $\beta = -1$, then

$$(v, q) \in \mathbb{X}_\gamma^\beta(a, \Omega)$$

constitutes a maximal regular solution to the Stokes equations $(S \mid a, \Omega)_{f,g,h,u_0}^{\alpha,\beta}$. On the other hand, if $\beta \in \{ 0, +1 \}$, then

$$(\bar{u}, \bar{p}) := (u - v, p - q)$$

solves the Stokes equations $(S \mid a, \Omega)_{0,0,\bar{h},0}^{\alpha,\beta}$ with

$$\bar{h} = Q_\Gamma(h - [v]_\Gamma), \text{ if } \beta = 0 \quad \text{respectively} \quad \bar{h} = -2\mu\, Q_\Gamma \partial_\nu v, \text{ if } \beta = +1.$$

Hence, $P_\Gamma \bar{h} = 0$ and $\bar{h}$ enjoys the stated regularity properties. $\qquad\square$

4.61 Note that the proof of Theorem 4.60 only shows, that the solution constructed in steps 1 and 2 has the desired regularity properties and solves the Stokes equations, where the normal part of the desired boundary condition may be violated if $\beta \in \{ 0, +1 \}$. In these cases the first two steps reduce the task to prove the existence of unique maximal regular solutions to the Stokes equations as claimed in Theorem 3.30 to the special case $f = 0$, $g = 0$, $P_\Gamma h = 0$ and $u_0 = 0$ with

$$Q_\Gamma h \in {}_0H_p^1((0, a), \dot{W}_p^{-1/p}(\Gamma, N\Gamma)) \cap \mathbb{N}_{h,-\infty}^0(a, \Gamma), \quad \text{if } \beta = 0,$$

respectively

$$Q_\Gamma h \in \mathbb{N}_{h,1/2-1/2p}^{+1}(a, \Gamma), \quad \text{if } \beta = +1.$$

Once this additional task is accomplished, Theorem 3.30 will be completely proved in these cases, see also 3.33. In particular, there is no need to consider the uniqueness of maximal regular solutions in the proof of Theorem 4.60.

4.62 On the other hand, if $\beta = -1$, the first two steps establish the existence of a maximal regular solution to the Stokes equations and it remains to prove its uniqueness to complete the proof of Theorem 3.30. However, if

$$(u, p) \in \mathbb{X}_\gamma^\beta(a, \Omega)$$

constitutes a maximal regular solution to the Stokes equations $(S \mid a, \Omega)_{0,0,0,0}^{\alpha,\beta}$ and $\beta = -1$, then

$$p \in L_p((0, a), {}_0\dot{H}_p^1(\Omega))$$

and the momentum equation delivers

$$-(\nabla p \mid \nabla \phi)_\Omega = 0, \qquad \phi \in C_0^\infty(\Omega),$$

which implies $p = 0$ by Proposition 4.4, and, hence, $u = 0$ by uniqueness of the solution to the corresponding parabolic problem, cf. Proposition 4.21. Therefore, the proof of Theorem 3.30 for $\beta = -1$ is complete and, again, there is no need to consider the uniqueness of maximal regular solutions in the proof of Theorem 4.60.

4.63 As has been shown in 4.61, the splitting scheme Theorem 4.60 allows the proof of Theorem 3.30 to be reduced to the special case

$$f = 0, \quad g = 0, \quad u_0 = 0 \quad \text{and}$$

$$P_\Gamma h = 0, \quad Q_\Gamma h \in {}_0H_p^1((0, a), \dot{W}_p^{-1/p}(\Gamma, N\Gamma)) \cap \mathbb{N}_{h,-\infty}^0(a, \Gamma)$$

for $\beta = 0$. However, if $(u, p) \in {}_0\mathbb{X}_{-\infty}^0(a, \Omega)$ should be a solution to $(S \mid a, \Omega)_{f,g,h,u_0}^{\alpha,\beta}$, where the data is subject to these additional conditions, then we may first obtain a maximal regular solution

$$\nabla\varphi \in {}_0H_p^1((0, a), L_p(\Omega, \mathbb{R}^n)) \cap L_p((0, a), H_p^2(\Omega, \mathbb{R}^n))$$

to the elliptic problem

$$-\Delta\varphi = 0 \qquad \text{in } \Omega,$$

$$\partial_\nu\varphi = Q_\Gamma h \cdot \nu_\Gamma \qquad \text{on } \partial\Omega$$

and set $(u, p) = (\bar{u}, \bar{p}) + (\nabla\varphi, \mu\Delta\varphi - \rho\partial_t\varphi)$, where

$$(\bar{u}, \bar{p}) \in {}_0\mathbb{X}_{-\infty}^0(a, \Omega)$$

solves the Stokes equations $(S \mid a, \Omega)_{0,0,\bar{h},0}^{\alpha,\beta}$ with

$$P_\Gamma\bar{h} = -P_\Gamma\mathcal{B}^\alpha(\nabla\varphi) \quad \text{and} \quad Q_\Gamma\bar{h} = 0.$$

Hence, in order to prove Theorem 3.30 for $\beta = 0$ it is sufficient to consider the special case

$$f = 0, \quad g = 0, \quad Q_\Gamma h = 0 \quad \text{and} \quad u_0 = 0.$$

4.6 Uniform Sobolev Type Embeddings

4.64 To treat the perturbations of the Stokes equations, which arise during the localisation procedure, we will make frequent use of a family of uniform Sobolev type embeddings, which allow for the exploitation of available additional time regularity. These types of embeddings are inspired by [PSS07, Proposition 6.2 (a)]. To derive the corresponding estimates, we employ the ideas used in [PSS07] and start with a result on uniform extension operators.

4.65 PROPOSITION. *Let X be a Banach space and let $1 < p < \infty$. Then there exists a family of bounded linear extension operators*

$$\mathcal{E}_a : L_p((0, a), X) \longrightarrow L_p((0, \infty), X), \qquad 0 < a < \infty$$

such that

$$\|\mathcal{E}_a\|_{\mathcal{B}({}_0H_p^\tau((0,a), X), {}_0H_p^\tau((0,\infty), X))} \leq C, \qquad 0 < a < \infty, \ \tau \in (0, 1]$$

with a constant $C > 0$, which is independent of $0 < a < \infty$ and $\tau \in (0, 1]$.

Proof. We define the operator $\mathcal{E}_a$ as in the proof of [PSS07, Proposition 6.1], i.e.

$$\mathcal{E}_a\phi(t) := \begin{cases} \phi(t), & t \in (0, a), \\ \phi(2a - t), & t \in (a, 2a), \\ 0, & t \in (2a, \infty), \end{cases} \qquad \phi \in L_p((0, a), X).$$

Then we have

$$\|\mathcal{E}_a\|_{\mathcal{B}(L_p((0,a),X),\, L_p((0,\infty),X))}, \; \|\mathcal{E}_a\|_{\mathcal{B}(_0W_p^1((0,a),X),\,_0W_p^1((0,\infty),X))} \leq 2^{1/p}.$$

Now, we denote by $\mathcal{E}_0 : L_p((0, \infty), X) \longrightarrow L_p(\mathbb{R}, X)$ the extension by zero and infer

$$\|\phi\|_{_0H_p^1((0,a),X)} \leq \|\mathcal{E}_0\mathcal{E}_a\phi\|_{_0H_p^1(\mathbb{R},X)} \leq M\|\mathcal{E}_0\mathcal{E}_a\phi\|_{_0W_p^1(\mathbb{R},X)} \leq 2^{1/p}M\|\phi\|_{_0W_p^1((0,a),X)}$$

for all $\phi \in {}_0W_p^1((0, a), X)$, the constant $M > 0$ being independent of $0 < a < \infty$. Conversely, if $\phi \in {}_0H_p^1((0, a), X)$ and $\varepsilon > 0$, then there exists $\psi_\varepsilon \in {}_0H_p^1(\mathbb{R}, X)$ such that $\mathcal{R}_a\psi_\varepsilon = \phi$ and

$$\|\phi\|_{_0W_p^1((0,a),X)} \leq \|\psi_\varepsilon\|_{_0W_p^1(\mathbb{R},X)} \leq M'\|\psi_\varepsilon\|_{_0H_p^1(\mathbb{R},X)} \leq M'\left(\|\phi\|_{_0H_p^1((0,a),X)} + \varepsilon\right),$$

the constant $M' > 0$ being independent of $0 < a < \infty$. Here

$$\mathcal{R}_a : L_p(\mathbb{R}, X) \longrightarrow L_p((0, a), X)$$

denotes the restriction to $(0, a)$. Since $\varepsilon > 0$ was arbitrary, we have proved the identity $_0H_p^1((0, a), X) \cong {}_0W_p^1((0, a), X)$ and

$$c^{-1}\|\phi\|_{_0H_p^1((0,a),X)} \leq \|\phi\|_{_0W_p^1((0,a),X)} \leq c\|\phi\|_{_0H_p^1((0,a),X)}, \qquad \phi \in {}_0H_p^1((0, a), X)$$

with $c := \max\{2^{1/p}M, M'\} > 0$ being independent of $0 < a < \infty$. Hence, the assertion first follows for $\tau = 1$ and then for all $\tau \in (0, 1]$ by complex interpolation. $\square$

4.66 The reason for insisting on the fact that the extension operators constructed in Proposition 4.65 are also bounded in L_p is the application to anisotropic spaces.

4.67 COROLLARY. *Let X and Y be Banach spaces and let $1 < p < \infty$. Then there exists a family of extension operators*

$$\mathcal{E}_a : {}_0H_p^\tau((0, a), X) \cap L_p((0, a), Y)$$

$$\longrightarrow {}_0H_p^\tau((0, \infty), X) \cap L_p((0, \infty), Y), \qquad 0 < a < \infty, \ \tau \in (0, 1]$$

which are uniformly bounded w. r. t. $0 < a < \infty$ and $\tau \in (0, 1]$. $\square$

4.68 Following the proof of [PSS07, Proposition 6.2 (a)] we also obtain the following result on uniform embeddings.

4.69 COROLLARY. *Let X be a Banach space and let $1 < p < \infty$. Moreover, let $\tau \in (0, 1]$. Then there exists a constant $c > 0$ and an exponent $\eta > 0$ such that*

$$_0H_p^\tau((0, a), X) \hookrightarrow L_p((0, a), X),$$

$$\|\phi\|_{L_p((0,a),X)} \leq ca^\eta\|\phi\|_{_0H_p^\tau((0,a),X)}, \qquad \phi \in {}_0H_p^\tau((0, a), X)$$

for all $0 < a < \infty$.

Proof. We first employ the uniform extension of $_0H_p^\tau((0, a), X)$ to $_0H_p^\tau((0, \infty), X)$ due to Proposition 4.65 and then the embedding

$$_0H_p^\tau((0, \infty), X) \hookrightarrow L_q((0, \infty), X)$$

with a suitable $p < q < \infty$. After restriction to $L_q((0, a), X)$ the embedding

$$L_q((0, a), X) \hookrightarrow L_p((0, a), X)$$

then completes the proof.   □

4.70 To obtain suitable estimates for the pressure in Chapter 7 it will be convenient to have a suitable norm at hand. Therefore, we close this chapter with another corollary of Proposition 4.65.

4.71 COROLLARY. *Let X be a Banach space and let $1 < p < \infty$. Moreover, let $0 < a < \infty$ and let $\tau \in (0, 1]$. Then the operator*

$$\partial_t : {}_0H_p^1((0, \bar{a}), X) \subseteq L_p((0, \bar{a}), X) \longrightarrow L_p((0, \bar{a}), X)$$

is sectorial and admits bounded imaginary powers for all $\bar{a} \in (0, a]$. The functional

$$\phi \mapsto \|\phi\|_{\bar{a},\tau,p} := \left(\|\phi\|_{L_p((0,\bar{a}),X)}^p + \|\partial_t^\tau \phi\|_{L_p((0,\bar{a}),X)}^p \right)^{1/p} : {}_0H_p^\tau((0, \bar{a}), X) \longrightarrow [0, \infty)$$

defines an equivalent norm in $_0H_p^\tau((0, \bar{a}), X)$ and

$$c^{-1}\|\phi\|_{\bar{a},\tau,p} \leq \|\phi\|_{{}_0H_p^\tau((0,\bar{a}),X)} \leq c\|\phi\|_{\bar{a},p,\tau}, \qquad \phi \in {}_0H_p^\tau((0, \bar{a}), X)$$

with some constant $c > 0$, which is independent of $\bar{a} \in (0, a]$.

Proof. For $\bar{a} \in (0, a]$ we denote by

$$G_{\bar{a}} : D(G_{\bar{a}}) \subseteq X_{\bar{a}} \longrightarrow X_{\bar{a}}, \qquad G_{\bar{a}} := \partial_t$$

the time derivative in $X_{\bar{a}} := L_p((0, \bar{a}), X)$ with domain $D(G_{\bar{a}}) := {}_0H_p^1((0, \bar{a}), X)$. It is well known, see e.g. [DSS08], that $G_{\bar{a}}$ is sectorial with spectral angle $\phi_{G_{\bar{a}}} = \frac{\pi}{2}$ and admits bounded imaginary powers. Thus, the characterisation of the domains of fractional powers due to R. T. SEELEY, cf. [See71], implies

$$X_{G_{\bar{a}}^\tau} \cong [X_{\bar{a}}, X_{G_{\bar{a}}}]_\tau \cong [L_p((0, \bar{a}), X), {}_0H_p^1((0, \bar{a}), X)]_\tau = {}_0H_p^\tau((0, \bar{a}, X), \qquad \tau \in (0, 1),$$

where $X_{G_{\bar{a}}}$ denotes the domain $D(G_{\bar{a}})$ of $G_{\bar{a}}$ equipped with the graph norm $\| \cdot \|_{\bar{a},1,p}$ and $X_{G_{\bar{a}}^\tau}$ denotes the domain $D(G_{\bar{a}}^\tau)$ of $G_{\bar{a}}^\tau$ equipped with the graph norm $\| \cdot \|_{\bar{a},\tau,p}$, see also [DHP03, Theorem 2.5]. Now, the assertion for $\tau = 1$ may be proved by the very same arguments as used in the proof of Proposition 4.65 and, thus, the second of the above topological identifications allows for an equivalence estimate of the involved norms with constants independent of $0 < \bar{a} < \infty$. Moreover, it is well-known that the proof of the first of the above topological identifications due to R. T. SEELEY may be modified to deduce an estimate

$$c^{-1}\|\phi\|_{\bar{a},\tau,p} \leq \|\phi\|_{[X_{\bar{a}}, X_{G_{\bar{a}}}]_\tau} \leq c\|\phi\|_{\bar{a},\tau,p}, \qquad \phi \in [X_{\bar{a}}, X_{G_{\bar{a}}}]_\tau$$

with some constant $c = c(M) > 0$, which only depends on the bound $M > 0$ of the imaginary powers of $G_{\bar{a}}$, see e.g. [Ama95, Section 2.9]. Therefore, it remains to prove an estimate

$$\|G_{\bar{a}}^{i\sigma}\|_{\mathcal{B}(X_{\bar{a}})} \leq M, \qquad \sigma \in [-1, 1]$$

with some constant $M > 0$, which is independent of $\bar{a} \in (0, a]$.

Now, we first have the relation

$$(\lambda + G_{\bar{a}})^{-1} = R_{\bar{a}}(\lambda + G_a)^{-1}E_{\bar{a}}, \quad \lambda \in \Sigma_{\frac{\pi}{2}} \cup \{0\}, \qquad \bar{a} \in (0, a]$$

for the resolvents of the $G_{\bar{a}}$, where the extension operators

$$E_{\bar{a}} : L_p((0, \bar{a}), X) \longrightarrow L_p((0, a), X), \qquad E_{\bar{a}} := \mathcal{R}_a \mathcal{E}_{\bar{a}}$$

are uniformly bounded w.r.t. $\bar{a} \in (0, a]$ thanks to Proposition 4.65 and the restriction operators $R_{\bar{a}} : L_p((0, a), X) \longrightarrow L_p((0, \bar{a}), X)$ are uniformly bounded w.r.t. $\bar{a} \in (0, a]$ by definition. Based on the definition of the functional calculus for sectorial operators as introduced in 4.44 we infer

$$f(G_{\bar{a}}) = R_{\bar{a}}f(G_a)E_{\bar{a}}, \quad f \in \mathcal{H}_0(\Sigma_\phi), \qquad \bar{a} \in (0, a],$$

provided $\frac{\pi}{2} < \phi < \pi$. However, to obtain an estimate for the imaginary powers $\pi_{i\sigma}(G_{\bar{a}})$, which are defined via the holomorphic functions $\pi_{i\sigma} \in \mathcal{H}_{1/2}(\Sigma_\pi)$ introduced in 4.44, we further need to study the extended functional calculus for the $G_{\bar{a}}$ as introduced in 4.45. We denote by $\psi \in \mathcal{H}_0(\Sigma_\pi)$ the holomorphic multiplier defined in 4.45 and observe that $\psi f \in \mathcal{H}_0(\Sigma_\phi)$ for all $f \in \mathcal{H}_\omega(\Sigma_\phi)$, provided $0 \leq \omega < 1$. Thus, we have

$$_0H_p^1((0, \bar{a}), X) = D(G_{\bar{a}}) \subseteq D(f(G_{\bar{a}})) = \left\{ \phi \in X_{\bar{a}} : (\psi f)(G_{\bar{a}})\phi \in D(G_{\bar{a}}) \cap R(G_{\bar{a}}) \right\}$$

for all $f \in \mathcal{H}_\omega(\Sigma_\phi)$, provided $\frac{\pi}{2} < \phi < \pi$ and $0 \leq \omega < 1$, and, since $E_{\bar{a}}D(G_{\bar{a}}) \subseteq D(G_a)$, we further infer

$$
\begin{aligned}
f(G_{\bar{a}})\phi &= \psi(G_{\bar{a}})^{-1}(\psi f)(G_{\bar{a}})\phi \\
&= R_{\bar{a}}(2 + G_a + G_a^{-1})E_{\bar{a}}R_{\bar{a}}(\psi f)(G_a)E_{\bar{a}}\phi \\
&= R_{\bar{a}}(2 + G_a + G_a^{-1})E_{\bar{a}}R_{\bar{a}}\psi(G_a)\psi(G_a)^{-1}(\psi f)(G_a)E_{\bar{a}}\phi \\
&=: R_{\bar{a}}F_{\bar{a}}f(G_a)E_{\bar{a}}\phi, \qquad \phi \in D(G_{\bar{a}})
\end{aligned}
$$

for all $f \in \mathcal{H}_\omega(\Sigma_\phi)$, provided $\frac{\pi}{2} < \phi < \pi$ and $0 \leq \omega < 1$. Now, the involved operators $F_{\bar{a}} : X_a \longrightarrow X_a$ are uniformly bounded w.r.t. $\bar{a} \in (0, a]$ and the desired estimate for the imaginary powers of the $G_{\bar{a}}$ immediately follows, since $_0H_p^1((0, \bar{a}), X)$ is a dense subspace of $_0H_p^\tau((0, \bar{a}), X)$ for $\tau \in (0, 1)$. $\qquad\square$

Remarks

4.72 Concerning the sectoriality and the $\mathcal{H}^\infty$-calculus of the operators G_s as defined in Section 4.4 for $s \in \mathbb{R}$, first note that it is well-known that the operator G_0 is sectorial, invertible and admits a bounded $\mathcal{H}^\infty$-calculus with angle $\phi_{G_0}^\infty = \frac{\pi}{2}$. A proof may be found e.g. in [DSS08]. Since

$$G_s = (-\Delta_\Gamma)^{-s/2} G_0 (-\Delta_\Gamma)^{s/2}, \qquad s \in \mathbb{R}$$

and

$$(-\Delta_\Gamma)^{s/2} : L_p((0, \infty), \dot{H}_p^s(\mathbb{R}^{n-1})) \longrightarrow L_p((0, \infty), L_p(\mathbb{R}^{n-1}))$$

is an isomorphism for all $s \in \mathbb{R}$ as follows from the definition of the homogeneous Bessel potential spaces, the properties of G_0 directly transfer to G_s for all $s \in \mathbb{R}$, cf. [DHP03, Proposition 2.11].

4.73 Concerning the sectoriality and the $\mathcal{H}^\infty$-calculus of the operators D_s as defined in Section 4.4 for $s \in \mathbb{R}$, an argumentation similar to that used in 4.72 may be used. It is well-known that the operator D_0 is sectorial and admits a bounded $\mathcal{H}^\infty$-calculus with angle $\phi_{D_0}^\infty = 0$. A proof may be found e.g. in [DSS08]. Since

$$D_s = (-\Delta_\Gamma)^{-s/2} D_0 (-\Delta_\Gamma)^{s/2}, \qquad s \in \mathbb{R},$$

these properties directly transfer to D_s for all $s \in \mathbb{R}$, cf. [DHP03, Proposition 2.11].

4.74 The pioneering work of G. DA PRATO and P. GRISVARD resulted in the following theorem on the sum of two sectorial operators, cf. [PG75]. If A and B are two sectorial operators in a Banach space X with domains $D(A)$ respectively $D(B)$ and spectral angles $\phi_A \geq 0$ and $\phi_B \geq 0$ that commute and satisfy the *parabolicity condition* $\phi_A + \phi_B < \pi$, then $A + B$ with natural domain $D(A) \cap D(B)$ is closable, its closure being sectorial with spectral angle of at most $\max\{\phi_A, \phi_B\}$. Note that no condition on the geometry of the underlying Banach space X is needed.

4.75 The result derived by G. DORE and A. VENNI on the sum of two sectorial operators in the extended version due to J. PRÜSS and H. SOHR, cf. [DV87, PS90], reads as follows. If under the conditions of 4.74 the operators A and B admit bounded imaginary powers with power angles $\theta_A \geq 0$ respectively $\theta_B \geq 0$ and satisfy the *strong parabolicity condition* $\theta_A + \theta_B < \pi$, then $A + B$ is closed, sectorial and admits bounded imaginary powers with power angle of at most $\max\{\theta_A, \theta_B\}$, provided the underlying Banach space X is a UMD-space. For details on this class of Banach spaces, we refer to [KW01, DHP03]. We only note that all L_p-based function spaces with $1 < p < \infty$ that are used Section 4.4 are well-known to be UMD-spaces.

4.76 The original result of G. DORE and A. VENNI, cf. [DV87], employed the stronger assumptions

$$[0, \infty) \subseteq \rho(-A), \, \rho(-B)$$

and

$$\|t(t + A)^{-1}\|_{B(X)}, \|t(t + B)^{-1}\|_{B(X)} \leq \frac{M}{1 + t}, \quad t \geq 0$$

and delivered the slightly weaker result $\theta_{A+B} = \max\{\theta_A, \theta_B\} + \varepsilon$, where $\varepsilon > 0$ may be chosen arbitrarily small.

4.77 The last cited result concerning the sum of two sectorial operators was derived by N. J. KALTON and L. WEIS cf. [KW01], and reads as follows. If under the conditions of 4.74 the operators A and B admit a bounded $\mathcal{H}^\infty$-calculus with $\mathcal{H}^\infty$ angles $\phi_A^\infty \geq 0$ respectively $\phi_B^\infty \geq 0$ and satisfy $\phi_A^\infty + \phi_B^\infty < \pi$, then $A + B$ is closed, sectorial and admits a bounded $\mathcal{H}^\infty$-calculus with $\mathcal{H}^\infty$-angle of at most max $\{\, \phi_A^\infty,\ \phi_B^\infty \,\}$, provided the underlying Banach space X satisfies a certain geometrical condition, which is called the *property* (α). For details on this class of Banach spaces, we refer to [KW01]. We only note that all L_p-based function spaces with $1 < p < \infty$ that are used in Section 4.4 are well-known to have this property. Moreover, the condition on the underlying Banach space may be removed, if at least one of the involved operators admits a more restrictive functional calculus. For details we refer again to [KW01].

4.78 An extension of the Kalton-Weis theorem to the case of non-commuting operators has recently been obtained by J. PRÜSS and G. SIMONETT. For this result as well as a more detailed overview on known results concerning the sum of two sectorial operators we refer to [PS07].

4.79 An analogous result as provided by Proposition 4.65 has been proved by J. PRÜSS, J. SAAL and G. SIMONETT for the Sobolev respectively Sobolev-Slobodeckij spaces. To be precise, [PSS07, Proposition 6.1] states that the extension operators introduced in the proof of Proposition 4.65 also act as uniformly bounded linear operators

$$\mathcal{E}_{\bar{a}} : {}_0W_p^\tau((0, \bar{a}), X) \longrightarrow {}_0W_p^\tau((0, \infty), X), \qquad \bar{a} \in (0, a], \ \tau \in (1/p, 1],$$

where X denotes an arbitrary Banach space and $0 < a < \infty$. However, in this case the norm of these operators may depends on $\tau \in (1/p, 1]$ and it is not possible to obtain a result for $\tau \in (0, 1)$ by interpolation, since

$${}_0W_p^\tau((0, \bar{a}), X) \cong (L_p((0, \bar{a}), X), {}_0W_p^1((0, \bar{a}), X))_{\tau,p}, \qquad \tau \in (0, 1)$$

but the equivalence estimates of the involved norms are not uniform w.r.t. $\bar{a} \in (0, a]$ for any $0 < a < \infty$. Therefore, the intrinsic norms of the Sobolev-Slobodeckij spaces as introduced in 3.11 have to be estimated, which is the reason for the restrictions $\bar{a} \in (0, a]$ and $\tau \in (1/p, 1]$. In these restricted cases, however, similar embedding results as provided by Corollary 4.69 are an immediate consequence also for the Sobolev respectively Sobolev-Slobodeckij spaces, cf. [PSS07, Proposition 6.2 (a)].

References

[Ama95] H. AMANN: Linear and Quasilinear Parabolic Problems. Volume I. Abstract Linear Theory, *Monographs in Mathematics*, vol. 89. Birkhäuser, 1995.

[DHP01] W. DESCH, M. HIEBER, and J. PRÜSS: *L^p-Theory of the Stokes Equation in a Half-Space*. J. Evol. Equ., 1, 115–142, 2001.

[DHP03] R. DENK, M. HIEBER, and J. PRÜSS: *$\mathcal{R}$-Boundedness, Fourier-Multipliers and Problems of Elliptic and Parabolic Type*, *Mem. Amer. Math. Soc.*, vol. 166. American Mathematical Society, 2003.

[DS11] R. DENK and J. SEILER: *On the Maximal L_p-Regularity of Parabolic Mixed Order Systems*. J. Evol. Equ., 11, 371–404, 2011.

[DSS08] R. DENK, J. SAAL, and J. SEILER: *Inhomogeneous Symbols, the Newton Polygon, and Maximal L_p-Regularity.* Russian J. Math. Phys., 15 (2), 171–192, 2008.

[DV87] G. DORE and A. VENNI: *On the Closedness of the Sum of Two Closed Operators.* Math. Z., 196, 189–201, 1987.

[HP98] M. HIEBER and J. PRÜSS: *Functional Calculi for Linear Operators in Vector-Valued L^p-Spaces via the Transference Principle.* Adv. Differential Equations, 3, 847–872, 1998.

[KW01] N. J. KALTON and L. WEIS: *The H^∞-Calculus and Sums of Closed Operators.* Math. Ann., 321, 319–345, 2001.

[Nec67] J. NEČAS: *Les Méthodes Directes en Théorie des Équations Elliptiques.* Academia, 1967.

[PG75] G. DA PRATO and P. GRISVARD: *Sommes d'Opératours Linéaires et équations Différentielles Opérationelles.* J. Math. Pures Appl., 54, 305–387, 1975.

[Prü03] J. PRÜSS: *Maximal Regularity for Evolution Equations in L_p-Spaces.* Conf. Sem. Mat. Univ. Bari, 285, 1–39, 2003.

[PS90] J. PRÜSS and H. SOHR: *On Operators with Bounded Imaginary Powers in Banach Spaces.* Math. Z., 203, 429–452, 1990.

[PS07] J. PRÜSS and G. SIMONETT: *H^∞-Calculus for the Sum of Non-Commuting Operators.* Trans. Amer. Math. Soc., 359, 3549–3565, 2007.

[PSS07] J. PRÜSS, J. SAAL, and G. SIMONETT: *Existence of Analytic Solutions for the Classical Stefan Problem.* Math. Ann., 338, 703–755, 2007.

[Rha60] G. DE RHAM: *Variétés Différentiables. Formes, Courants, Formes Harmoniques.* III, *Actualites Scientifiques et Industrielles,* vol. 1222. Publications de lInstitut de Mathématique de lUniversité de Nancago. Hermann & Cie., 2nd ed., 1960.

[See71] R. T. SEELEY: *Norms and Domains of the Complex Powers A_b^z.* Amer. J. Math., 93, 299–309, 1971.

[SS92] C. G. SIMADER and H. SOHR: *A New Approach to the Helmholtz Decomposition and the Neumann Problem in L_q-Spaces for Bounded and Exterior Domains.* In: Mathematical Problems Relating to the Navier-Stokes Equations, Ser. Adv. Math. Appl. Sci., vol. 11 (G. P. GALDI, ed.), (1–35), World Scientific Publishing, 1992.

[SS96] C. G. SIMADER and H. SOHR: *The Dirichlet Problem for the Laplacian in Bounded and Unbounded Domains,* *Pitman Research Notes in Mathematics,* vol. 360. Longman, 1996.

Chapter 5

Maximal L_p-Regularity in a Halfspace

This chapter is devoted to the study of the Stokes equations in a halfspace subject to one of the energy preserving respectively artificial boundary conditions introduced 2.19, 2.22 and 2.23. We prove an analogous result as formulated in Theorem 3.30 for a bounded domain $\Omega \subseteq \mathbb{R}^n$, i.e. we establish a maximal L_p-regularity result for the Stokes equations for all $1 < p < \infty$ with $p \neq \frac{3}{2}$, 3.

Since our analysis is based on the derivation of solution formulas obtained via Laplace and Fourier transformations, we want to focus on the unbounded time interval $(0, \infty)$. However, we expect the Stokes equations to have the property of maximal L_p-regularity on bounded time intervals only. Therefore, we study the shifted equations on the unbounded time interval.

Given $\rho, \mu > 0$ and parameters $\alpha, \beta \in \{-1, 0, +1\}$, which define the boundary condition, as well as a parameter

- $\gamma = -\infty$, if $\beta = 0$;

- $\gamma \in \{-\infty\} \cup [0, 1/2 - 1/2p]$, if $\beta = +1$;

- $\gamma \in \{-\infty\} \cup [0, \infty)$, if $\beta = -1$,

which defines the regularity of the pressure trace, we prove the existence of a unique maximal regular solution

$$(u, p) \in \mathbb{X}_\gamma^\beta(\infty, \mathbb{R}_+^n)$$

to the shifted Stokes equations

$$(S \,|\, \infty, \mathbb{R}_+^n)_{f,g,h,u_0}^{\alpha,\beta,\varepsilon}$$

$$\begin{aligned}
\rho\varepsilon u + \rho\partial_t u - \mu\Delta u + \nabla p &= \rho f && \text{in } (0, \infty) \times \mathbb{R}_+^n, \\
\operatorname{div} u &= g && \text{in } (0, \infty) \times \mathbb{R}_+^n, \\
\mathcal{B}^{\alpha,\beta}(u, p) &= h && \text{on } (0, \infty) \times \partial\mathbb{R}_+^n, \\
u(0) &= u_0 && \text{in } \mathbb{R}_+^n,
\end{aligned}$$

whenever the data is subject to the regularity and compatibility conditions

$$(f, g, h, u_0) \in \mathbb{Y}_\gamma^{\alpha,\beta}(\infty, \mathbb{R}_+^n).$$

The bounded linear operators $\mathcal{B}^{\alpha,\beta}$ realise the desired boundary condition. Recall that they were defined in Chapter 3 as

$$P_\Gamma \mathcal{B}^{0,\beta}(u,\,p) := P_\Gamma[u]_\Gamma, \qquad P_\Gamma \mathcal{B}^{\pm 1,\beta}(u,\,p) := \mu P_\Gamma[\nabla u \pm \nabla u^\mathsf{T}]_\Gamma\, \nu_\Gamma$$

for $\beta \in \{-1,\,0,\,+1\}$ and

$$Q_\Gamma \mathcal{B}^{\alpha,0}(u,\,p) \cdot \nu_\Gamma := [u]_\Gamma \cdot \nu_\Gamma,$$

$$Q_\Gamma \mathcal{B}^{\alpha,+1}(u,\,p) \cdot \nu_\Gamma := 2\mu\,\partial_\nu u \cdot \nu_\Gamma - [p]_\Gamma, \qquad Q_\Gamma \mathcal{B}^{\alpha,-1}(u,\,p) \cdot \nu_\Gamma := -[p]_\Gamma$$

for $\alpha \in \{-1,\,0,\,+1\}$. Also recall that $[\,\cdot\,]_\Gamma$ denotes the trace of a function defined in $\mathbb{R}^n_+$ on the boundary $\Gamma := \partial \mathbb{R}^n_+$. The normal derivative has to be understood as $\partial_\nu = [\nabla \cdot^\mathsf{T}]_\Gamma\, \nu_\Gamma$. Moreover, $P_\Gamma = P_\Gamma(y) := 1 - \nu_\Gamma(y) \otimes \nu_\Gamma(y)$ denotes the projection onto the tangent space of Γ at a point $y \in \Gamma$. Furthermore, $Q_\Gamma := 1 - P_\Gamma$ denotes the projection onto the normal bundle of Γ and $\nu_\Gamma : \Gamma \longrightarrow \mathbb{R}^n$ denotes the outer unit normal field of $\mathbb{R}^n_+$.

The solution spaces have been defined in 3.1, 3.21, 3.22, 3.23, 3.28 and 3.59 as

$$\mathbb{X}^\beta_\gamma(\infty,\,\mathbb{R}^n_+) = \mathbb{X}_u(\infty,\,\mathbb{R}^n_+) \times \mathbb{X}^\beta_{p,\gamma}(\infty,\,\mathbb{R}^n_+)$$

with

$$\mathbb{X}_u(\infty,\,\mathbb{R}^n_+) = H^1_p((0,\,\infty),\,L_p(\mathbb{R}^n_+,\,\mathbb{R}^n)) \cap L_p((0,\,\infty),\,H^2_p(\mathbb{R}^n_+,\,\mathbb{R}^n)),$$

$$\mathbb{X}^0_{p,-\infty}(\infty,\,\mathbb{R}^n_+) = L_p((0,\,\infty),\,\dot{H}^1_p(\mathbb{R}^n_+)), \qquad \mathbb{X}^{\pm 1}_{p,-\infty}(\infty,\,\mathbb{R}^n_+) = L_p((0,\,\infty),\,\dot{H}^1_p(\mathbb{R}^n_+))$$

and

$$\mathbb{X}^{\pm 1}_{p,\gamma}(\infty,\,\mathbb{R}^n_+) = \left\{ q \in L_p((0,\,\infty),\,\dot{H}^1_p(\mathbb{R}^n_+)) \; : \; \begin{array}{l} [q]_\Gamma \in W^\gamma_p((0,\,a),\,L_p(\Gamma)) \\ \qquad \cap\, L_p((0,\,a),\,W^{1-1/p}_p(\Gamma)) \end{array} \right\}$$

for $\gamma \geq 0$. The data spaces have been defined in 3.1, 3.12, 3.19, 3.23 and 3.59 as

$$\mathbb{Y}_f(\infty,\,\mathbb{R}^n_+) = L_p((0,\,\infty) \times \mathbb{R}^n_+,\,\mathbb{R}^n),$$

$$\mathbb{Y}_g(\infty,\,\mathbb{R}^n_+) = H^{1/2}_p((0,\,\infty),\,H^1_p(\mathbb{R}^n_+)) \cap L_p((0,\,\infty),\,H^1_p(\mathbb{R}^n_+)),$$

$$\mathbb{Y}_0(\mathbb{R}^n_+) = W^{2-2/p}_p(\mathbb{R}^n_+,\,\mathbb{R}^n)$$

and

$$\mathbb{Y}^{\alpha,\beta}_{h,\gamma}(\infty,\,\Gamma) = \left\{ \eta \in \mathbb{Y}_h(\infty,\,\Gamma) \; : \; P_\Gamma \eta \in \mathbb{T}^\alpha_h(\infty,\,\Gamma),\; Q_\Gamma \eta \in \mathbb{N}^\beta_{h,\gamma}(\infty,\,\Gamma) \right\},$$

where $\mathbb{Y}_h(\infty,\,\Gamma) = L_p((0,\,\infty),\,L_{p,loc}(\Gamma,\,\mathbb{R}^n))$ has been defined in 3.13 and

$$\mathbb{T}^0_h(\infty,\,\Gamma) = W^{1-1/2p}_p((0,\,\infty),\,L_p(\Gamma,\,T\Gamma)) \cap L_p((0,\,\infty),\,W^{2-1/p}_p(\Gamma,\,T\Gamma)),$$

$$\mathbb{T}^{\pm 1}_h(\infty,\,\Gamma) = W^{1/2-1/2p}_p((0,\,\infty),\,L_p(\Gamma,\,T\Gamma)) \cap L_p((0,\,\infty),\,W^{1-1/p}_p(\Gamma,\,T\Gamma)),$$

$$\mathbb{N}^0_{h,-\infty}(\infty,\,\Gamma) = W^{1-1/2p}_p((0,\,\infty),\,L_p(\Gamma,\,N\Gamma)) \cap L_p((0,\,\infty),\,W^{2-1/p}_p(\Gamma,\,N\Gamma)),$$

$$\mathbb{N}^{\pm 1}_{h,-\infty}(\infty,\,\Gamma) = L_p((0,\,\infty),\,\dot{W}^{1-1/p}_p(\Gamma,\,N\Gamma))$$

have been defined in 3.18. Moreover, for $\gamma \geq 0$ the spaces

$$\mathbb{N}^{\pm 1}_{h,\gamma}(\infty,\,\Gamma) := W^\gamma_p((0,\,\infty),\,L_p(\Gamma,\,N\Gamma)) \cap L_p((0,\,\infty),\,W^{1-1/p}_p(\Gamma,\,N\Gamma))$$

have been defined in 3.21 and 3.22 and $\mathbb{Y}_\gamma^{\alpha,\beta}(\infty, \mathbb{R}_+^n)$ has been defined in 3.28 and 3.59 to consist of all

$$(f, g, h, u_0) \in \mathbb{Y}_f(\infty, \mathbb{R}_+^n) \times \mathbb{Y}_g(\infty, \mathbb{R}_+^n) \times \mathbb{Y}_{h,\gamma}^{\alpha,\beta}(\infty, \Gamma) \times \mathbb{Y}_0(\mathbb{R}_+^n)$$

that satisfy the compatibility condition

$(C_1)_{g,u_0}$ $\qquad\qquad\qquad\qquad \operatorname{div} u_0 = g(0),$

which stems from the divergence constraint, the compatibility condition

$(C_2)_{h,u_0}^\alpha$ $\qquad\qquad \begin{aligned} P_\Gamma[u_0]_\Gamma &= P_\Gamma h(0), \text{ if } \alpha = 0 \quad \text{and } p > \tfrac{3}{2}, \\ \mu P_\Gamma[\nabla u_0 \pm \nabla u_0^\mathsf{T}]_\Gamma \, \nu_\Gamma &= P_\Gamma h(0), \text{ if } \alpha = \pm 1 \text{ and } p > 3, \end{aligned}$

which stems from the tangential boundary condition, and the condition

$(C_3)_{g,h,u_0}^\beta$
$$\begin{aligned} &\quad Q_\Gamma[u_0]_\Gamma = Q_\Gamma h, \quad \text{if } p > \tfrac{3}{2}, \\ &\text{and} \quad (\, g, Q_\Gamma h\,) \in H_p^1((0, \infty), \, {}_0\dot{H}_p^{-1}(\mathbb{R}_+^n)), \end{aligned} \quad \text{if } \beta = 0,$$
$$\begin{aligned} &\text{there exists } \eta \in \mathbb{N}_{h,-\infty}^0(\infty, \Gamma) \text{ such that} \\ &\quad Q_\Gamma[u_0]_\Gamma = \eta, \quad \text{if } p > \tfrac{3}{2}, \qquad\qquad \text{if } \beta = \pm 1, \\ &\text{and} \quad (\, g, \eta\,) \in H_p^1((0, \infty), \, {}_0\dot{H}_p^{-1}(\mathbb{R}_+^n)), \end{aligned}$$

which stems from the divergence constraint and the normal boundary condition. For a proof of the necessity of these conditions see Section 3.3 and 3.59.

5.1 Strategy

5.1 The case $\beta = -1$ has been completely treated by Theorem 4.60, see 4.62. Note that Theorem 4.60 has only been formulated for bounded time intervals. However, in the halfspace case the shifted version of the parabolic problem $(P \,|\, \infty, \mathbb{R}_+^n)_{f,h,u_0}^\alpha$ has the property of maximal L_p-regularity, provided $\varepsilon > 0$ is chosen to be sufficiently large, i.e. $\varepsilon > \varepsilon_0$. This can be seen as in 4.28 by applying the theory of abstract parabolic evolution equations. Therefore, based on 4.61 we may restrict our considerations to the cases

- $\beta = 0$, $\gamma = -\infty$, $f = 0$, $g = 0$, $u_0 = 0$ and
$$P_\Gamma h = 0, \quad Q_\Gamma h \in {}_0 H_p^1((0, \infty), \, \dot{W}_p^{-1/p}(\Gamma, \, N\Gamma)) \cap \mathbb{N}_{h,-\infty}^0(\infty, \Gamma);$$

- $\beta = +1$, $\gamma = 1/2 - 1/2p$, $f = 0$, $g = 0$, $P_\Gamma h = 0$ and $u_0 = 0$.

Once these two cases have been treated we have proved our main result on the Stokes equations in a halfspace.

5.2 THEOREM. *Let $1 < p < \infty$ with $p \neq \frac{3}{2}, 3$ and let $\varepsilon > 0$. Let $\rho, \mu > 0$ and let $\alpha, \beta \in \{-1, 0, +1\}$. Moreover,*

- *let $\gamma = -\infty$, if $\beta = 0$;*
- *let $\gamma \in \{-\infty\} \cup [0, 1/2 - 1/2p]$, if $\beta = +1$;*
- *let $\gamma \in \{-\infty\} \cup [0, \infty)$, if $\beta = -1$.*

Then there exists a unique maximal regular solution

$$(u, p) \in \mathbb{X}_\gamma^\beta(\infty, \mathbb{R}_+^n)$$

to the shifted Stokes equations $(S \mid \infty, \mathbb{R}_+^n)_{f,g,h,u_0}^{\alpha,\beta,\varepsilon}$, if and only if the data satisfies

$$(f, g, h, u_0) \in \mathbb{Y}_\gamma^{\alpha,\beta}(\infty, \mathbb{R}_+^n).$$

Furthermore, the solutions depend continuously on the data.

5.3 The continuous dependence of the solutions on the data is a consequence of the open mapping principle, cf. 3.33. However, the scaling invariance of a halfspace $\mathbb{R}_+^n$ allows for a more refined estimate. Note that the proof of the following corollary also reveals, that it is sufficient to prove Theorem 5.2 for some $\varepsilon > 0$. Indeed, we may restrict the proof of Theorem 5.2 to the case $\varepsilon > \varepsilon_0$ as introduced in 5.1 and obtain the assertions for every $\varepsilon > 0$ by using the time-space rescaling employed in the proof of the following corollary.

5.4 COROLLARY. *Let $1 < p < \infty$ with $p \neq \frac{3}{2}, 3$ and let $\varepsilon \geq 1$. Let $\rho, \mu > 0$ and let $\alpha, \beta \in \{-1, 0, +1\}$. Let $f \in \mathbb{Y}_f(\infty, \mathbb{R}_+^n)$. Then the unique maximal regular solution*

$$(u, p) \in \mathbb{X}_\gamma^\beta(\infty, \mathbb{R}_+^n)$$

to the shifted Stokes equations $(S \mid \infty, \mathbb{R}_+^n)_{f,0,0,0}^{\alpha,\beta,\varepsilon}$ satisfies the estimate

$$\|u\|_{{}_0\mathbb{X}_u(\infty, \mathbb{R}_+^n)} \leq \varepsilon^{1/p-1} C \|f\|_{\mathbb{Y}_f(\infty, \mathbb{R}_+^n)},$$
$$\|p\|_{\mathbb{X}_{p,\gamma}^\beta(\infty, \mathbb{R}_+^n)} \leq \varepsilon^{1/2p-1/2} C \|f\|_{\mathbb{Y}_f(\infty, \mathbb{R}_+^n)},$$

where the constant $C > 0$ is independent of $\varepsilon \geq 1$.

Proof. We rescale the solution and the data as

$$\bar{u}(t, x) := u(\varepsilon^{-1}t, \varepsilon^{-1/2}x), \quad t \geq 0, \ x \in \mathbb{R}_+^n,$$
$$\bar{p}(t, x) := \varepsilon^{-1/2}p(\varepsilon^{-1}t, \varepsilon^{-1/2}x), \quad t \geq 0, \ x \in \mathbb{R}_+^n,$$
$$\bar{f}(t, x) := \varepsilon^{-1}f(\varepsilon^{-1}t, \varepsilon^{-1/2}x), \quad t \geq 0, \ x \in \mathbb{R}_+^n$$

to obtain a solution $(\bar{u}, \bar{p}) \in \mathbb{X}_\gamma^\beta(\infty, \mathbb{R}_+^n)$ to the shifted Stokes equations $(S \mid \infty, \mathbb{R}_+^n)_{\bar{f},0,0,0}^{\alpha,\beta,1}$. Indeed, the regularity of all three functions as well as the divergence constraint on the velocity field, the homogeneous boundary condition and the homogeneous initial condition are preserved by this rescaling. Moreover, the momentum balance reads

$$\rho\bar{u} + \rho\partial_t\bar{u} - \mu\Delta\bar{u} + \nabla\bar{p}$$
$$= \varepsilon^{-1}\{\rho\varepsilon u + \rho\partial_t u - \mu\Delta u + \nabla p\} \circ \Theta_\varepsilon = \rho\bar{f} \quad \text{in } (0, \infty) \times \mathbb{R}_+^n,$$

where $\Theta_\varepsilon : (0, \infty) \times \mathbb{R}_+^n \longrightarrow (0, \infty) \times \mathbb{R}_+^n$ denotes the employed time-space rescaling $(t, x) \mapsto (\varepsilon^{-1}t, \varepsilon^{-1/2}x)$. Now, we may employ the identification

$$\|\phi\|_{H_p^1(\mathbb{R}_+, L_p(\mathbb{R}_+^n)) \cap L_p(\mathbb{R}_+, H_p^2(\mathbb{R}_+^n))}^p \cong \|\partial_t \phi\|_{L_p(\mathbb{R}_+ \times \mathbb{R}_+^n)}^p + \|\phi\|_{L_p(\mathbb{R}_+ \times \mathbb{R}_+^n)}^p$$
$$+ \|\nabla \phi\|_{L_p(\mathbb{R}_+ \times \mathbb{R}_+^n, \mathbb{R}^n)}^p + \|\nabla^2 \phi\|_{L_p(\mathbb{R}_+ \times \mathbb{R}_+^n, \mathbb{R}^{n \times n})}^p,$$

which is valid for $\phi \in H_p^1(\mathbb{R}_+, L_p(\mathbb{R}_+^n)) \cap L_p(\mathbb{R}_+, H_p^2(\mathbb{R}_+^n))$ up to equivalence of norms, and obtain the estimates

$$\|u\|_{{}_0\mathbb{X}_u(\infty, \mathbb{R}_+^n)} \leq \varepsilon^{-n/2p} \|\bar{u}\|_{{}_0\mathbb{X}_u(\infty, \mathbb{R}_+^n)},$$
$$\|p\|_{\mathbb{X}_{p,\gamma}^\beta(\infty, \mathbb{R}_+^n)} \leq \varepsilon^{1/2 - (n+1)/2p} \|\bar{p}\|_{\mathbb{X}_{p,\gamma}^\beta(\infty, \mathbb{R}_+^n)},$$
$$\|\bar{f}\|_{\mathbb{Y}_f(\infty, \mathbb{R}_+^n)} \leq \varepsilon^{(n+2)/2p - 1} \|f\|_{\mathbb{Y}_f(\infty, \mathbb{R}_+^n)}.$$

Thus, the proof is complete, if we denote by $C > 0$ the norm of the solution operator to the shifted Stokes equations $(S \mid \infty, \mathbb{R}_+^n)_{\cdot,0,0,0}^{\alpha,\beta,1}$. $\qquad\square$

5.5 As an immediate consequence, we obtain maximal L_p-regularity for the Stokes equations on bounded time intervals.

5.6 COROLLARY. *Let $a > 0$, let $1 < p < \infty$ with $p \neq \frac{3}{2}, 3$. Let $\rho, \mu > 0$ and let $\alpha, \beta \in \{-1, 0, +1\}$. Moreover,*

- *let $\gamma = -\infty$, if $\beta = 0$;*
- *let $\gamma \in \{-\infty\} \cup [0, 1/2 - 1/2p]$, if $\beta = +1$;*
- *let $\gamma \in \{-\infty\} \cup [0, \infty)$, if $\beta = -1$.*

Then there exists a unique maximal regular solution

$$(u, p) \in \mathbb{X}_\gamma^\beta(a, \mathbb{R}_+^n)$$

to the Stokes equations $(S \mid a, \mathbb{R}_+^n)_{f,g,h,u_0}^{\alpha,\beta}$, if and only if the data satisfies

$$(f, g, h, u_0) \in \mathbb{Y}_\gamma^{\alpha,\beta}(a, \mathbb{R}_+^n).$$

Furthermore, the solutions depend continuously on the data.

Proof. Given data $(f, g, h, u_0) \in \mathbb{Y}_\gamma^{\alpha,\beta}(a, \mathbb{R}_+^n)$, we define

$$(\bar{f}, \bar{g}, \bar{h}) \in \mathbb{Y}_f(\infty, \mathbb{R}_+^n) \times \mathbb{Y}_g(\infty, \mathbb{R}_+^n) \times \mathbb{Y}_{h,\gamma}^{\alpha,\beta}(\infty, \partial\mathbb{R}_+^n)$$

to be extensions to $e^{-\varepsilon t}(f, g, h)$ with an $\varepsilon \geq 1$. Now, Theorem 5.2 ensures the existence of a unique maximal regular solution

$$(\bar{u}, \bar{p}) \in \mathbb{X}_\gamma^\beta(\infty, \mathbb{R}_+^n)$$

to the shifted Stokes equations $(S \mid \infty, \mathbb{R}_+^n)_{\bar{f},\bar{g},\bar{h},u_0}^{\alpha,\beta,\varepsilon}$. Thus, if we define

$$(u, p) \in \mathbb{X}_\gamma^\beta(a, \mathbb{R}_+^n)$$

to be the restriction of $e^{\varepsilon t}(\bar{u}, \bar{p})$ to the interval $(0, a)$, we obtain the desired maximal regular solution to the Stokes equations $(S \,|\, a, \, \mathbb{R}^n_+)^{\alpha,\beta}_{f,g,h,u_0}$.

Conversely, if

$$(u, p) \in {}_0\mathbb{X}^\beta_\gamma(a, \, \mathbb{R}^n_+)$$

is a solution to the fully homogeneous Stokes equations $(S \,|\, a, \, \mathbb{R}^n_+)^{\alpha,\beta}_{0,0,0,0}$, then

$$(\bar{u}, \bar{p}) := \mathcal{E}_a(u, p) \in {}_0\mathbb{X}^\beta_\gamma(\infty, \, \mathbb{R}^n_+)$$

satisfies the Stokes equations $(S \,|\, \infty, \, \mathbb{R}^n_+)^{\alpha,\beta}_{\bar{f},0,0,0}$. Here $\mathcal{E}_a$ denotes the extension operator employed in the proof of Proposition 4.65, see also Corollary 4.67, and

$$\bar{f} := \partial_t \mathcal{E}_a u - \mathcal{E}_a \partial_t u \in \mathbb{Y}_f(\infty, \, \mathbb{R}^n_+)$$

is supported in the time interval $(a, 2a)$. Now, for $\varepsilon \geq 1$ we set

$$(\bar{u}_\varepsilon, \bar{p}_\varepsilon) := e^{-\varepsilon t}(\bar{u}, \bar{p}) \in {}_0\mathbb{X}^\beta_\gamma(\infty, \, \mathbb{R}^n_+), \qquad \bar{f}_\varepsilon := e^{-\varepsilon t}\bar{f} \in \mathbb{Y}_f(\infty, \, \mathbb{R}^n_+)$$

and note that $(\bar{u}_\varepsilon, \bar{p}_\varepsilon)$ constitutes the unique maximal regular solution to the shifted Stokes equations $(S \,|\, \infty, \, \mathbb{R}^n_+)^{\alpha,\beta,\varepsilon}_{\bar{f},0,0,0}$ delivered by Theorem 5.2. Thus, Corollary 5.4 yields

$$\begin{aligned}
\|u\|_{L_p((0,a)\times\mathbb{R}^n_+,\mathbb{R}^n)} &= \|\bar{u}\|_{L_p((0,a)\times\mathbb{R}^n_+,\mathbb{R}^n)} \leq e^{a\varepsilon}\|\bar{u}_\varepsilon\|_{L_p((0,a)\times\mathbb{R}^n_+,\mathbb{R}^n)} \\
&\leq e^{a\varepsilon}\|\bar{u}_\varepsilon\|_{{}_0\mathbb{X}_u(\infty,\mathbb{R}^n_+)} \leq \varepsilon^{1/p-1}Ce^{a\varepsilon}\|\bar{f}_\varepsilon\|_{\mathbb{Y}_f(\infty,\mathbb{R}^n_+)} \\
&= \varepsilon^{1/p-1}Ce^{a\varepsilon}\|\bar{f}_\varepsilon\|_{L_p((a,2a)\times\mathbb{R}^n_+)} \leq \varepsilon^{1/p-1}C\|\bar{f}\|_{L_p((a,2a)\times\mathbb{R}^n_+)}
\end{aligned}$$

and, since $\varepsilon \geq 1$ was arbitrary, we obtain $u = 0$ in $L_p((0, a) \times \mathbb{R}^n_+, \, \mathbb{R}^n)$. Hence, solutions in the maximal regularity class $\mathbb{X}^\beta_\gamma(a, \, \mathbb{R}^n_+)$ are unique.

Finally, the continuous dependence of the solutions on the data is a consequence of the open mapping principle, see also 3.33.   □

5.7 To prove Theorem 5.2 in the two special cases derived in 5.1, we exploit the simple geometry of a halfspace and employ a Laplace transformation in time and a Fourier transformation in the tangential spatial variables. Recall that a similar approach has also been used in Section 4.3 to derive solution formulas for the shifted parabolic problems $(P \,|\, \infty, \, \mathbb{R}^n_+)^{\alpha,\varepsilon}_{\cdot,0,0}$. The derivation of solution formulas for the shifted Stokes equations and the proof of Theorem 5.2 will be carried out in the next sections. From this point on, we will always assume one of the simplified situations derived in 5.1.

5.2 The Transformed Stokes Equations

5.8 We split the spatial variable into a tangential part $x \in \mathbb{R}^{n-1}$ and a normal part $y > 0$. Moreover, we may split the velocity field as $u = (v, w)$ into a tangential part

$v : [0, \infty) \times \mathbb{R}_+^n \longrightarrow \mathbb{R}^{n-1}$ and a normal part $w : [0, \infty) \times \mathbb{R}_+^n \longrightarrow \mathbb{R}$ to obtain the system of interior partial differential equations

$$\rho\varepsilon v + \rho\partial_t v - \mu\Delta_x v - \mu\partial_y^2 v + \nabla_x p = 0 \quad \text{in } (0, \infty) \times \mathbb{R}_+^n,$$
$$\rho\varepsilon w + \rho\partial_t w - \mu\Delta_x w - \mu\partial_y^2 w + \partial_y p = 0 \quad \text{in } (0, \infty) \times \mathbb{R}_+^n,$$
$$\nabla_x \cdot v + \partial_y w = 0 \quad \text{in } (0, \infty) \times \mathbb{R}_+^n,$$

which have to be complemented by the given initial and boundary conditions.

5.9 Due to $v(0) = 0$ and $w(0) = 0$ we may employ a Laplace transformation in time and a Fourier transformation in the tangential spatial variable to obtain the transformed system

$$\omega^2 \hat{v} - \mu\partial_y^2 \hat{v} + i\xi\hat{p} = 0 \quad \text{Re}\,\lambda \geq 0, \, \xi \in \mathbb{R}^{n-1}, \, y > 0,$$
$$\omega^2 \hat{w} - \mu\partial_y^2 \hat{w} + \partial_y\hat{p} = 0 \quad \text{Re}\,\lambda \geq 0, \, \xi \in \mathbb{R}^{n-1}, \, y > 0,$$
$$i\xi^\mathsf{T}\hat{v} + \partial_y\hat{w} = 0 \quad \text{Re}\,\lambda \geq 0, \, \xi \in \mathbb{R}^{n-1}, \, y > 0,$$

where $\hat{v}$, $\hat{w}$ and $\hat{p}$ denote the Laplace-Fourier transforms of the components of the solution, $\lambda \in \mathbb{C}$ denotes the Laplace covariable of t and $\xi \in \mathbb{R}^{n-1}$ denotes the Fourier covariable of x. Moreover, we used the abbreviation

$$\omega := \sqrt{\rho\lambda_\varepsilon + \mu|\xi|^2},$$

where we have set $\lambda_\varepsilon := \varepsilon + \lambda$. Now, an exponential ansatz leads to

$$\begin{bmatrix} \hat{v}(\lambda, \xi, y) \\ \hat{w}(\lambda, \xi, y) \\ \hat{p}(\lambda, \xi, y) \end{bmatrix} = \begin{bmatrix} \omega & -i\zeta \\ i\zeta^\mathsf{T} & |\zeta| \\ 0 & \kappa\lambda_\varepsilon \end{bmatrix} \begin{bmatrix} \hat{z}_v(\lambda, \xi)e^{-\frac{\omega}{\sqrt{\mu}}y} \\ \hat{z}_w(\lambda, \xi)e^{-|\xi|y} \end{bmatrix}, \quad \begin{array}{l} \text{Re}\,\lambda \geq 0, \\ \xi \in \mathbb{R}^{n-1}, \, y > 0, \end{array}$$

where we have set $\zeta := \sqrt{\mu}\,\xi$ and $\kappa := \rho\sqrt{\mu}$ and where $\hat{z}_v$ and $\hat{z}_w$ denote the transformed components of a function $z = (z_v, z_w) : (0, \infty) \times \mathbb{R}^{n-1} \longrightarrow \mathbb{R}^n$, which has to be determined via the boundary conditions. We will treat the various boundary conditions under consideration separately in the next sections.

5.10 Note that the normal component of the boundary datum h is the only non-zero part of the data. In the sequel we will denote it by

$$h_w(x) := -h(x, 0) \cdot \nu_\Gamma(x, 0) = h_n(x, 0), \qquad x \in \mathbb{R}^{n-1}.$$

Also note that every boundary condition will lead to a linear system

$$\hat{\mathcal{B}}^{\alpha,\beta}(\lambda, \xi) \begin{bmatrix} \hat{z}_v(\lambda, \xi) \\ \hat{z}_w(\lambda, \xi) \end{bmatrix} = \begin{bmatrix} 0 \\ \hat{h}_w(\lambda, \xi) \end{bmatrix}, \qquad \text{Re}\,\lambda > 0, \, \xi \in \mathbb{R}^{n-1},$$

with a family of linear operators $\hat{\mathcal{B}}^{\alpha,\beta}(\lambda, \xi)$, which will be shown to uniquely determine $\hat{z}(\lambda, \xi)$ for all $\text{Re}\,\lambda \geq 0$ and $\xi \in \mathbb{R}^{n-1}$.

5.3 Boundary Conditions Involving the Normal Velocity

5.11 We start with the boundary conditions that involve the normal velocity, i. e. $\beta = 0$. In this case we have

$$h_w \in {}_0H_p^1((0, \infty), \dot{W}_p^{-1/p}(\mathbb{R}^{n-1})) \cap L_p((0, \infty), W_p^{2-1/p}(\mathbb{R}^{n-1})).$$

5.12 Now, suppose $\alpha = 0$. In this case we have to treat the plain Dirichlet conditions

$$[v]_y = 0 \quad \text{and} \quad [w]_y = h_w \quad \text{on } (0, \infty) \times \partial\mathbb{R}_+^n.$$

These lead to

$$\hat{\mathcal{B}}^{0,0}(\lambda, \xi) = \begin{bmatrix} \omega & -i\zeta \\ i\zeta^\mathsf{T} & |\zeta| \end{bmatrix} = \begin{bmatrix} \omega & 0 \\ 0 & \omega \end{bmatrix} \begin{bmatrix} 1 & -\frac{i\zeta}{\omega} \\ \frac{i\zeta^\mathsf{T}}{\omega} & \frac{|\zeta|}{\omega} \end{bmatrix}.$$

Now, we have

$$\begin{bmatrix} 1 & -\frac{i\zeta}{\omega} \\ \frac{i\zeta^\mathsf{T}}{\omega} & \frac{|\zeta|}{\omega} \end{bmatrix}^{-1} = \left\{ \left(1 - \frac{|\zeta|}{\omega}\right) \frac{|\zeta|}{\omega} \right\}^{-1} \begin{bmatrix} \left(1 - \frac{|\zeta|}{\omega}\right)\frac{|\zeta|}{\omega} - \frac{i\zeta \otimes i\zeta}{\omega^2} & \frac{i\zeta}{\omega} \\ -\frac{i\zeta^\mathsf{T}}{\omega} & 1 \end{bmatrix}$$

and, hence,

$$\hat{z}_w = \left\{ \left(1 - \frac{|\zeta|}{\omega}\right) \frac{|\zeta|}{\omega} \right\}^{-1} \omega^{-1} \hat{h}_w = \left(1 - \frac{|\zeta|}{\omega}\right)^{-1} |\zeta|^{-1} \hat{h}_w.$$

This implies

$$\widehat{\partial_\nu p} = -[\partial_y \hat{p}]_y = \frac{\kappa}{\sqrt{\mu}} \lambda_\varepsilon |\zeta| \hat{z}_w = \rho\lambda_\varepsilon \left(1 - \frac{|\zeta|}{\omega}\right)^{-1} \hat{h}_w$$

$$= \rho\lambda_\varepsilon \frac{1 + \frac{|\zeta|}{\omega}}{1 - \frac{|\zeta|^2}{\omega^2}} \hat{h}_w = \omega(\omega + |\zeta|)\hat{h}_w$$

and, therefore, the desired solution may be obtained by solving the splitting scheme

$$\begin{aligned} \rho\varepsilon u + \rho\partial_t u - \mu\Delta u &= -\nabla p &&\text{in } (0, \infty) \times \mathbb{R}_+^n, \\ P_\Gamma[u]_\Gamma &= 0 &&\text{on } (0, \infty) \times \partial\mathbb{R}_+^n, \\ [\operatorname{div} u]_\Gamma &= 0 &&\text{on } (0, \infty) \times \partial\mathbb{R}_+^n, \\ u(0) &= 0 &&\text{in } \mathbb{R}_+^n, \end{aligned}$$

together with

$$\begin{aligned} -\Delta p &= 0 &&\text{in } (0, \infty) \times \mathbb{R}_+^n, \\ \partial_\nu p &= \mathcal{M}^{0,0} h_w &&\text{on } (0, \infty) \times \partial\mathbb{R}_+^n. \end{aligned}$$

The linear operator

$$\mathcal{M}^{0,0} : {}_0H_p^1((0, \infty), \dot{W}_p^{-1/p}(\mathbb{R}^{n-1})) \cap L_p((0, \infty), W_p^{2-1/p}(\mathbb{R}^{n-1}))$$

$$\subseteq L_p((0, \infty), \dot{W}_p^{-1/p}(\mathbb{R}^{n-1})) \longrightarrow L_p((0, \infty), \dot{W}_p^{-1/p}(\mathbb{R}^{n-1}))$$

is defined via its Laplace-Fourier symbol

$$m^{0,0}(\lambda, \xi) = \omega(\omega + |\zeta|),$$

and it has been shown in Section 4.4 that $\mathcal{M}^{0,0}$ is closed. Therefore, the above splitting scheme admits a unique maximal regular solution

$$(u, p) \in {}_0\mathbb{X}^0_{-\infty}(\infty, \mathbb{R}^n_+)$$

for every

$$h_w \in {}_0\dot{H}^1_p((0, \infty), \dot{W}^{-1/p}_p(\mathbb{R}^{n-1})) \cap L_p((0, \infty), W^{2-2/p}_p(\mathbb{R}^{n-1}))$$

and the relation $(\mathrm{M})^{0,0}$ derived in 4.32 shows that (u, p) constitutes a maximal regular solution to the shifted Stokes equations. Conversely, any solution to the shifted Stokes equations with $h_w = 0$ satisfies the equations of the above splitting scheme with a boundary condition

$$\partial_\nu p = h_p \quad \text{on } (0, \infty) \times \partial\mathbb{R}^n_+$$

for some $h_p \in L_p((0, a), \dot{W}^{-1/p}_p(\mathbb{R}^{n-1}))$. Employing relation $(\mathrm{M})^{0,0}$ once again we infer $h_p = \mathcal{M}^{0,0}[w]_y = \mathcal{M}^{0,0}h_w = 0$. Therefore, p is constant and $u = 0$.

5.13 Now, suppose $\alpha = \pm 1$. In this case the boundary conditions read

$$\mp\mu[\partial_y v]_y - \mu\nabla_x[w]_y = 0 \quad \text{and} \quad [w]_y = h_w \quad \text{on } (0, \infty) \times \partial\mathbb{R}^n_+.$$

These lead to

$$
\mathcal{B}^{\pm 1,0}(\lambda, \xi) \;=\;
\begin{bmatrix}
\pm\sqrt{\mu}\omega^2 - \sqrt{\mu}(i\zeta \otimes i\zeta) & -\sqrt{\mu}(i\zeta|\zeta| \pm i\zeta|\zeta|) \\
i\zeta^\mathsf{T} & |\zeta|
\end{bmatrix}
$$

$$
\;=\;
\begin{bmatrix}
\pm\sqrt{\mu}\omega^2 & 0 \\
0 & \omega
\end{bmatrix}
\begin{bmatrix}
1 \mp \frac{i\zeta \otimes i\zeta}{\omega^2} & -\left(\frac{i\zeta}{\omega}\frac{|\zeta|}{\omega} \pm \frac{i\zeta}{\omega}\frac{|\zeta|}{\omega}\right) \\
\frac{i\zeta^\mathsf{T}}{\omega} & \frac{|\zeta|}{\omega}
\end{bmatrix}.
$$

Now, we have

$$
\begin{bmatrix}
1 \mp \frac{i\zeta \otimes i\zeta}{\omega^2} & -\left(\frac{i\zeta}{\omega}\frac{|\zeta|}{\omega} \pm \frac{i\zeta}{\omega}\frac{|\zeta|}{\omega}\right) \\
\frac{i\zeta^\mathsf{T}}{\omega} & \frac{|\zeta|}{\omega}
\end{bmatrix}^{-1}
$$

$$
= \left\{\left(1 - \frac{|\zeta|^2}{\omega^2}\right)\frac{|\zeta|}{\omega}\right\}^{-1}
\begin{bmatrix}
\left(1 - \frac{|\zeta|^2}{\omega^2}\right)\frac{|\zeta|}{\omega} - \frac{i\zeta \otimes i\zeta}{\omega^2}\frac{|\zeta|}{\omega} & \frac{i\zeta}{\omega}\frac{|\zeta|}{\omega} \pm \frac{i\zeta}{\omega}\frac{|\zeta|}{\omega} \\
-\frac{i\zeta^\mathsf{T}}{\omega} & 1 \pm \frac{|\zeta|^2}{\omega^2}
\end{bmatrix}
$$

and, hence,

$$
\hat{z}_w = \left\{\left(1 - \frac{|\zeta|^2}{\omega^2}\right)\frac{|\zeta|}{\omega}\right\}^{-1}\left(1 \pm \frac{|\zeta|^2}{\omega^2}\right)\omega^{-1}\hat{h}_w
$$

$$
= \left(1 - \frac{|\zeta|^2}{\omega^2}\right)^{-1}\left(1 \pm \frac{|\zeta|^2}{\omega^2}\right)|\zeta|^{-1}\hat{h}_w.
$$

This implies

$$\widehat{\partial_\nu p} = -[\partial_y \hat{p}]_y = \tfrac{\kappa}{\sqrt{\mu}} \lambda_\varepsilon |\zeta| \hat{z}_w = \rho \lambda_\varepsilon \left(1 - \tfrac{|\zeta|^2}{\omega^2}\right)^{-1} \left(1 \pm \tfrac{|\zeta|^2}{\omega^2}\right) \hat{h}_w$$

$$= \rho \lambda_\varepsilon \frac{1 \pm \frac{|\zeta|^2}{\omega^2}}{1 - \frac{|\zeta|^2}{\omega^2}} \hat{h}_w = (\omega^2 \pm |\zeta|^2) \hat{h}_w$$

and, therefore, the desired solution may be obtained by solving the splitting scheme

$$\rho \varepsilon u + \rho \partial_t u - \mu \Delta u = -\nabla p \quad \text{in } (0, \infty) \times \mathbb{R}^n_+,$$

$$\mu P_\Gamma [\nabla u \pm \nabla u^\mathsf{T}]_\Gamma \nu_\Gamma = 0 \quad \text{on } (0, \infty) \times \partial \mathbb{R}^n_+,$$

$$[\operatorname{div} u]_\Gamma = 0 \quad \text{on } (0, \infty) \times \partial \mathbb{R}^n_+,$$

$$u(0) = 0 \quad \text{in } \mathbb{R}^n_+,$$

together with

$$-\Delta p = 0 \quad \text{in } (0, \infty) \times \mathbb{R}^n_+,$$

$$\partial_\nu p = \mathcal{M}^{\pm 1,0} h_w \quad \text{on } (0, \infty) \times \partial \mathbb{R}^n_+.$$

The linear operators

$$\mathcal{M}^{\pm 1,0} : {}_0 H^1_p((0, \infty), \dot{W}^{-1/p}_p(\mathbb{R}^{n-1})) \cap L_p((0, \infty), W^{2-1/p}_p(\mathbb{R}^{n-1}))$$

$$\subseteq L_p((0, \infty), \dot{W}^{-1/p}_p(\mathbb{R}^{n-1})) \longrightarrow L_p((0, \infty), \dot{W}^{-1/p}_p(\mathbb{R}^{n-1}))$$

are defined via their Laplace-Fourier symbols

$$m^{\pm 1,0}(\lambda, \xi) = \omega^2 \pm |\zeta|^2,$$

and it has been shown in Section 4.4 that $\mathcal{M}^{\pm 1,0}$ is closed. Therefore, the above splitting scheme admits a unique maximal regular solution

$$(u, p) \in {}_0 \mathbb{X}^0_{-\infty}(\infty, \mathbb{R}^n_+)$$

for every

$$h_w \in {}_0 \dot{H}^1_p((0, \infty), \dot{W}^{-1/p}_p(\mathbb{R}^{n-1})) \cap L_p((0, \infty), W^{2-2/p}_p(\mathbb{R}^{n-1}))$$

and the relation (M)$^{\pm 1,0}$ derived in 4.34 shows that (u, p) constitutes a maximal regular solution to the shifted Stokes equations. Conversely, any solution to the shifted Stokes equations with $h_w = 0$ satisfies the equations of the above splitting scheme with a boundary condition

$$\partial_\nu p = h_p \quad \text{on } (0, \infty) \times \partial \mathbb{R}^n_+$$

for some $h_p \in L_p((0, a), \dot{W}^{-1/p}_p(\mathbb{R}^{n-1}))$. Employing relation (M)$^{\pm 1,0}$ once again we infer $h_p = \mathcal{M}^{\pm 1,0}[w]_y = \mathcal{M}^{\pm 1,0} h_w = 0$. Therefore, p is constant and $u = 0$.

5.4 Boundary Conditions Involving the Pressure

5.14 Now, we consider the boundary conditions that involve the pressure, i. e. $\beta = 1$. In this case we have

$$h_w \in {}_0W_p^{1/2-1/2p}((0, \infty), L_p(\mathbb{R}^{n-1})) \cap L_p((0, \infty), W_p^{1-1/p}(\mathbb{R}^{n-1})).$$

5.15 First, suppose $\alpha = 0$. In this case the boundary conditions read

$$[v]_y = 0 \quad \text{and} \quad -2\mu[\partial_y w]_y + [p]_y = h_w \quad \text{on } (0, \infty) \times \partial\mathbb{R}_+^n.$$

These lead to

$$
\hat{\mathcal{B}}^{0,1}(\lambda, \xi) = \begin{bmatrix} \omega & -i\zeta \\ 2\sqrt{\mu}\omega i\zeta^{\mathsf{T}} & \kappa\lambda_\varepsilon + 2\sqrt{\mu}|\zeta|^2 \end{bmatrix}
$$

$$
= \begin{bmatrix} \omega & 0 \\ 0 & 2\sqrt{\mu}\omega^2 \end{bmatrix} \begin{bmatrix} 1 & -\frac{i\zeta}{\omega} \\ \frac{i\zeta^{\mathsf{T}}}{\omega} & \frac{1}{2} + \frac{1}{2}\frac{|\zeta|^2}{\omega^2} \end{bmatrix}.
$$

Now, we have

$$
\begin{bmatrix} 1 & -\frac{i\zeta}{\omega} \\ \frac{i\zeta^{\mathsf{T}}}{\omega} & \frac{1}{2} + \frac{1}{2}\frac{|\zeta|^2}{\omega^2} \end{bmatrix}^{-1} = \left\{\frac{1}{2}\left(1 - \frac{|\zeta|^2}{\omega^2}\right)\right\}^{-1} \begin{bmatrix} \frac{1}{2}\left(1 - \frac{|\zeta|^2}{\omega^2}\right) - \frac{i\zeta \otimes i\zeta}{\omega^2} & \frac{i\zeta}{\omega} \\ -\frac{i\zeta^{\mathsf{T}}}{\omega} & 1 \end{bmatrix}
$$

and, hence,

$$
\hat{z}_w = \left\{\frac{1}{2}\left(1 - \frac{|\zeta|^2}{\omega^2}\right)\right\}^{-1} \frac{1}{2\sqrt{\mu}}\omega^{-2}\hat{h}_w = \left(1 - \frac{|\zeta|^2}{\omega^2}\right)^{-1} \frac{1}{\sqrt{\mu}}\omega^{-2}\hat{h}_w.
$$

This implies

$$
[\hat{p}]_y = \kappa\lambda_\varepsilon \hat{z}_w = \rho\lambda_\varepsilon \left(1 - \frac{|\zeta|^2}{\omega^2}\right)^{-1} \omega^{-2}\hat{h}_w
$$

$$
= \rho\lambda_\varepsilon \frac{1}{1 - \frac{|\zeta|^2}{\omega^2}} \omega^{-2}\hat{h}_w = \hat{h}_w
$$

and, therefore, the desired solution may be obtained by solving the splitting scheme

$$
\begin{aligned}
\rho\varepsilon u + \rho\partial_t u - \mu\Delta u &= -\nabla p && \text{in } (0, \infty) \times \mathbb{R}_+^n, \\
P_\Gamma[u]_\Gamma &= 0 && \text{on } (0, \infty) \times \partial\mathbb{R}_+^n, \\
[\operatorname{div} u]_\Gamma &= 0 && \text{on } (0, \infty) \times \partial\mathbb{R}_+^n, \\
u(0) &= 0 && \text{in } \mathbb{R}_+^n,
\end{aligned}
$$

together with

$$
\begin{aligned}
-\Delta p &= 0 && \text{in } (0, \infty) \times \mathbb{R}_+^n, \\
[p]_\Gamma &= h_w && \text{on } (0, \infty) \times \partial\mathbb{R}_+^n.
\end{aligned}
$$

Obviously, a solution to this splitting scheme solves the shifted Stokes equations. On the other hand, any solution to the shifted Stokes equations with $h_w = 0$ satisfies the equations of the above splitting scheme with boundary condition

$$[p]_\Gamma = 0 \quad \text{on } (0, \infty) \times \partial\mathbb{R}^n_+$$

and we infer $p = 0$ as well as $u = 0$.

5.16 Now, suppose $\alpha = \pm 1$. In this case the boundary conditions read

$$\mp\mu[\partial_y v]_y - \mu\nabla_x[w]_y = 0 \quad \text{and} \quad -2\mu[\partial_y w]_y + [p]_y = h_w \quad \text{on } (0, \infty) \times \partial\mathbb{R}^n_+.$$

These lead to

$$
\hat{\mathcal{B}}^{\pm 1,1}(\lambda, \xi) = \begin{bmatrix} \pm\sqrt{\mu}\omega^2 - \sqrt{\mu}(i\zeta \otimes i\zeta) & -\sqrt{\mu}(i\zeta|\zeta| \pm i\zeta|\zeta|) \\ 2\sqrt{\mu}\omega i\zeta^\mathsf{T} & \kappa\lambda_\varepsilon + 2\sqrt{\mu}|\zeta|^2 \end{bmatrix}
$$

$$
= \begin{bmatrix} \pm\sqrt{\mu}\omega^2 & 0 \\ 0 & 2\sqrt{\mu}\omega^2 \end{bmatrix} \begin{bmatrix} 1 \mp \frac{i\zeta\otimes i\zeta}{\omega^2} & -\left(\frac{i\zeta}{\omega}\frac{|\zeta|}{\omega} \pm \frac{i\zeta}{\omega}\frac{|\zeta|}{\omega}\right) \\ \frac{i\zeta^\mathsf{T}}{\omega} & \frac{1}{2} + \frac{1}{2}\frac{|\zeta|^2}{\omega^2} \end{bmatrix}.
$$

Now, we have

$$
\begin{bmatrix} 1 \mp \frac{i\zeta\otimes i\zeta}{\omega^2} & -\left(\frac{i\zeta}{\omega}\frac{|\zeta|}{\omega} \pm \frac{i\zeta}{\omega}\frac{|\zeta|}{\omega}\right) \\ \frac{i\zeta^\mathsf{T}}{\omega} & \frac{1}{2} + \frac{1}{2}\frac{|\zeta|^2}{\omega^2} \end{bmatrix}^{-1}
$$

$$
= \delta^{-1}\begin{bmatrix} \delta \pm \left\{\frac{1}{2}\left(1 + \frac{|\zeta|^2}{\omega^2}\right) \mp \left(\frac{|\zeta|}{\omega} \pm \frac{|\zeta|}{\omega}\right)\right\}\frac{i\zeta\otimes i\zeta}{\omega^2} & \frac{i\zeta}{\omega}\frac{|\zeta|}{\omega} \pm \frac{i\zeta}{\omega}\frac{|\zeta|}{\omega} \\ -\frac{i\zeta^\mathsf{T}}{\omega} & 1 \pm \frac{|\zeta|^2}{\omega^2} \end{bmatrix}
$$

with

$$
\delta = \frac{1}{2}\left(1 + \frac{|\zeta|^2}{\omega^2}\right)\left(1 \pm \frac{|\zeta|^2}{\omega^2}\right) - \left(\frac{|\zeta|^3}{\omega^3} \pm \frac{|\zeta|^3}{\omega^3}\right)
$$

and, hence,

$$
\hat{z}_w = \left\{\frac{1}{2}\left(1 + \frac{|\zeta|^2}{\omega^2}\right)\left(1 \pm \frac{|\zeta|^2}{\omega^2}\right) - \left(\frac{|\zeta|^3}{\omega^3} \pm \frac{|\zeta|^3}{\omega^3}\right)\right\}^{-1}\left(1 \pm \frac{|\zeta|^2}{\omega^2}\right)\frac{1}{2\sqrt{\mu}}\omega^{-2}\hat{h}_w
$$

$$
= \begin{cases} \dfrac{1}{\sqrt{\mu}}\dfrac{\omega^2+|\zeta|^2}{(\omega^2+|\zeta|^2)^2 - 4\omega|\zeta|^3}\hat{h}_w, & \text{if } \alpha = +1, \\[2ex] \dfrac{1}{\sqrt{\mu}}\dfrac{1}{\omega^2+|\zeta|^2}\hat{h}_w, & \text{if } \alpha = -1. \end{cases}
$$

This implies

$$
[\hat{p}]_y = \kappa\lambda_\varepsilon\hat{z}_w = \frac{\kappa}{\sqrt{\mu}}\lambda_\varepsilon\frac{\omega^2+|\zeta|^2}{(\omega^2+|\zeta|^2)^2 - 4\omega|\zeta|^3}\hat{h}_w = \rho\lambda_\varepsilon\frac{\omega^2+|\zeta|^2}{((\omega^2-|\zeta|^2)+2|\zeta|^2)^2 - 4\omega|\zeta|^3}\hat{h}_w
$$

$$
= (\omega^2 - |\zeta|^2)\frac{\omega^2+|\zeta|^2}{(\omega^2-|\zeta|^2)^2 + 4(\omega^2-|\zeta|^2)|\zeta|^2 + 4|\zeta|^4 - 4\omega|\zeta|^3}\hat{h}_w
$$

$$
= \frac{\omega^2+|\zeta|^2}{(\omega^2-|\zeta|^2) + 4\frac{\omega}{\omega+|\zeta|}|\zeta|^2}\hat{h}_w
$$

for $\alpha = +1$ and

$$[\hat{p}]_y = \kappa \lambda_\varepsilon \hat{z}_w = \frac{\kappa}{\sqrt{\mu}} \lambda_\varepsilon \frac{1}{\omega^2 + |\zeta|^2} \hat{h}_w = \frac{\omega^2 - |\zeta|^2}{\omega^2 + |\zeta|^2} \hat{h}_w$$

for $\alpha = -1$. Therefore, a solution may be obtained by solving the splitting scheme

$$\rho \varepsilon u + \rho \partial_t u - \mu \Delta u = -\nabla p \quad \text{in } (0, \infty) \times \mathbb{R}^n_+,$$
$$\mu P_\Gamma [\nabla u \pm \nabla u^\mathsf{T}]_\Gamma \nu_\Gamma = 0 \quad \text{on } (0, \infty) \times \partial \mathbb{R}^n_+,$$
$$[\operatorname{div} u]_\Gamma = 0 \quad \text{on } (0, \infty) \times \partial \mathbb{R}^n_+,$$
$$u(0) = 0 \quad \text{in } \mathbb{R}^n_+,$$

together with

$$-\Delta p = 0 \quad \text{in } (0, \infty) \times \mathbb{R}^n_+,$$
$$[p]_\Gamma = \mathcal{M}^{\pm 1, +1} h_w \quad \text{on } (0, \infty) \times \partial \mathbb{R}^n_+.$$

The linear operators

$$\mathcal{M}^{\pm 1, +1} : L_p((0, \infty), \dot{W}_p^{1-1/p}(\mathbb{R}^{n-1})) \longrightarrow L_p((0, \infty), \dot{W}_p^{1-1/p}(\mathbb{R}^{n-1}))$$

are defined via their Laplace-Fourier symbols

$$m^{\pm 1, +1}(\lambda, \xi) = \frac{\omega^2 \pm |\zeta|^2}{(\omega^2 \mp |\zeta|^2) + 2\frac{\omega}{\omega + |\zeta|}(|\zeta|^2 \pm |\zeta|^2)},$$

and it has been shown in Section 4.4, that $\mathcal{M}^{\pm 1, +1}$ is bounded. Therefore, the above splitting scheme admits a unique maximal regular solution

$$(u, p) \in {}_0\mathbb{X}_{-\infty}^{\pm 1}(\infty, \mathbb{R}^n_+)$$

for every

$$h_w \in {}_0\dot{W}_p^{1/2-1/2p}((0, \infty), L_p(\mathbb{R}^{n-1})) \cap L_p((0, \infty), W_p^{1-1/p}(\mathbb{R}^{n-1}))$$

and the relation $(\mathrm{M})^{\pm 1, +1}$ derived in 4.35 shows that (u, p) constitutes a maximal regular solution to the shifted Stokes equations. Moreover, the boundary condition

$$[p]_\Gamma = 2\mu \, \partial_\nu u \cdot \nu_\Gamma - h \cdot \nu_\Gamma \quad \text{on } (0, a) \times \Gamma$$

implies $p \in \mathbb{X}_{p, 1/2-1/2p}^{\pm 1}(\infty, \mathbb{R}^n_+)$. Conversely, any solution to the shifted Stokes equations with $h_w = 0$ satisfies the equations of the above splitting scheme with a boundary condition

$$[p]_\Gamma = h_p \quad \text{on } (0, \infty) \times \partial \mathbb{R}^n_+$$

for some $h_p \in L_p((0, \infty), \dot{W}_p^{1-1/p}(\mathbb{R}^{n-1}))$. Employing relation $(\mathrm{M})^{\pm 1, +1}$ once again we infer $h_p = \mathcal{M}^{\pm 1, +1}(-2\mu[\partial_y w]_y - [p]_y) = \mathcal{M}^{\pm 1, +1} h_w = 0$. Therefore, $p = 0$ and $u = 0$.

Remarks

5.17 An alternative approach to the Stokes equations in a halfspace, which also relies on the Laplace-Fourier transformation, the theory of sectorial operators and the $\mathcal{H}^\infty$-calculus may be found in the work of D. BOTHE and J. PRÜSS, cf. [BP07]. The approach presented there is applicable to the more complex equations for a class of non-Newtonian fluids.

5.18 A maximal L_p-regularity result in a halfspace for some of the boundary conditions considered here has also been obtained by the author employing a slightly different method, which makes direct use of *Fourier multiplier theorems*, cf. [Köh07].

5.19 A large class of parabolic problems is considered by R. DENK, M. HIEBER and J. PRÜSS, cf. [DHP03]. The approach presented there also relies on the Laplace-Fourier transformation, the theory of sectorial operators and the $\mathcal{H}^\infty$-calculus. However, the equations are allowed to have variable coefficients, which requires an additional localisation procedure, even, if the whole space $\mathbb{R}^n$ or a halfspace is considered.

References

[BP07]   D. BOTHE and J. PRÜSS: L_p-Theory for a Class of Non-Newtonian Fluids. SIAM J. Math. Anal., 39, 379–421, 2007.

[DHP03] R. DENK, M. HIEBER, and J. PRÜSS: $\mathcal{R}$-Boundedness, Fourier-Multipliers and Problems of Elliptic and Parabolic Type, *Mem. Amer. Math. Soc.*, vol. 166. American Mathematical Society, 2003.

[Köh07] M. KÖHNE: Zur Analysis und Numerik der Navier-Stokes-Gleichungen in Gebieten mit künstlichen Rändern. Diplomarbeit, Universität Paderborn, 2007.

Chapter 6

Maximal L_p-Regularity in a Bent Halfspace

This chapter is devoted to the study of the Stokes equations in a bent halfspace subject to one of the energy preserving respectively artificial boundary conditions introduced 2.19, 2.22 and 2.23. We prove an analogous result as formulated in Theorem 3.30 for a bounded domain $\Omega \subseteq \mathbb{R}^n$, i.e. we establish maximal L_p-regularity for all $1 < p < \infty$ with $p \neq \frac{3}{2}, 3$.

To be precise, we will consider a bent halfspace

$$\mathbb{R}^n_\omega := \left\{ (x, y) \in \mathbb{R}^{n-1} \times \mathbb{R} : y > \omega(x) \right\},$$

which is defined by a sufficiently smooth and flat function $\omega \in BUC^{3-}(\mathbb{R}^{n-1})$.

Given $a > 0$ as well as $\rho, \mu > 0$ and parameters $\alpha, \beta \in \{-1, 0, +1\}$, which define the boundary condition, as well as a parameter

- $\gamma = -\infty$, if $\beta = 0$;
- $\gamma \in \{-\infty\} \cup [0, 1/2 - 1/2p]$, if $\beta = +1$;
- $\gamma \in \{-\infty\} \cup [0, \infty)$, if $\beta = -1$,

which defines the regularity of the pressure trace, we prove the existence of a unique maximal regular solution

$$(u, p) \in \mathbb{X}^\beta_\gamma(a, \mathbb{R}^n_\omega)$$

to the Stokes equations

$$
\begin{aligned}
\rho \partial_t u - \mu \Delta u + \nabla p &= \rho f &&\text{in } (0, a) \times \mathbb{R}^n_\omega, \\
\operatorname{div} u &= g &&\text{in } (0, a) \times \mathbb{R}^n_\omega, \\
\mathcal{B}^{\alpha,\beta}(u, p) &= h &&\text{on } (0, a) \times \partial\mathbb{R}^n_\omega, \\
u(0) &= u_0 &&\text{in } \mathbb{R}^n_\omega,
\end{aligned}
$$

$(S \,|\, a, \, \mathbb{R}^n_\omega)^{\alpha,\beta}_{f,g,h,u_0}$

whenever the data is subject to the regularity and compatibility conditions

$$(f, g, h, u_0) \in \mathbb{Y}^{\alpha,\beta}_\gamma(a, \mathbb{R}^n_\omega).$$

The bounded linear operators $\mathcal{B}^{\alpha,\beta}$ realise the desired boundary condition. Recall that they were defined in Chapter 3 as

$$P_\Gamma \mathcal{B}^{0,\beta}(u,\,p) := P_\Gamma[u]_\Gamma, \qquad P_\Gamma \mathcal{B}^{\pm 1,\beta}(u,\,p) := \mu P_\Gamma[\nabla u \pm \nabla u^\mathsf{T}]_\Gamma \, \nu_\Gamma$$

for $\beta \in \{-1,\,0,\,+1\}$ and

$$Q_\Gamma \mathcal{B}^{\alpha,0}(u,\,p) \cdot \nu_\Gamma := [u]_\Gamma \cdot \nu_\Gamma,$$

$$Q_\Gamma \mathcal{B}^{\alpha,+1}(u,\,p) \cdot \nu_\Gamma := 2\mu\,\partial_\nu u \cdot \nu_\Gamma - [p]_\Gamma, \qquad Q_\Gamma \mathcal{B}^{\alpha,-1}(u,\,p) \cdot \nu_\Gamma := -[p]_\Gamma$$

for $\alpha \in \{-1,\,0,\,+1\}$. Also recall that $[\,\cdot\,]_\Gamma$ denotes the trace of a function defined in $\mathbb{R}^n_\omega$ on the boundary $\Gamma := \partial\mathbb{R}^n_\omega$. The normal derivative has to be understood as $\partial_\nu = [\nabla\cdot^\mathsf{T}]_\Gamma\,\nu_\Gamma$. Moreover, $P_\Gamma = P_\Gamma(y) := 1 - \nu_\Gamma(y) \otimes \nu_\Gamma(y)$ denotes the projection onto the tangent space of Γ at a point $y \in \Gamma$. Furthermore, $Q_\Gamma := 1 - P_\Gamma$ denotes the projection onto the normal bundle of Γ and $\nu_\Gamma : \Gamma \longrightarrow \mathbb{R}^n$ denotes the outer unit normal field of $\mathbb{R}^n_\omega$.

The solution spaces have been defined in 3.1, 3.21, 3.22, 3.23, 3.28 and 3.59 as

$$\mathbb{X}^\beta_\gamma(a,\,\mathbb{R}^n_\omega) = \mathbb{X}_u(a,\,\mathbb{R}^n_\omega) \times \mathbb{X}^\beta_{p,\gamma}(a,\,\mathbb{R}^n_\omega)$$

with

$$\mathbb{X}_u(a,\,\mathbb{R}^n_\omega) = H^1_p((0,\,a),\,L_p(\mathbb{R}^n_\omega,\,\mathbb{R}^n)) \cap L_p((0,\,a),\,H^2_p(\mathbb{R}^n_\omega,\,\mathbb{R}^n)),$$

$$\mathbb{X}^0_{p,-\infty}(a,\,\mathbb{R}^n_\omega) = L_p((0,\,a),\,\hat{H}^1_p(\mathbb{R}^n_\omega)), \quad \mathbb{X}^{\pm 1}_{p,-\infty}(a,\,\mathbb{R}^n_\omega) = L_p((0,\,a),\,\dot{H}^1_p(\mathbb{R}^n_\omega))$$

and

$$\mathbb{X}^{\pm 1}_{p,\gamma}(a,\,\mathbb{R}^n_\omega) = \left\{ q \in L_p((0,\,a),\,\dot{H}^1_p(\mathbb{R}^n_\omega)) : \begin{array}{l} [q]_\Gamma \in W^\gamma_p((0,\,a),\,L_p(\Gamma)) \\ \qquad \cap L_p((0,\,a),\,W^{1-1/p}_p(\Gamma)) \end{array} \right\}$$

for $\gamma \geq 0$. The data spaces have been defined in 3.1, 3.12, 3.19, 3.23 and 3.59 as

$$\mathbb{Y}_f(a,\,\mathbb{R}^n_\omega) = L_p((0,\,a) \times \mathbb{R}^n_\omega,\,\mathbb{R}^n),$$

$$\mathbb{Y}_g(a,\,\mathbb{R}^n_\omega) = H^{1/2}_p((0,\,a),\,H^1_p(\mathbb{R}^n_\omega)) \cap L_p((0,\,a),\,H^1_p(\mathbb{R}^n_\omega)),$$

$$\mathbb{Y}_0(\mathbb{R}^n_\omega) = W^{2-2/p}_p(\mathbb{R}^n_\omega,\,\mathbb{R}^n)$$

and

$$\mathbb{Y}^{\alpha,\beta}_{h,\gamma}(a,\,\Gamma) = \left\{ \eta \in \mathbb{Y}_h(a,\,\Gamma) : P_\Gamma \eta \in \mathbb{T}^\alpha_h(a,\,\Gamma),\, Q_\Gamma \eta \in \mathbb{N}^\beta_{h,\gamma}(a,\,\Gamma) \right\},$$

where $\mathbb{Y}_h(a,\,\Gamma) = L_p((0,\,a),\,L_{p,loc}(\Gamma,\,\mathbb{R}^n))$ has been defined in 3.13 and

$$\mathbb{T}^0_h(a,\,\Gamma) = W^{1-1/2p}_p((0,\,a),\,L_p(\Gamma,\,T\Gamma)) \cap L_p((0,\,a),\,W^{2-1/p}_p(\Gamma,\,T\Gamma)),$$

$$\mathbb{T}^{\pm 1}_h(a,\,\Gamma) = W^{1/2-1/2p}_p((0,\,a),\,L_p(\Gamma,\,T\Gamma)) \cap L_p((0,\,a),\,W^{1-1/p}_p(\Gamma,\,T\Gamma)),$$

$$\mathbb{N}^0_{h,-\infty}(a,\,\Gamma) = W^{1-1/2p}_p((0,\,a),\,L_p(\Gamma,\,N\Gamma)) \cap L_p((0,\,a),\,W^{2-1/p}_p(\Gamma,\,N\Gamma)),$$

$$\mathbb{N}^{\pm 1}_{h,-\infty}(a,\,\Gamma) = L_p((0,\,a),\,\dot{W}^{1-1/p}_p(\Gamma,\,N\Gamma))$$

have been defined in 3.18 and 3.23. Moreover, for $\gamma \geq 0$ the spaces

$$\mathbb{N}^{\pm 1}_{h,\gamma}(a,\,\Gamma) := W^\gamma_p((0,\,a),\,L_p(\Gamma,\,N\Gamma)) \cap L_p((0,\,a),\,W^{1-1/p}_p(\Gamma,\,N\Gamma))$$

have been defined in 3.21 and 3.22 and $\mathbb{Y}_\gamma^{\alpha,\beta}(a, \mathbb{R}_\omega^n)$ has been defined in 3.28 and 3.59 to consist of all

$$(f, g, h, u_0) \in \mathbb{Y}_f(a, \mathbb{R}_\omega^n) \times \mathbb{Y}_g(a, \mathbb{R}_\omega^n) \times \mathbb{Y}_{h,\gamma}^{\alpha,\beta}(a, \Gamma) \times \mathbb{Y}_0(\mathbb{R}_\omega^n)$$

that satisfy the compatibility condition

$(C_1)_{g,u_0}$
$$\operatorname{div} u_0 = g(0),$$

which stems from the divergence constraint, the compatibility condition

$(C_2)_{h,u_0}^\alpha$
$$P_\Gamma[u_0]_\Gamma = P_\Gamma h(0), \text{ if } \alpha = 0 \quad \text{and } p > \tfrac{3}{2},$$
$$\mu P_\Gamma[\nabla u_0 \pm \nabla u_0^\mathsf{T}]_\Gamma \nu_\Gamma = P_\Gamma h(0), \text{ if } \alpha = \pm 1 \text{ and } p > 3,$$

which stems from the tangential boundary condition, and the condition

$(C_3)_{g,h,u_0}^\beta$

$$Q_\Gamma[u_0]_\Gamma = Q_\Gamma h, \quad \text{if } p > \tfrac{3}{2},$$
$$\text{and} \quad (g, Q_\Gamma h) \in H_p^1((0, a), {}_0\dot{H}_p^{-1}(\mathbb{R}_\omega^n)), \qquad \text{if } \beta = 0,$$

$$\text{there exists } \eta \in \mathbb{N}_{h,-a}^0(a, \Gamma) \text{ such that}$$
$$Q_\Gamma[u_0]_\Gamma = \eta, \quad \text{if } p > \tfrac{3}{2}, \qquad \text{if } \beta = \pm 1,$$
$$\text{and} \quad (g, \eta) \in H_p^1((0, a), {}_0\dot{H}_p^{-1}(\mathbb{R}_\omega^n)),$$

which stems from the divergence constraint and the normal boundary condition. For a proof of the necessity of these conditions see Section 3.3 and 3.59.

6.1 Strategy

6.1 The case $\beta = -1$ has been completely treated by Theorem 4.60, see 4.62. Therefore, based on 4.61 we may restrict our considerations to the cases

- $\beta = 0$, $\gamma = -\infty$, $f = 0$, $g = 0$, $P_\Gamma h = 0$ and $u_0 = 0$;
- $\beta = +1$, $\gamma = 1/2 - 1/2p$, $f = 0$, $g = 0$, $P_\Gamma h = 0$ and $u_0 = 0$.

A careful examination of the argumentation in the next sections reveals that the restriction to nearly homogeneous data is not necessary. However, it serves to reduce the notational effort.

6.2 The boundary of the considered bent halfspaces has to exhibit a minimal regularity. Therefore, we will assume the function $\omega : \mathbb{R}^{n-1} \longrightarrow \mathbb{R}$ that defines the bent halfspace $\mathbb{R}_\omega^n$ to be at least of class BUC^{3-}. Here BUC^{m-} with $m \in \mathbb{N}$ denotes the subspace of BUC^{m-1} that consists of the functions with Lipschitz continuous partial derivatives of order $m - 1$ only.

6.3 Also note that our proof requires the bent halfspace to be sufficiently flat. As the next sections will reveal, there exists some $\delta > 0$ such that the following theorem is valid.

6.4 THEOREM. *Let $a > 0$ and let $\omega \in BUC^{3-}(\mathbb{R}^{n-1})$ with*

$$\|\nabla\omega\|_{L_\infty(\mathbb{R}^{n-1})} < \delta.$$

Let $1 < p < \infty$ with $p \neq \frac{3}{2}, 3$, let $\rho, \mu > 0$ and let $\alpha, \beta \in \{-1, 0, +1\}$. Moreover,

- *let $\gamma = -\infty$, if $\beta = 0$;*
- *let $\gamma \in \{-\infty\} \cup [0, 1/2 - 1/2p]$, if $\beta = +1$;*
- *let $\gamma \in \{-\infty\} \cup [0, \infty)$, if $\beta = -1$.*

Then there exists a unique maximal regular solution

$$(u, p) \in \mathbb{X}_\gamma^\beta(a, \mathbb{R}_\omega^n)$$

to the Stokes equations $(S \,|\, a, \, \mathbb{R}_\omega^n)_{f,g,h,u_0}^{\alpha,\beta}$, if and only if the data satisfies

$$(f, g, h, u_0) \in \mathbb{Y}_\gamma^{\alpha,\beta}(a, \mathbb{R}_\omega^n).$$

Furthermore, the solutions depend continuously on the data.

6.5 To prove Theorem 6.4 in the two special cases derived in 6.1 we transform the given bent halfspace into a halfspace and employ the maximal L_p-regularity of the Stokes equations in a halfspace, cf. Corollary 5.6, and a fixed point argument. To be precise, we will first show for all $\bar{a} > 0$ that

$$(u, p) \in {}_0\mathbb{X}_\gamma^\beta(\bar{a}, \mathbb{R}_\omega^n)$$

is a solution to the Stokes equations $(S \,|\, \bar{a}, \, \mathbb{R}_\omega^n)_{0,0,h,0}^{\alpha,\beta}$, if and only if

$$(6.1) \qquad {}_0L_{\bar{a},\gamma}^{\alpha,\beta}(u \circ \Theta_\omega, p \circ \Theta_\omega) = R_{\bar{a},\gamma}^{\alpha,\beta}(u \circ \Theta_\omega, p \circ \Theta_\omega) + T_{\bar{a},\gamma}^\beta(h)$$

with $(u \circ \Theta_\omega, p \circ \Theta_\omega) \in {}_0\mathbb{X}_\gamma^\beta(\bar{a}, \mathbb{R}_+^n)$. Here

$${}_0L_{\bar{a},\gamma}^{\alpha,\beta} : {}_0\mathbb{X}_\gamma^\beta(\bar{a}, \mathbb{R}_+^n) \longrightarrow {}_0\mathbb{Y}_\gamma^{\alpha,\beta}(\bar{a}, \mathbb{R}_+^n)$$

denotes the bounded linear operator induced by the left hand side of the Stokes equations $(S \,|\, \bar{a}, \, \mathbb{R}_+^n)$ without the initial condition. Note that this operator is an isomorphism thanks to Corollary 5.6, cf. 3.31 and Corollary 3.32. Moreover,

$$R_{\bar{a},\gamma}^{\alpha,\beta} : {}_0\mathbb{X}_\gamma^\beta(\bar{a}, \mathbb{R}_+^n) \longrightarrow {}_0\mathbb{Y}_\gamma^{\alpha,\beta}(\bar{a}, \mathbb{R}_+^n)$$

denotes the right hand side, which arises due to the transformation of the equations via the isomorphism

$$(6.2) \qquad \Theta_\omega : \mathbb{R}_+^n \longrightarrow \mathbb{R}_\omega^n, \qquad \Theta_\omega(x, y) := (x, y + \omega(x)), \quad x \in \mathbb{R}^{n-1}, y > 0$$

and

$$T_{\bar{a},\gamma}^\beta : {}_\nu\mathbb{Y}_{h,\gamma}^\beta(\bar{a}, \Gamma) \longrightarrow {}_0\mathbb{Y}_\gamma^{\alpha,\beta}(\bar{a}, \mathbb{R}_+^n)$$

denotes the transformed right hand side. Recall that the solution space

$${}_0\mathbb{X}_\gamma^\beta(\bar{a}, \Omega) = \left\{ (v, q) \in \mathbb{X}_\gamma^\beta(\bar{a}, \Omega) : v(0) = 0 \right\}$$

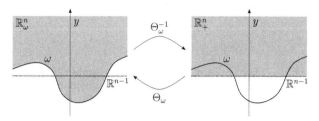

– Transformation of a Bent Halfspace –

and the data space ${}_0\mathbb{Y}_\gamma^{\alpha,\beta}(\bar{a}, \Omega)$, which consists of all

$$(f, g, h) \in \mathbb{Y}_f(\bar{a}, \Omega) \times \mathbb{Y}_g(\bar{a}, \Omega) \times \mathbb{Y}_{h,\gamma}^{\alpha,\beta}(\bar{a}, \Gamma)$$

that satisfy the compatibility conditions $(C_1)_{g,0}$, $(C_2)_{h,0}^\alpha$ and $(C_3)_{g,h,0}^\beta$, have been defined in 3.31 and 3.59 for $\Omega = \mathbb{R}_+^n$ respectively $\Omega = \mathbb{R}_\omega^n$. Moreover, we use

$${}_\nu\mathbb{Y}_{h,\gamma}^0(\bar{a}, \Gamma) := \left\{ \eta \in \mathbb{Y}_h(\bar{a}, \Gamma) : \begin{array}{l} P_\Gamma \eta = 0, \\ Q_\Gamma \eta \in {}_0H_p^1((0, \bar{a}), \dot{W}_p^{-1/p}(\Gamma, N\Gamma)) \\ \qquad \cap L_p((0, \bar{a}), W_p^{2-1/p}(\Gamma, N\Gamma)) \end{array} \right\}$$

as well as

$${}_\nu\mathbb{Y}_{h,\gamma}^{+1}(\bar{a}, \Gamma) := \left\{ \eta \in \mathbb{Y}_h(\bar{a}, \Gamma) : \begin{array}{l} P_\Gamma \eta = 0, \\ Q_\Gamma \eta \in {}_0W_p^\gamma((0, \bar{a}), L_p(\Gamma, N\Gamma)) \\ \qquad \cap L_p((0, \bar{a}), W_p^{1-1/p}(\Gamma, N\Gamma)) \end{array} \right\}$$

to denote the spaces of boundary data that match the reduced setting derived in 6.1.

6.6 To complete the proof Theorem 6.4 we will then show that the estimate

$$(6.3) \qquad \|R_{\bar{a},\gamma}^{\alpha,\beta}\|_{\mathcal{B}({}_0\mathbb{X}_\gamma^\beta(\bar{a}, \mathbb{R}_+^n), {}_0\mathbb{Y}_\gamma^{\alpha,\beta}(\bar{a}, \mathbb{R}_+^n))} < c \left(\bar{a}^\eta \|\omega\|_{BUC^{3-}(\mathbb{R}^{n-1})} + \|\nabla\omega\|_{L_\infty(\mathbb{R}^{n-1})} \right)$$

is valid for all $\bar{a} \in (0, 1]$ with some constant $c > 0$ and some exponent $\eta > 0$, which are independent of $\bar{a}$. Hence, there exists some $\delta \in (0, 1]$, which is independent of ω, and some $a^* = a^*(\omega) \in (0, 1]$ such that the linear operator

$$1 - {}_0S_{\bar{a},\gamma}^{\alpha,\beta} R_{\bar{a},\gamma}^{\alpha,\beta} : {}_0\mathbb{X}_\gamma^\beta(\bar{a}, \mathbb{R}_+^n) \longrightarrow {}_0\mathbb{X}_\gamma^\beta(\bar{a}, \mathbb{R}_+^n)$$

is invertible by a Neumann series, provided the flatness and smallness constraints

$$\|\nabla\omega\|_{L_\infty(\mathbb{R}^{n-1})} < \delta \qquad \text{and} \qquad \bar{a} < a^*(\omega)$$

are satisfied. Indeed, we then obtain a unique maximal regular solution

$$(\bar{u}, \bar{p}) \in {}_0\mathbb{X}_\gamma^\beta(\bar{a}, \mathbb{R}_+^n)$$

to (6.1) as

$$(\bar{u}, \bar{p}) = \left\{ 1 - {}_0S_{\bar{a},\gamma}^{\alpha,\beta} R_{\bar{a},\gamma}^{\alpha,\beta} \right\}^{-1} {}_0S_{\bar{a},\gamma}^{\alpha,\beta} T_{\bar{a},\gamma}^{\beta}(h).$$

As usual

$${}_0S_{\bar{a},\gamma}^{\alpha,\beta} : {}_0\mathbb{Y}_\gamma^{\alpha,\beta}(\bar{a}, \mathbb{R}_+^n) \longrightarrow {}_0\mathbb{X}_\gamma^{\beta}(\bar{a}, \mathbb{R}_+^n)$$

denotes the bounded linear inverse of ${}_0L_{\bar{a},\gamma}^{\alpha,\beta}$, which exists thanks to Corollary 5.6. Setting $(u, p) := (\bar{u} \circ \Theta_\omega^{-1}, \bar{p} \circ \Theta_\omega^{-1})$ we obtain the unique maximal regular solution to the Stokes equations in a bent halfspace on a small time interval $(0, \bar{a})$. However, since $\delta \in (0, 1]$ and $a^*(\omega) \in (0, 1]$ are independent of the data, the assertion of Theorem 6.4 is valid for all $a > 0$. Indeed, for fixed $\omega \in BUC^{3-}(\mathbb{R}^{n-1})$ subject to the above flatness condition we may choose $\bar{a} > 0$ such that

$$k\bar{a} = a \quad \text{for some } k \in \mathbb{N} \quad \text{and} \quad \bar{a} < a^*(\omega)$$

and successively solve the Stokes equations in a bent halfspace on the time intervals

$$(0, \bar{a}), \ (\bar{a}, 2\bar{a}), \ \ldots, \ ((k-1)\bar{a}, k\bar{a}).$$

This way we obtain the unique maximal regular solution on the time interval $(0, a)$. Thus, the proof of Theorem 6.4 will be complete.

6.2 Transformation to a Halfspace

6.7 As explained in 6.5, we transform the Stokes equations $(S \,|\, \bar{a}, \mathbb{R}_\omega^n)$ on a bent halfspace via (6.2) to a perturbed variant of the Stokes equations in a halfspace. Let $h \in {}_\nu\mathbb{Y}_{h,\gamma}^{\beta}(\bar{a}, \Gamma)$ and assume

$$(u, p) \in {}_0\mathbb{X}_\gamma^{\beta}(\bar{a}, \mathbb{R}_\omega^n)$$

to be a maximal regular solution to the Stokes equations $(S \,|\, \bar{a}, \mathbb{R}_+^n)_{0,0,h,0}^{\alpha,\beta}$. If

$$(\bar{u}, \bar{p}) := (u \circ \Theta_\omega, p \circ \Theta_\omega)$$

denotes the transformed solution, we have

$$(\bar{u}, \bar{p}) \in {}_0\mathbb{X}_\gamma^{\beta}(\bar{a}, \mathbb{R}_+^n).$$

Moreover,

$$\partial_t u = (\partial_t \bar{u}) \circ \Theta_\omega^{-1},$$

$$\partial_k u = (\partial_k \bar{u}) \circ \Theta_\omega^{-1} - (\partial_k \omega) \left\{ (\partial_y \bar{u}) \circ \Theta_\omega^{-1} \right\},$$

$$\partial_k^2 u = (\partial_k^2 \bar{u}) \circ \Theta_\omega^{-1} - 2(\partial_k \omega) \left\{ (\partial_k \partial_y \bar{u}) \circ \Theta_\omega^{-1} \right\},$$

$$\qquad + (\partial_k \omega)^2 \left\{ (\partial_y^2 \bar{u}) \circ \Theta_\omega^{-1} \right\} - (\partial_k^2 \omega) \left\{ (\partial_y \bar{u}) \circ \Theta_\omega^{-1} \right\},$$

$$\partial_y u = (\partial_y \bar{u}) \circ \Theta_\omega^{-1},$$

$$\partial_y^2 u = (\partial_y^2 \bar{u}) \circ \Theta_\omega^{-1}$$

for $k = 1, 2, \ldots, n - 1$.

6.8 To simplify our notation, we split $\bar{u} = (\bar{v}, \bar{w})$ according to the halfspace geometry into a tangential part $\bar{v} : (0, \bar{a}) \times \mathbb{R}_+^n \longrightarrow \mathbb{R}^{n-1}$ and a normal part $\bar{w} : (0, \bar{a}) \times \mathbb{R}_+^n \longrightarrow \mathbb{R}$. This way we obtain

$$\operatorname{div} u = (\operatorname{div} \bar{u}) \circ \Theta_\omega^{-1} - \nabla \omega \cdot \left\{ (\partial_y \bar{v}) \circ \Theta_\omega^{-1} \right\},$$

$$\Delta u = (\Delta \bar{u}) \circ \Theta_\omega^{-1} - 2 \nabla \omega \cdot \left\{ (\partial_y \nabla_x \bar{u}) \circ \Theta_\omega^{-1} \right\}$$
$$+ |\nabla \omega|^2 \left\{ (\partial_y^2 \bar{u}) \circ \Theta_\omega^{-1} \right\} - (\Delta \omega) \left\{ (\partial_y \bar{u}) \circ \Theta_\omega^{-1} \right\}.$$

Hence, the interior partial differential equations transform as

$$\rho \partial_t \bar{v} - \mu \Delta \bar{v} + \nabla_x \bar{p} = \rho F_v(\bar{v}, \bar{p}) \quad \text{in } (0, \bar{a}) \times \mathbb{R}_+^n,$$
$$\rho \partial_t \bar{w} - \mu \Delta \bar{w} + \partial_y \bar{p} = \rho F_w(\bar{w}) \quad \text{in } (0, \bar{a}) \times \mathbb{R}_+^n,$$
$$\operatorname{div}_x \bar{v} + \partial_y \bar{w} = G(\bar{v}) \quad \text{in } (0, \bar{a}) \times \mathbb{R}_+^n,$$

where

$$\rho F_v(\bar{v}, \bar{p}) := -2\mu (\nabla \omega \cdot \nabla_x) \partial_y \bar{v} + \mu |\nabla \omega|^2 \partial_y^2 \bar{v} - \mu (\Delta \omega) \partial_y \bar{v} + (\nabla \omega) \partial_y \bar{p},$$
(6.4) $$\rho F_w(\bar{w}) := -2\mu (\nabla \omega \cdot \nabla_x) \partial_y \bar{w} + \mu |\nabla \omega|^2 \partial_y^2 \bar{w} - \mu (\Delta \omega) \partial_y \bar{w},$$
$$G(\bar{v}) := \nabla \omega \cdot \partial_y \bar{v}.$$

6.9 Concerning the transformation of the boundary conditions first observe that the outer unit normal field of $\mathbb{R}_\omega^n$ is given as

$$\nu_\Gamma(x, \omega(x)) = \frac{1}{\sqrt{1 + |\nabla \omega|^2}} \begin{bmatrix} \nabla \omega(x) \\ -1 \end{bmatrix}, \qquad x \in \mathbb{R}^{n-1}.$$

Thus, we infer

$$\kappa_\omega \left\{ \nu_\Gamma \circ \Theta_\omega \right\} = \nu_\Sigma + \begin{bmatrix} \nabla \omega \\ 0 \end{bmatrix} =: \nu_\Sigma + N(\nabla \omega),$$

where we have set $\kappa_\omega := \sqrt{1 + |\nabla \omega|^2}$ and ν_Σ denotes the outer unit normal field of $\mathbb{R}_+^n$ on $\Sigma := \partial \mathbb{R}_+^n$. Moreover,

$$P_\Gamma(x, \omega(x)) = 1 - \frac{1}{1 + |\nabla \omega|^2} \begin{bmatrix} \nabla \omega(x) \otimes \nabla \omega(x) & -\nabla \omega(x) \\ -\nabla \omega(x)^\top & 1 \end{bmatrix}, \qquad x \in \mathbb{R}^{n-1}$$

and we infer

$$P_\Gamma \circ \Theta_\omega = P_\Sigma - \frac{1}{1 + |\nabla \omega|^2} \begin{bmatrix} \nabla \omega \otimes \nabla \omega & -\nabla \omega \\ -\nabla \omega^\top & -|\nabla \omega|^2 \end{bmatrix} =: P_\Sigma - L(\nabla \omega)$$

6.10 Now, for $\alpha = 0$, the tangential boundary condition $P_\Gamma[u]_\Gamma = 0$ is equivalent to

$$P_\Sigma[\bar{u}]_\Sigma - L(\nabla \omega)[\bar{u}]_\Sigma = (P_\Gamma[u]_\Gamma) \circ \Theta_\omega = 0,$$

which may be rewritten as

$$[\bar{v}]_\Sigma = L_v(\nabla\omega)[\bar{u}]_\Sigma, \qquad L_w(\nabla\omega)[\bar{u}]_\Sigma = 0,$$

where L_v denotes the first $n-1$ rows of L and L_w denotes the last row of L. However, the first equation already implies the second one, which is not surprising, since a tangential boundary condition on $\Gamma = \partial\mathbb{R}^n_\omega$ may not lead to n linearly independent transformed boundary conditions on $\Sigma = \partial\mathbb{R}^n_+$. The original boundary condition in a bent halfspace is therefore equivalent to

$$[\bar{v}]_\Sigma = H^0_v(\bar{v},\,\bar{w}) \quad \text{on } (0,\,\bar{a}) \times \partial\mathbb{R}^n_+,$$

where

(6.5a)
$$H^0_v(\bar{v},\,\bar{w}) := \frac{\nabla\omega}{1 + |\nabla\omega|^2}\left\{\,\nabla\omega\cdot[\bar{v}]_\Sigma - [\bar{w}]_\Sigma\,\right\}.$$

6.11 On the other hand, for $\alpha = \pm 1$, we have to treat the terms

$$[\nabla u \pm \nabla u^\mathsf{T}]_\Gamma \circ \Theta_\omega$$

$$= [\nabla\bar{u} \pm \nabla\bar{u}^\mathsf{T}]_\Sigma - \left[\begin{array}{cc} \nabla\omega\otimes[\partial_y\bar{v}]_\Sigma \pm [\partial_y\bar{v}]_\Sigma\otimes\nabla\omega & (\nabla\omega)\,[\partial_y\bar{w}]_\Sigma \\ \pm\nabla\omega\cdot[\partial_y\bar{v}]_\Sigma & 0 \end{array}\right]$$

$$=: [\nabla\bar{u} \pm \nabla\bar{u}^\mathsf{T}]_\Sigma - M(\nabla\omega,\,[\partial_y\bar{v}]_\Sigma,\,[\partial_y\bar{w}]_\Sigma)$$

and the tangential boundary condition $\mu P_\Gamma[\nabla u \pm \nabla u^\mathsf{T}]_\Gamma\,\nu_\Gamma = 0$ is equivalent to

$$\mu P_\Sigma[\nabla\bar{u} \pm \nabla\bar{u}^\mathsf{T}]_\Sigma\,\nu_\Sigma - \mu L(\nabla\omega)\left\{\,[\nabla u \pm \nabla u^\mathsf{T}]_\Gamma \circ \Theta_\omega\,\right\}\nu_\Sigma$$
$$- \mu P_\Sigma M(\nabla\omega,\,[\partial_y\bar{v}]_\Sigma,\,[\partial_y\bar{w}]_\Sigma)\nu_\Sigma$$
$$+ \mu\left\{\,P_\Gamma[\nabla u \pm \nabla u^\mathsf{T}]_\Gamma \circ \Theta_\omega\,\right\}N(\nabla\omega)$$
$$= \kappa_\omega\mu\left\{\,P_\Gamma[\nabla u \pm \nabla u^\mathsf{T}]_\Gamma\,\nu_\Gamma \circ \Theta_\omega\,\right\}$$
$$= 0,$$

which may be rewritten as

$$\mp\mu[\partial_y v]_\Sigma - \mu\nabla_x[w]_\Sigma = \mu L_v(\nabla\omega)\left\{\,[\nabla u \pm \nabla u^\mathsf{T}]_\Gamma \circ \Theta_\omega\,\right\}\nu_\Sigma$$
$$+ \mu M_v(\nabla\omega,\,[\partial_y\bar{v}]_\Sigma,\,[\partial_y\bar{w}]_\Sigma)\nu_\Sigma$$
$$- \mu\left\{\,P^v_\Gamma[\nabla u \pm \nabla u^\mathsf{T}]_\Gamma \circ \Theta_\omega\,\right\}N(\nabla\omega)$$

and

$$\mu L_w(\nabla\omega)\left\{\,[\nabla u \pm \nabla u^\mathsf{T}]_\Gamma \circ \Theta_\omega\,\right\}\nu_\Sigma = \mu\left\{\,P^w_\Gamma[\nabla u \pm \nabla u^\mathsf{T}]_\Gamma \circ \Theta_\omega\,\right\}N(\nabla\omega),$$

where M_v denotes the first $n-1$ rows of $P_\Sigma M$, P^v_Γ denotes the first $n-1$ rows of P_Γ and P^w_Γ denotes the last row of P_Γ. Again, the first equation already implies the second one and the original boundary condition in the bent halfspace is therefore equivalent to

$$\mp\mu[\partial_y v]_\Sigma - \mu\nabla_x[w]_\Sigma = H^{\pm 1}_v(\bar{v},\,\bar{w}) \quad \text{on } (0,\,\bar{a}) \times \partial\mathbb{R}^n_+,$$

where

$$(6.5b) \quad H_v^{\pm 1}(\bar{v}, \bar{w}) := \mu L_v(\nabla \omega) \left\{ [\nabla u \pm \nabla u^\mathsf{T}]_\Gamma \circ \Theta_\omega \right\} \nu_\Sigma$$
$$+ \mu M_v(\nabla \omega, [\partial_y \bar{v}]_\Sigma, [\partial_y \bar{w}]_\Sigma) \nu_\Sigma$$
$$- \mu \left\{ P_\Gamma^v [\nabla u \pm \nabla u^\mathsf{T}]_\Gamma \circ \Theta_\omega \right\} N(\nabla \omega).$$

6.12 Concerning the normal boundary condition, we start with the case $\beta = 0$. In this case the normal boundary condition is equivalent to

$$-[\bar{w}]_\Sigma + \nabla \omega \cdot [\bar{v}]_\Sigma = [\bar{u}]_\Sigma \cdot \nu_\Sigma + [\bar{u}]_\Sigma \cdot N(\nabla \omega) = \kappa_\omega \left\{ ([u]_\Gamma \cdot \nu_\Gamma) \circ \Theta_\omega \right\} = \kappa_\omega \bar{h}_w$$

with $\bar{h}_w := (Q_\Gamma h \cdot \nu_\Gamma) \circ \Theta_\omega$. Therefore, the original boundary condition in the bent halfspace is equivalent to

$$(6.6a) \quad -[\bar{w}]_\Sigma = -\nabla \omega \cdot [\bar{v}]_\Sigma + \kappa_\omega \bar{h}_w =: H_w^0(\bar{v}) + \kappa_\omega \bar{h}_w \quad \text{on } (0, \bar{a}) \times \partial \mathbb{R}_+^n.$$

6.13 Finally, if $\beta = 1$, the normal boundary condition is equivalent to

$$2\mu \, \partial_\nu \bar{u} \cdot \nu_\Sigma - [\bar{p}]_\Sigma - 2\mu \, \bar{M}(\nabla \omega, [\partial_y \bar{v}]_\Sigma, [\partial_y \bar{w}]_\Sigma) \nu_\Sigma \cdot \nu_\Sigma$$
$$+ 2\mu \left\{ [\nabla u]_\Gamma \circ \Theta_\omega \right\} N(\nabla \omega) \cdot \nu_\Sigma$$
$$+ 2\mu \left\{ [\nabla u]_\Gamma \nu_\Gamma \circ \Theta_\omega \right\} \cdot \kappa_\omega N(\nabla \omega)$$
$$- |\nabla \omega|^2 [\bar{p}]_\Sigma$$
$$= \kappa_\omega^2 \left\{ (2\mu \, \partial_\nu \cdot \nu_\Gamma - [p]_\Gamma) \circ \Theta_\omega \right\}$$
$$= \kappa_\omega^2 \bar{h}_w$$

with $\bar{h}_w$ as in 6.13 and

$$[\nabla u]_\Gamma \circ \Theta_\omega$$
$$= [\nabla \bar{u}]_\Sigma - \begin{bmatrix} \nabla \omega \otimes [\partial_y \bar{v}]_\Sigma & (\nabla \omega) \, [\partial_y \bar{w}]_\Sigma \\ 0 & 0 \end{bmatrix}$$
$$=: [\nabla \bar{u}]_\Sigma - \bar{M}(\nabla \omega, [\partial_y \bar{v}]_\Sigma, [\partial_y \bar{w}]_\Sigma).$$

Therefore, the original boundary condition in a bent halfspace is equivalent to

$$(6.6b) \quad 2\mu \, \partial_\nu \bar{u} \cdot \nu_\Sigma - [\bar{p}]_\Sigma = 2\mu \, \bar{M}(\nabla \omega, [\partial_y \bar{v}]_\Sigma, [\partial_y \bar{w}]_\Sigma) \nu_\Sigma \cdot \nu_\Sigma$$
$$- 2\mu \left\{ [\nabla u]_\Gamma \circ \Theta_\omega \right\} N(\nabla \omega) \cdot \nu_\Sigma$$
$$- 2\mu \left\{ [\nabla u]_\Gamma \nu_\Gamma \circ \Theta_\omega \right\} \cdot \kappa_\omega N(\nabla \omega)$$
$$+ |\nabla \omega|^2 [\bar{p}]_\Sigma + \kappa_\omega^2 \bar{h}_w$$
$$=: H_w^{+1}(\bar{v}, \bar{w}, \bar{p}) + \kappa_\omega^2 \bar{h}_w \quad \text{on } (0, \bar{a}) \times \partial \mathbb{R}_+^n.$$

6.14 In summary, for $\alpha \in \{-1, 0, +1\}$, $\beta \in \{0, +1\}$ and $\gamma = -\infty$ for $\beta = 0$ respectively $\gamma = 1/2 - 1/2p$ for $\beta = +1$ as well as

$$h \in {}_\nu \mathbb{Y}_{h,\gamma}^{\alpha,\beta}(\bar{a}, \Gamma)$$

the Stokes equations $(S \mid \bar{a}, \mathbb{R}^n_\omega)^{\alpha,\beta}_{0,0,h,0}$ in a bent halfspace are equivalent to the perturbed halfspace Stokes equations

$$\rho \partial_t \bar{v} - \mu \Delta \bar{v} + \nabla_x \bar{p} = \rho F_v(\bar{v}, \bar{p}) \qquad \text{in } (0, \bar{a}) \times \mathbb{R}^n_+,$$

$$\rho \partial_t \bar{w} - \mu \Delta \bar{w} + \partial_y \bar{p} = \rho F_w(\bar{w}) \qquad \text{in } (0, \bar{a}) \times \mathbb{R}^n_+,$$

$$\text{div}_x \bar{v} + \partial_y \bar{w} = G(\bar{v}) \qquad \text{in } (0, \bar{a}) \times \mathbb{R}^n_+,$$

$$P_\Sigma \mathcal{B}^\alpha(\bar{v}, \bar{w}) = H^\alpha_v(\bar{v}, \bar{w}) \qquad \text{on } (0, \bar{a}) \times \partial \mathbb{R}^n_+,$$

$$Q_\Sigma \mathcal{B}^\beta(\bar{v}, \bar{w}, \bar{p}) \cdot \nu_\Sigma = H^\beta_w(\bar{v}, \bar{w}, \bar{p}) + T^\beta_{\bar{a},h,\gamma}(h) \qquad \text{on } (0, \bar{a}) \times \partial \mathbb{R}^n_+,$$

$$(\bar{v}, \bar{w})(0) = 0 \qquad \text{in } \mathbb{R}^n_+,$$

where F_v, F_w and G are given by (6.4), H^α_v is given by (6.5) and H^β_w is given by (6.6). Furthermore,

$$T^0_{\bar{a},h,\gamma}(h) = \kappa_\omega (Q_\Gamma h \cdot \nu_\Gamma) \circ \Theta_\omega \qquad \text{and} \qquad T^{+1}_{\bar{a},h,\gamma}(h) = \kappa^2_\omega (Q_\Gamma h \cdot \nu_\Gamma) \circ \Theta_\omega$$

Setting

$$T^\beta_{\bar{a},\gamma}(h) = (0, 0, T^\beta_{\bar{a},h,\gamma}(h)), \qquad \beta \in \{0, +1\}$$

we obtain the desired formulation (6.1). Thus, to complete the proof of Theorem 6.4 it remains to derive the estimate (6.3). This will be done separately for the different boundary conditions in the next sections.

6.3 Boundary Conditions Involving the Normal Velocity

6.15 We start with the case $\beta = 0$ and $\gamma = -\infty$. First, we derive the necessary estimates for F_v and F_w. Based on (6.4) we have

$$\|\rho F_v(\bar{v}, \bar{p})\|_{L_p((0,\bar{a}) \times \mathbb{R}^n_+, \mathbb{R}^{n-1})} \leq \mu \|\omega\|_{BUC^3-(\mathbb{R}^{n-1})} \cdot \|\bar{v}\|_{L_p((0,\bar{a}), H^1_p(\mathbb{R}^n_+, \mathbb{R}^{n-1}))}$$
$$+ 3\mu\delta \|\bar{v}\|_{L_p((0,\bar{a}), H^2_p(\mathbb{R}^n_+, \mathbb{R}^{n-1}))}$$
$$+ \delta \|\nabla \bar{p}\|_{L_p((0,\bar{a}) \times \mathbb{R}^n_+, \mathbb{R}^n)},$$

where we assumed

$$\|\nabla \omega\|_{L_\infty(\mathbb{R}^{n-1})} < \delta \leq 1.$$

Now, due to Propositions 3.9 and 4.69 we have an estimate

$$\begin{aligned}
(6.7) \qquad & \|\bar{v}\|_{L_p((0,\bar{a}), H^1_p(\mathbb{R}^n_+, \mathbb{R}^{n-1}))} \\
& \leq c\bar{a}^\eta \|\bar{v}\|_{{}_0 H^{1/2}_p((0,\bar{a}), H^1_p(\mathbb{R}^n_+, \mathbb{R}^{n-1}))} \\
& \leq c\bar{a}^\eta \|\bar{v}\|_{{}_0 H^1_p((0,\bar{a}), L_p(\mathbb{R}^n_+, \mathbb{R}^{n-1})) \cap L_p((0,\bar{a}), H^2_p(\mathbb{R}^n_+, \mathbb{R}^{n-1}))}
\end{aligned}$$

at hand with some constant $c > 0$ and some exponent $\eta > 0$, which are independent of $\bar{a} \in (0, 1]$. Since analogous estimates are valid for F_w, we obtain

$$\|(\rho F_v(\bar{v}, \bar{p}), \rho F_w(\bar{w}))\|_{\mathbb{Y}_f(\bar{a}, \mathbb{R}^n_+)} \leq c \left(\bar{a}^\eta \|\omega\|_{BUC^3-(\mathbb{R}^{n-1})} + \delta \right) \|(\bar{v}, \bar{w}, \bar{p})\|_{{}_0 \mathbb{X}^\beta_\gamma(\bar{a}, \mathbb{R}^n_+)}.$$

6.16 Concerning the right hand side of the divergence equation as defined by (6.4), we have

$$\left\|G(\bar{v})\right\|_{{}_0 H_p^{1/2}((0,\bar{a}),\,L_p(\mathbb{R}_+^n))} \le \delta\|\bar{v}\|_{{}_0 H_p^{1/2}((0,\bar{a}),\,L_p(\mathbb{R}_+^n))}$$

as well as

$$\left\|G(\bar{v})\right\|_{L_p((0,\bar{a}),\,H_p^1(\mathbb{R}_+^n))}^p$$

$$= \left\|\nabla\omega \cdot \partial_y\bar{v}\right\|_{L_p((0,\bar{a})\times\mathbb{R}_+^n)}^p + \sum_{k=1}^{n-1}\left\|\partial_k\nabla\omega \cdot \partial_y\bar{v}\right\|_{L_p((0,\bar{a})\times\mathbb{R}_+^n)}^p$$

$$+ \sum_{k=1}^{n-1}\left\|\nabla\omega \cdot \partial_k\partial_y\bar{v}\right\|_{L_p((0,\bar{a})\times\mathbb{R}_+^n)}^p + \left\|\nabla\omega \cdot \partial_y^2\bar{v}\right\|_{L_p((0,\bar{a})\times\mathbb{R}_+^n)}^p$$

$$\le \|\omega\|_{BUC^{3-}(\mathbb{R}^{n-1})}^p \cdot \left\|\partial_y\bar{v}\right\|_{L_p((0,\bar{a})\times\mathbb{R}_+^n)}^p + \delta^p\left\|\partial_y\bar{v}\right\|_{L_p((0,\bar{a}),\,H_p^1(\mathbb{R}_+^n,\mathbb{R}^{n-1}))}^p.$$

Thus, estimate (6.7) leads to

$$\left\|G(\bar{v}))\right\|_{\mathbb{Y}_g(\bar{a},\mathbb{R}_+^n)} \le c\left(\bar{a}^\eta\|\omega\|_{BUC^{3-}(\mathbb{R}^{n-1})} + \delta\right)\|(\bar{v},\,\bar{w})\|_{{}_0\mathbb{X}_u(\bar{a},\mathbb{R}_+^n)}.$$

Again, $c > 0$ and $\eta > 0$ are independent of $\bar{a} \in (0,\,1]$.

6.17 Analogously, the right hand sides H_v^α as defined by (6.5) are composed of terms, which either carry a factor $\nabla\omega$ or are of lower order. Arguing as in 6.15 and 6.16 we obtain

$$\left\|H_v^\alpha(\bar{v},\,\bar{w})\right\|_{\mathbb{T}_h^\alpha(\bar{a},\Sigma)} \le c\left(\bar{a}^\eta\|\omega\|_{BUC^{3-}(\mathbb{R}^{n-1})} + \delta\right)\|(\bar{v},\,\bar{w})\|_{{}_0\mathbb{X}_u(\bar{a},\mathbb{R}_+^n)}$$

and the same is true for H_w^0 as defined by (6.6), i.e.

$$\left\|H_w^0(\bar{v})\right\|_{\mathbb{N}_{h,\gamma}^0(\bar{a},\Sigma)} \le c\left(\bar{a}^\eta\|\omega\|_{BUC^{3-}(\mathbb{R}^{n-1})} + \delta\right)\|(\bar{v},\,\bar{w})\|_{{}_0\mathbb{X}_u(\bar{a},\mathbb{R}_+^n)}.$$

As in 6.15 and 6.16 the constants $c > 0$ and $\eta > 0$ are both independent of $\bar{a} \in (0,\,1]$.

6.18 It remains to check the compatibility condition, which stems from the divergence constraint and the normal boundary condition. Indeed, we have

$$\langle\phi\,|\,(G(\bar{v}),\,H_w^0(\bar{v})\nu_\Sigma)\rangle = \int_\Sigma [\phi]_\Sigma(\nabla\omega\cdot[\bar{v}]_\Sigma)\,\mathrm{d}\mathcal{H}^2 - \int_{\mathbb{R}_+^n}\phi\,(\nabla\omega\cdot\partial_y\bar{v})\,\mathrm{d}\mathcal{H}^3$$

$$= \int_{\mathbb{R}_+^n}(\partial_y\phi)\,(\nabla\omega\cdot\bar{v})\,\mathrm{d}\mathcal{H}^3, \qquad \phi \in H_{p'}^1(\mathbb{R}_+^n),$$

where $1/p + 1/p' = 1$. Hence,

$$\left\|(G(\bar{v}),\,H_w^0(\bar{v})\nu_\Sigma)\right\|_{{}_0 H_p^1((0,\bar{a}),\,{}_0\dot{H}_p^{-1}(\mathbb{R}_+^n))}$$

$$\le \left\|\nabla\omega\cdot\bar{v}\right\|_{{}_0 H_p^1((0,\bar{a}),\,L_p(\mathbb{R}_+^n))}$$

$$\le \delta\|(\bar{v},\,\bar{w})\|_{{}_0\mathbb{X}_u(\bar{a},\mathbb{R}_+^n)}$$

and the proof of (6.3) for $\beta = 0$ is complete.

6.4 Boundary Conditions Involving the Pressure

6.19 Now, assume $\beta = +1$ and $\gamma = 1/2 - 1/2p$. The necessary estimates for F, G and H_v^α have been derived in 6.15, 6.16 and 6.17. Concerning the right hand side H_w^{+1} of the normal boundary condition as defined in (6.6) we observe that again all terms either carry a factor $\nabla\omega$ or are of lower order. Arguing as in 6.15, 6.16 and 6.17 we obtain

$$\|H_w^0(\bar{v})\|_{\mathbb{N}_{h,\gamma}^{+1}(\bar{a},\,\Sigma)} \le c\left(\bar{a}^\eta\|\omega\|_{BUC^3-(\mathbb{R}^{n-1})} + \delta\right)\|(\bar{v},\,\bar{w},\,\bar{p})\|_{{}_0\mathbb{X}_\gamma^\beta(\bar{a},\mathbb{R}_+^n)}$$

with some constants $c > 0$ and $\eta > 0$, which are independent of $\bar{a} \in (0, 1]$.

6.20 Finally, it remains to check the compatibility condition, which stems from the divergence constraint and the normal boundary condition. Setting

$$\eta(\bar{v}) := (\nabla\omega \cdot [\bar{v}]_\Sigma)\nu_\Sigma$$

we obtain as in 6.18 by partial integration

$$(\,G(\bar{v}),\,\eta(\bar{v})\,) \in {}_0H_p^1((0,\,\bar{a}),\,{}_0\dot{H}_p^{-1}(\mathbb{R}_+^n))$$

with

$$\|(\,G(\bar{v}),\,\eta(\bar{v})\,)\|_{{}_0H_p^1((0,\bar{a}),\,{}_0\dot{H}_p^{-1}(\mathbb{R}_+^n))} \le \delta\|(\bar{v},\,\bar{w})\|_{{}_0\mathbb{X}_u(\bar{a},\mathbb{R}_+^n)}$$

and, therefore, the proof of (6.3) for $\beta = +1$ and, hence, of Theorem 6.4 is complete.

Remarks

6.21 The approach presented in this section to reduce the problem in a bent halfspace by a suitable transformation to a problem in a halfspace is well-known. In particular, it is applicable to a large class of parabolic problems, see the monograph [DHP03] by R. DENK, M. HIEBER and J. PRÜSS, and to the more complex *generalised Stokes equations*, which occur for a class of non-Newtonian fluids, see the work [BP07] of D. BOTHE and J. PRÜSS.

References

[BP07]   D. BOTHE and J. PRÜSS: L_p-Theory for a Class of Non-Newtonian Fluids. SIAM J. Math. Anal., 39, 379–421, 2007.

[DHP03] R. DENK, M. HIEBER, and J. PRÜSS: $\mathcal{R}$-Boundedness, Fourier-Multipliers and Problems of Elliptic and Parabolic Type, *Mem. Amer. Math. Soc.*, vol. 166. American Mathematical Society, 2003.

Chapter 7

Maximal L_p-Regularity in a Bounded Smooth Domain

This chapter is devoted to the study of the Stokes equations subject to one of the energy preserving respectively artificial boundary conditions introduced 2.19, 2.22 and 2.23 in a bounded domain $\Omega \subseteq \mathbb{R}^n$ with boundary $\Gamma := \partial\Omega$ of class C^{3-}, i.e. we prove Theorem 3.30.

Given $a > 0$ as well as $\rho, \mu > 0$ and parameters $\alpha, \beta \in \{-1, 0, +1\}$, which define the boundary condition, as well as a parameter

- $\gamma = -\infty$, if $\beta = 0$;
- $\gamma \in \{-\infty\} \cup [0, 1/2 - 1/2p]$, if $\beta = +1$;
- $\gamma \in \{-\infty\} \cup [0, \infty)$, if $\beta = -1$,

which defines the regularity of the pressure trace, we prove the existence of a unique maximal regular solution

$$(u, p) \in \mathbb{X}_\gamma^\beta(a, \Omega)$$

to the Stokes equations

$$(\mathrm{S}\,|\,a,\,\Omega)_{f,g,h,u_0}^{\alpha,\beta} \qquad \begin{aligned} \rho\partial_t u - \mu\Delta u + \nabla p &= \rho f && \text{in } (0, a) \times \Omega, \\ \operatorname{div} u &= g && \text{in } (0, a) \times \Omega, \\ \mathcal{B}^{\alpha,\beta}(u, p) &= h && \text{on } (0, a) \times \partial\Omega, \\ u(0) &= u_0 && \text{in } \Omega, \end{aligned}$$

whenever the data is subject to the regularity and compatibility conditions

$$(f, g, h, u_0) \in \mathbb{Y}_\gamma^{\alpha,\beta}(a, \Omega).$$

The bounded linear operators $\mathcal{B}^{\alpha,\beta}$ realise the desired boundary condition. Recall that they were defined in Chapter 3 as

$$P_\Gamma \mathcal{B}^{0,\beta}(u, p) = P_\Gamma[u]_\Gamma, \qquad P_\Gamma \mathcal{B}^{\pm 1,\beta}(u, p) = \mu P_\Gamma[\nabla u \pm \nabla u^\top]_\Gamma \nu_\Gamma$$

for $\beta \in \{-1, 0, +1\}$ and

$$Q_\Gamma \mathcal{B}^{\alpha,0}(u, p) \cdot \nu_\Gamma = [u]_\Gamma \cdot \nu_\Gamma,$$

$$Q_\Gamma \mathcal{B}^{\alpha,+1}(u, p) \cdot \nu_\Gamma = 2\mu\,\partial_\nu u \cdot \nu_\Gamma - [p]_\Gamma, \qquad Q_\Gamma \mathcal{B}^{\alpha,-1}(u, p) \cdot \nu_\Gamma = -[p]_\Gamma$$

for $\alpha \in \{-1, 0, +1\}$. Also recall that $[\,\cdot\,]_\Gamma$ denotes the trace of a function defined in Ω on the boundary $\Gamma := \partial\Omega$. The normal derivative has to be understood as $\partial_\nu = [\nabla \cdot {}^\mathsf{T}]_\Gamma \nu_\Gamma$. Moreover, $P_\Gamma = P_\Gamma(y) := 1 - \nu_\Gamma(y) \otimes \nu_\Gamma(y)$ denotes the projection onto the tangent space $T_y\Gamma$ of Γ at a point $y \in \Gamma$. Furthermore, $Q_\Gamma := 1 - P_\Gamma$ denotes the projection onto the normal bundle of Γ and $\nu_\Gamma : \Gamma \longrightarrow \mathbb{R}^n$ denotes the outer unit normal field of Ω.

The solution spaces have been defined in 3.1, 3.21, 3.22, 3.23 and 3.28 as

$$\mathbb{X}_\gamma^\beta(a,\,\Omega) = \mathbb{X}_u(a,\,\Omega) \times \mathbb{X}_{p,\gamma}^\beta(a,\,\Omega)$$

with

$$\mathbb{X}_u(a,\,\Omega) = H_p^1((0,\,a),\,L_p(\Omega,\,\mathbb{R}^n)) \cap L_p((0,\,a),\,H_p^2(\Omega,\,\mathbb{R}^n)),$$

$$\mathbb{X}_{p,-\infty}^0(a,\,\Omega) = \left\{ q \in L_p((0,\,a),\,\dot{H}_p^1(\Omega)) : (q)_\Omega = 0 \right\},$$

$$\mathbb{X}_{p,-\infty}^{\pm 1}(a,\,\Omega) = L_p((0,\,a),\,\dot{H}_p^1(\Omega))$$

and

$$\mathbb{X}_{p,\gamma}^{\pm 1}(a,\,\Omega) = \left\{ q \in L_p((0,\,a),\,\dot{H}_p^1(\Omega)) : \begin{array}{l} [q]_\Gamma \in W_p^\gamma((0,\,a),\,L_p(\Gamma)) \\[4pt] \qquad \cap\, L_p((0,\,a),\,W_p^{1-1/p}(\Gamma)) \end{array} \right\}$$

for $\gamma \geq 0$. The data spaces have been defined in 3.1, 3.12, 3.19 and 3.23 as

$$\mathbb{Y}_f(a,\,\Omega) = L_p((0,\,a) \times \Omega,\,\mathbb{R}^n),$$

$$\mathbb{Y}_g(a,\,\Omega) = H_p^{1/2}((0,\,a),\,H_p^1(\Omega)) \cap L_p((0,\,a),\,H_p^1(\Omega)),$$

$$\mathbb{Y}_0(\Omega) = W_p^{2-2/p}(\Omega,\,\mathbb{R}^n)$$

and

$$\mathbb{Y}_{h,\gamma}^{\alpha,\beta}(a,\,\Gamma) = \left\{ \eta \in \mathbb{Y}_h(a,\,\Gamma) : P_\Gamma\eta \in \mathbb{T}_h^\alpha(a,\,\Gamma),\, Q_\Gamma\eta \in \mathbb{N}_{h,\gamma}^\beta(a,\,\Gamma) \right\},$$

where $\mathbb{Y}_h(a,\,\Gamma) = L_p((0,\,a),\,L_{p,loc}(\Gamma,\,\mathbb{R}^n))$ has been defined in 3.13 and

$$\mathbb{T}_h^0(a,\,\Gamma) = W_p^{1-1/2p}((0,\,a),\,L_p(\Gamma,\,T\Gamma)) \cap L_p((0,\,a),\,W_p^{2-1/p}(\Gamma,\,T\Gamma)),$$

$$\mathbb{T}_h^{\pm 1}(a,\,\Gamma) = W_p^{1/2-1/2p}((0,\,a),\,L_p(\Gamma,\,T\Gamma)) \cap L_p((0,\,a),\,W_p^{1-1/p}(\Gamma,\,T\Gamma)),$$

$$\mathbb{N}_{h,-\infty}^0(a,\,\Gamma) = W_p^{1-1/2p}((0,\,a),\,L_p(\Gamma,\,N\Gamma)) \cap L_p((0,\,a),\,W_p^{2-1/p}(\Gamma,\,N\Gamma)),$$

$$\mathbb{N}_{h,-\infty}^{\pm 1}(a,\,\Gamma) = L_p((0,\,a),\,\dot{W}_p^{1-1/p}(\Gamma,\,N\Gamma))$$

have been defined in 3.18 and 3.23. Moreover, for $\gamma \geq 0$ the spaces

$$\mathbb{N}_{h,\gamma}^{\pm 1}(a,\,\Gamma) := W_p^\gamma((0,\,a),\,L_p(\Gamma,\,N\Gamma)) \cap L_p((0,\,a),\,W_p^{1-1/p}(\Gamma,\,N\Gamma))$$

have been defined in 3.21 and 3.22 and $\mathbb{Y}_\gamma^{\alpha,\beta}(a,\,\Omega)$ has been defined in 3.28 to consist of all

$$(f,\,g,\,h,\,u_0) \in \mathbb{Y}_f(a,\,\Omega) \times \mathbb{Y}_g(a,\,\Omega) \times \mathbb{Y}_{h,\gamma}^{\alpha,\beta}(a,\,\Gamma) \times \mathbb{Y}_0(\Omega)$$

that satisfy the compatibility condition

$$(\mathrm{C}_1)_{g,u_0} \qquad\qquad\qquad\qquad \mathrm{div}\, u_0 = g(0),$$

which stems from the divergence constraint, the compatibility condition

$(C_2)^\alpha_{h,u_0}$
$$P_\Gamma[u_0]_\Gamma = P_\Gamma h(0), \text{ if } \alpha = 0 \quad \text{and } p > \tfrac{3}{2},$$
$$\mu P_\Gamma[\nabla u_0 \pm \nabla u_0^\mathsf{T}]_\Gamma \, \nu_\Gamma = P_\Gamma h(0), \text{ if } \alpha = \pm 1 \text{ and } p > 3,$$

which stems from the tangential boundary condition, and the condition

$$Q_\Gamma[u_0]_\Gamma = Q_\Gamma h, \quad \text{if } p > \tfrac{3}{2},$$
$$\text{and} \quad (\, g, \, Q_\Gamma h \,) \in H^1_p((0, \, a), \, {}_0\dot{H}^{-1}_p(\Omega)), \qquad \text{if } \beta = 0,$$

$(C_3)^\beta_{g,h,u_0}$
$$\text{there exists } \eta \in \mathbb{N}^0_{h,-\infty}(a, \, \Gamma) \text{ such that}$$
$$Q_\Gamma[u_0]_\Gamma = \eta, \quad \text{if } p > \tfrac{3}{2}, \qquad \text{if } \beta = \pm 1,$$
$$\text{and} \quad (\, g, \, \eta \,) \in H^1_p((0, \, a), \, {}_0\dot{H}^{-1}_p(\Omega)),$$

which stems from the divergence constraint and the normal boundary condition. For a proof of the necessity of these conditions see Section 3.3.

7.1 Strategy

7.1 The case $\beta = -1$ has been completely treated by Theorem 4.60, see 4.62, and based on 4.61 we may restrict our considerations to the cases

- $\beta = 0$, $\gamma = -\infty$ and $u_0 = 0$;

- $\beta = +1$, $\gamma = 1/2 - 1/2p$ and $u_0 = 0$.

Hence, it is sufficient to show that the bounded linear operator

$${}_0L^{\alpha,\beta}_{a,\gamma} : {}_0\mathbb{X}^\beta_\gamma(a, \, \Omega) \longrightarrow {}_0\mathbb{Y}^{\alpha,\beta}_\gamma(a, \, \Omega),$$

which is defined by the left hand side of the Stokes equations $(S \,|\, a, \, \Omega)^{\alpha,\beta}_{f,g,h,0}$ without the initial condition, constitutes an isomorphism between the solution space

$${}_0\mathbb{X}^\beta_\gamma(a, \, \Omega) = \left\{ \, (v, \, q) \in \mathbb{X}^\beta_\gamma(a, \, \Omega) \, : \, v(0) = 0 \, \right\}$$

and the data space ${}_0\mathbb{Y}^{\alpha,\beta}_\gamma(a, \, \Omega)$, which consists of all

$$(f, \, g, \, h) \in \mathbb{Y}_f(a, \, \Omega) \times \mathbb{Y}_g(a, \, \Omega) \times \mathbb{Y}^{\alpha,\beta}_{h,\gamma}(a, \, \Gamma),$$

which satisfy the compatibility conditions $(C_1)_{g,0}$, $(C_2)^\alpha_{h,0}$ and $(C_3)^\beta_{g,h,0}$, cf. 3.31 and Corollary 3.32. To do this we fix $\gamma = -\infty$, if $\beta = 0$, and $\gamma = 1/2 - 1/2p$, if $\beta = +1$. These two cases will be treated separately in the next sections.

7.2 To establish the desired mapping properties of ${}_0L^{\alpha,\beta}_{a,\gamma}$ for $\beta = 0$ we will first show that any solution

$$(u, \, p) \in {}_0\mathbb{X}^\beta_\gamma(a, \, \Omega)$$

to the Stokes equations $(S \mid a, \Omega)_{0,0,h,0}^{\alpha,\beta}$ with data $h \in {}_\tau \mathbb{Y}_{h,\gamma}^\alpha(a, \Gamma)$, where

$$_\tau \mathbb{Y}_{h,\gamma}^\alpha(a, \Gamma) := \Big\{ \eta \in \mathbb{Y}_h(a, \Gamma) \,:\, P_\Gamma \eta \in \mathbb{T}_h^\alpha(a, \Gamma), \ P_\Gamma \eta(0) = 0, \ Q_\Gamma \eta = 0 \Big\},$$

satisfies an estimate

(7.1a)
$$\|(u, p)\|_{_0\mathbb{X}_\gamma^\beta(a, \Omega)} \le ca^\eta \|u\|_{_0\mathbb{X}_u(a, \Omega)} + c' \|h\|_{\mathbb{Y}_{h,\gamma}^{\alpha,\beta}(a,\Gamma)}$$

with some constant $c > 0$ and some exponent $\eta > 0$, which are independent of $a \in (0, 1]$. This then implies that $_0 L_{a,\gamma}^{\alpha,\beta}$ is injective with closed range, provided we choose $a \in (0, 1]$ to be sufficiently small.

7.3 Analogously, for $\beta = +1$ we will show that any solution

$$(u, p) \in {}_0\mathbb{X}_\gamma^\beta(a, \Omega)$$

to the Stokes equations $(S \mid a, \Omega)_{0,0,h,0}^{\alpha,\beta}$ with data $h \in {}_\nu \mathbb{Y}_{h,\gamma}^{+1}(a, \Gamma)$, where

$$_\nu \mathbb{Y}_{h,\gamma}^{+1}(a, \Gamma) := \left\{ \eta \in \mathbb{Y}_h(a, \Gamma) \,:\, \begin{array}{c} P_\Gamma \eta = 0, \\ Q_\Gamma \eta \in {}_0 H_p^\gamma((0, a), L_p(\Gamma, N\Gamma)) \\ \cap\ L_p((0, a), W_p^{1-1/p}(\Gamma, N\Gamma)) \end{array} \right\},$$

satisfies an estimate

(7.1b)
$$\|(u, p)\|_{_0\mathbb{X}_\gamma^\beta(a, \Omega)} \le ca^\eta \|u\|_{_0\mathbb{X}_u(a, \Omega)} + c' \|h\|_{\mathbb{Y}_{h,\gamma}^{\alpha,\beta}(a,\Gamma)}$$

with some constant $c > 0$ and some exponent $\eta > 0$, which are again independent of $a \in (0, 1]$. Again, this then implies that $_0 L_{a,\gamma}^{\alpha,\beta}$ is injective with closed range, provided we choose $a \in (0, 1]$ to be sufficiently small.

7.4 Finally, we will construct bounded linear operators

$$S_{a,\gamma}^{\alpha,\beta} : {}_0\mathbb{Y}_\gamma^{\alpha,\beta}(a, \Omega) \longrightarrow {}_0\mathbb{X}_\gamma^\beta(a, \Omega)$$

for $\beta \in \{0, +1\}$ with the property

(7.2)
$$_0 L_{a,\gamma}^{\alpha,\beta} S_{a,\gamma}^{\alpha,\beta} = 1 - R_{a,\gamma}^{\alpha,\beta},$$

where the perturbation $R_{a,\gamma}^{\alpha,\beta} \in \mathcal{B}({}_0\mathbb{Y}_\gamma^{\alpha,\beta}(a, \Omega))$ satisfies an estimate

(7.3)
$$\|R_{a,\gamma}^{\alpha,\beta}\|_{\mathcal{B}({}_0\mathbb{Y}_\gamma^{\alpha,\beta}(a, \Omega))} \le ca^\eta$$

with some constants $c > 0$ and $\tau > 0$, which are independent of $a \in (0, 1]$. Thus, if we choose $a \in (0, 1]$ to be sufficiently small, the operator $1 - R_{a,\gamma}^{\alpha,\beta}$ is invertible by a Neumann series and delivers the right inverse operator

$$_0 S_{a,\gamma}^{\alpha,\beta} = S_{a,\gamma}^{\alpha,\beta} (1 - R_{a,\gamma}^{\alpha,\beta})^{-1}$$

for $_0 L_{a,\gamma}^{\alpha,\beta}$. Hence, Theorem 3.30 will be proved for small time intervals. However, since the admitted length of the time intervals, which allow the above argument to work, does not depend on the data, we may solve the Stokes equations $(S \mid a, \Omega)_{f,g,h,u_0}^{\alpha,\beta}$ for any given time interval by successively solving them on arbitrarily small time intervals of fixed length, cf. 6.6, where a similar argumentation has been used for the bent halfspace case.

7.5 The argumentation in the next sections will make frequent use of the exploitation of some extra time regularity of the solutions and the data based on Corollary 4.69. Given $1 < p < \infty$ we will denote in Sections 7.4 and 7.5 by $0 < \tau < 1/2 - 1/2p$ a fixed regularity index and choose constants $c > 0$ and $\eta > 0$ such that the embeddings

$$_0H_p^s((0, a), X) \hookrightarrow L_p((0, a), X)$$

together with the estimates

$$\|\phi\|_{L_p((0,a),X)} \leq ca^\eta \|\phi\|_{_0H_p^s((0,a),X)}, \qquad \phi \in {}_0H_p^s((0, a), X)$$

are valid uniformly in $a \in (0, 1]$ for $s \in \{\tau/2, \tau, 1/2\}$ and the finitely many Banach spaces X that are involved in Sections 7.4 and 7.5. The various constants that appear in the estimates derived in these sections may differ from line to line but will always be denoted by c, $c' > 0$, where $c > 0$ will always be independent of $a \in (0, 1]$.

7.2 Partition of the Domain

7.6 The derivation of the estimates (7.1) and (7.3) as well as the construction of the operator $S_{a,\gamma}^{\alpha,\beta}$ will rely on a localisation argument, which allows to locally transform the domain Ω into a bent halfspace respectively $\mathbb{R}^n$. This way we will be able to use the results for the Stokes equations in $\mathbb{R}^n$ and in a bent halfspace to construct approximate solutions to the Stokes equations in Ω.

7.7 By assumption, $\Gamma = \partial\Omega$ is of class C^{3-}. Thus, if $y \in \Gamma$, there exists an $\bar{r} = \bar{r}(y) > 0$ with the following properties:

(i) for all $0 < r < \bar{r}$ there exists a BUC^{3-}-function $\bar{\omega} : \Sigma \longrightarrow \mathbb{R}$, where $\Sigma \subseteq T_y\Gamma$ denotes the projection of $\Gamma \cap B_r(y)$ onto the tangent plane $T_y\Gamma$, such that

$$\Gamma \cap B_r(y) = \left\{ \eta - \bar{\omega}(\eta)\nu_\Gamma(y) : \eta \in \Sigma \right\},$$

i. e. $\Gamma \cap B_r(y)$ may be parametrised over its projection onto the tangent plane;

(ii) the above functions $\bar{\omega}$ may be extended to BUC^{3-}-functions $\omega : T_y\Gamma \longrightarrow \mathbb{R}$ with compact support, which by definition satisfy

$$\omega(y) = 0, \qquad \nabla_{T_y\Gamma}\omega(y) = 0;$$

(iii) the bent halfspaces

$$\mho := \left\{ \eta - \theta\nu_\Gamma(y) : \eta \in T_y\Gamma, \ \theta > \omega(\eta) \right\}$$

satisfy

$$V := \Omega \cap B_r(y) = \mho \cap B_r(y).$$

7.8 To simplify our notation we set $U := B_r(y)$. Due to the second constraint on the parametrizations, we may choose $0 < r < \bar{r}$, such that

$$|\nabla_{\mathcal{T}_y\Gamma}\omega(\eta)| < \delta, \quad \eta \in \mathcal{T}_y\Gamma,$$

where $\delta > 0$ is defined as in Chapter 6, and, therefore, the corresponding bent halfspace $\mho$ satisfies the requirements of Theorem 6.4 and the Stokes equations

$$(S \,|\, a, \, \mho)_{f,g,h,u_0}^{\alpha,\beta} \qquad \begin{aligned} \rho\partial_t u - \mu\Delta u + \nabla p &= \rho f && \text{in } (0, a) \times \mho, \\ \operatorname{div} u &= g && \text{in } (0, a) \times \mho, \\ \mathcal{B}^{\alpha,\beta}(u, \, p) &= h && \text{on } (0, a) \times \partial\mho, \\ u(0) &= u_0 && \text{in } \mho \end{aligned}$$

admit unique maximal regular solutions in the L_p-setting as described by Theorem 6.4. To establish a mapping between functions defined on Ω and functions defined on $\mho$ we denote by $\phi \in C_0^\infty(\mathbb{R}^n)$ a smooth function with compact support in U and by $\psi \in C_0^\infty(\mathbb{R}^n)$ a smooth function with compact support in U, which satisfies $\psi \equiv 1$ on spt ϕ. The mapping

$$\Phi : C^\infty(\bar{\Omega}) \longrightarrow C^\infty(\bar{\mho}), \qquad \Phi\varphi := E_\mho\phi R_V\varphi, \quad \varphi \in C^\infty(\bar{\Omega}),$$

where R_V denotes the restriction to V and $E_\mho$ denotes the extension by zero to $\mho$, may then be extended to a bounded linear operator to all L_p-based function spaces under consideration and we will denote these extensions for simplicity also by Φ. Analogously, the mapping

$$\Psi : C^\infty(\bar{\mho}) \longrightarrow C^\infty(\bar{\Omega}), \qquad \Psi\varphi := E_\Omega\psi R_V\varphi, \quad \varphi \in C^\infty(\bar{\mho}),$$

where E_Ω denotes the extension by zero to Ω, may be extended to a bounded linear operator to all L_p-based function spaces under consideration and we will denote these extensions for simplicity also by Ψ. Note that

$$\Psi\Phi\varphi = \phi\varphi, \quad \varphi \in C^\infty(\bar{\Omega}),$$

since spt $\phi \subseteq U$ and $\psi \equiv 1$ on spt ϕ.

7.9 Since $\Gamma = \partial\Omega$ is compact, there exist finitely many points $y_1, y_2, \ldots, y_N \in \Gamma$ such that

$$\Gamma \subseteq \bigcup_{k=1}^N B_{r_k}(y_k),$$

where the radii $r_k > 0$ may be chosen sufficiently small to ensure

$$|\nabla_{T_k}\omega_k(\eta)| < \delta, \quad \eta \in T_k,$$

where $\delta > 0$ is defined as in Chapter 6. Here, $T_k = T_{y_k}\Gamma$ denotes the tangent plane of Γ at the point y_k and ω_k denotes the local BUC^{3-}-parametrization of $\Gamma \cap B_{r_k}(y_k)$ constructed in 7.7 for $y = y_k$. Moreover, we set $U_k := B_{r_k}(y_k)$ and denote by Ω_k the resulting bent halfspace, which equals the bent halfspace $\mho$ constructed in 7.7 for $y = y_k$. Finally, we set $V_k := \Omega \cap U_k$ and note in passing that $V_k = \Omega_k \cap U_k$ by construction.

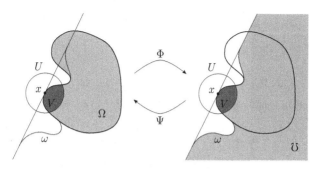

– Local Transformation to a Bent Halfspace –

7.10 Since Ω is compact, there exists a bounded, open set $U_0 \subseteq \Omega$, such that the covering of Γ as constructed in 7.9 is completed to a covering of Ω, i.e.

$$\Omega \subseteq \bigcup_{k=0}^{N} U_k.$$

To simplify our notation, we set $V_0 := U_0$. Now, there exist compactly supported, smooth functions $\phi_0, \phi_1, \ldots, \phi_N \in C_0^\infty(\mathbb{R}^n)$, which form a partition of unity subordinate to the above covering, i.e.

- $0 \le \phi_k \le 1$ for $k = 0, 1, \ldots, N$;

- $\operatorname{spt} \phi_k \subseteq U_k$ for $k = 0, 1, \ldots, N$;

- $\sum_{k=0}^{N} \phi_k \equiv 1$ on $\bar{\Omega}$.

Based on these functions we choose cut-off functions $\psi_0, \psi_1, \ldots, \psi_N \in C_0^\infty(\mathbb{R}^n)$ with $\operatorname{spt} \psi_k \subseteq U_k$ and $\psi_k \equiv 1$ on $\operatorname{spt} \phi_k$ for $k = 0, 1, \ldots, N$. Finally, we denote by Φ_k and Ψ_k for $k = 0, 1, \ldots, N$ the mappings constructed in 7.7, which act as bounded linear operators between the L_p-based function spaces defined on Ω and those defined on Ω_k.

7.11 To develop a first impression how a solution to the Stokes equations in Ω could possibly be constructed based on the localisation defined in 7.7 – 7.10, we now assume $(u, p) \in \mathbb{X}_\gamma^\beta(a, \Omega)$ to be a solution to $(S \mid a, \Omega)_{f,g,h,u_0}^{\alpha,\beta}$ for given data

$$(f, g, h, u_0) \in \mathbb{Y}_\gamma^{\alpha,\beta}(a, \Omega).$$

If we set $u_k := \Phi_k u$ and $p_k := \Phi_k p$ for $k = 0, 1, \ldots, N$, then the decomposition

$$(u, p) = \sum_{k=0}^{N}(\phi_k u, \phi_k p) = \sum_{k=0}^{N}(\Psi_k \Phi_k u, \Psi_k \Phi_k p) = \sum_{k=0}^{N}(\Psi_k u_k, \Psi_k p_k)$$

is valid, $(u_0, p_0) \in \mathbb{X}_\gamma^\beta(a, \mathbb{R}^n)$ is a solution to the Stokes equations

$$
\begin{aligned}
\rho \partial_t u_0 - \mu \Delta u_0 + \nabla p_0 &= \rho F_0(u, p) + \rho \Phi_0 f && \text{in } (0, a) \times \mathbb{R}^n, \\
\operatorname{div} u_0 &= G_0(u) + \Phi_0 g && \text{in } (0, a) \times \mathbb{R}^n, \\
u_0(0) &= \Phi_0 u_0 && \text{in } \mathbb{R}^n
\end{aligned}
$$

and $(u_k, p_k) \in \mathbb{X}_\gamma^\beta(a, \Omega_k)$ are solutions to the Stokes equations

$$
\begin{aligned}
\rho \partial_t u_k - \mu \Delta u_k + \nabla p_k &= \rho F_k(u, p) + \rho \Phi_k f && \text{in } (0, a) \times \Omega_k, \\
\operatorname{div} u_k &= G_k(u) + \Phi_k g && \text{in } (0, a) \times \Omega_k, \\
P_{\Gamma_k} \mathcal{B}^\alpha(u_k) &= P_{\Gamma_k} H_k^\alpha(u) + \Phi_k P_\Gamma h && \text{on } (0, a) \times \partial\Omega_k, \\
Q_{\Gamma_k} \mathcal{B}^\beta(u_k, p_k) &= Q_{\Gamma_k} H_k^\beta(u) + \Phi_k Q_\Gamma h && \text{on } (0, a) \times \partial\Omega_k, \\
u_k(0) &= \Phi_k u_0 && \text{in } \Omega_k,
\end{aligned}
$$

for $k = 1, 2, \dots, N$. The linear operators on the right hand sides are formed by the corresponding commutators, i. e.

$$
\begin{aligned}
\rho F_k(u, p) &:= -\mu[\Delta, \Phi_k]u + [\nabla, \Phi_k]p \\
&= -\mu E_k(\Delta\phi_k) R_k u - 2\mu E_k(\nabla\phi_k \cdot \nabla) R_k u + E_k(\nabla\phi_k) R_k p, \\
G_k(u) &:= [\operatorname{div}, \Phi_k]u \\
&= E_k(\nabla\phi_k \cdot R_k u)
\end{aligned}
$$

for $k = 0, 1, \dots, N$, where R_k denotes the restriction to V_k and E_k denotes the extension by zero to Ω_k. Moreover, the commutators of the boundary conditions are given as

$$
H_k^{\alpha,\beta}(u) := [\mathcal{B}^{\alpha,\beta}, \Phi_k]u = [P_\Gamma \mathcal{B}^\alpha, \Phi_k]u + [Q_\Gamma \mathcal{B}^\beta, \Phi_k]u =: P_{\Gamma_k} H_k^\alpha(u) + Q_{\Gamma_k} H_k^\beta(u)
$$

with

$$
\begin{aligned}
P_{\Gamma_k} H_k^0(u) &= 0, \\
P_{\Gamma_k} H_k^{\pm 1}(u) &= \mu E_k P_\Gamma([\nabla\phi_k \otimes R_k u]_\Gamma \pm [R_k u \otimes \nabla\phi_k]_\Gamma)\nu_\Gamma, \\
Q_{\Gamma_k} H_k^0(u) &= 0, \\
Q_{\Gamma_k} H_k^1(u) &= 2\mu E_k Q_\Gamma[\nabla\phi_k \otimes R_k u]_\Gamma \nu_\Gamma
\end{aligned}
$$

for $k = 1, 2, \dots, N$.

7.12 Note that the commutators involving the velocity u are all of lower order and, therefore, carry additional time regularity, which could be exploited to obtain (7.1) and (7.3). However, the right hand sides $\rho F_k(u, p)$ also contain a commutator involving the pressure p, which is of lower order but does – in general – not carry additional time regularity. Therefore, the nowadays standard localisation procedure for parabolic problems as presented e. g. in [DHP03] may not directly be transferred to the Stokes equations. In fact, as will be shown in the next sections, the presence of the pressure requires a more refined approach, which enables one to simultaneously deal with the pressure and the incompressibility constraint. This is one of the reasons, why we only want to deal with the special cases mentioned in 7.2 and 7.3. Indeed, in these cases we may establish an additional regularity property at least for a part of the pressure p. This is the subject of the next section.

7.3 The Regularity of the Pressure

7.13 The argumentation in the next sections will make frequent use of some extra time regularity, which is available for the pressure under some additional assumptions on the data. Indeed, we have the following result concerning the pressure.

7.14 PROPOSITION. *Let $a \in (0, 1]$, let $\alpha \in \{-1, 0, +1\}$ and let $\beta \in \{0, +1\}$. Let $(u, p) \in {}_0\mathbb{X}_\gamma^\beta(a, \Omega)$ be a solution to the Stokes equations $(S \,|\, a, \Omega)_{f,0,h,0}^{\alpha,\beta}$, where*

$$(f, 0, h) \in {}_0\mathbb{Y}_\gamma^{\alpha,\beta}(a, \Omega)$$

satisfy the additional regularity property

$$f \in {}_0H_p^\tau((0, a), L_p(\Omega, \mathbb{R}^n))$$

for some $\tau \in (0, 1/2 - 1/2p)$ and $\gamma = 1/2 - 1/2p$, if $\beta = +1$. Then

$$p \in {}_0H_p^\tau((0, a), L_p(\Omega))$$

and the estimate

$$\|p\|_{{}_0H_p^\tau((0,a),\,L_p(\Omega))}$$
$$\leq c \left(\|u\|_{{}_0\mathbb{X}_u(a,\Omega)} + \|f\|_{{}_0H_p^\tau((0,a),\,L_p(\Omega,\mathbb{R}^n))} + \beta\|Q_\Gamma h\|_{{}_0\mathbb{N}_{h,\gamma}^\beta(a,\Omega)} \right)$$

is valid with a constant $c > 0$, which is independent of a.

Proof. Given $\psi \in L_{p'}(\Omega)$, where $1/p + 1/p' = 1$, we choose $\phi \in H_{p'}^2(\Omega)$ to be a solution to the elliptic problem

$$-\Delta\phi = \psi_0 \quad \text{in } \Omega,$$
$$\partial_\nu \phi = 0 \quad \text{on } \partial\Omega,$$

if $\beta = 0$, where $\psi_0 := \psi - (\psi)_\Omega$, respectively

$$-\Delta\phi = \psi \quad \text{in } \Omega,$$
$$[\phi]_\Gamma = 0 \quad \text{on } \partial\Omega,$$

if $\beta = +1$. In the first case we have

$$(p \,|\, \psi)_\Omega = (p \,|\, \psi_0)_\Omega = -(p \,|\, \Delta\phi)_\Omega = (\nabla p \,|\, \nabla\phi)_\Omega$$

thanks to the fact that the pressure is assumed to have mean value zero and to the boundary condition on ϕ. In the second case we have

$$(p \,|\, \psi)_\Omega = -(p \,|\, \Delta\phi)_\Omega = (\nabla p \,|\, \nabla\phi)_\Omega - ([p]_\Gamma \,|\, \partial_\nu\phi)_\Gamma$$
$$= (\nabla p \,|\, \nabla\phi)_\Omega - (2\mu\, \partial_\nu u \cdot \nu_\Gamma - h \cdot \nu_\Gamma \,|\, \partial_\nu\phi)_\Gamma$$

thanks to the boundary condition on the pressure. Using the momentum equation we infer

$$(\nabla p \,|\, \nabla\phi)_\Omega = (\rho f - \rho\partial_t u + \mu\Delta u \,|\, \nabla\phi)_\Omega$$

and an integration by parts yields

$$(\rho \partial_t u \,|\, \nabla \phi)_\Omega = \rho \partial_t (\, [u]_\Gamma \cdot \nu_\Gamma \,|\, [\phi]_\Gamma \,)_\Gamma - \rho (\operatorname{div} u \,|\, \phi)_\Omega = 0$$

thanks to the incompressibility condition and the boundary condition on the velocity field respectively ϕ. Moreover, we have

$$(\mu \Delta u \,|\, \nabla \phi)_\Omega = \mu (\, \partial_\nu u \,|\, [\nabla \phi]_\Gamma \,)_\Gamma - \mu (\nabla u \,|\, \nabla^2 \phi)_\Omega$$

and we arrive at

$$(p \,|\, \psi)_\Omega = \mu (\, \partial_\nu u \,|\, [\nabla \phi]_\Gamma \,)_\Gamma - \mu (\nabla u \,|\, \nabla^2 \phi)_\Omega + \rho (f \,|\, \nabla \phi)_\Omega$$
$$- \beta (\, 2 \mu \, \partial_\nu u \cdot \nu_\Gamma - h \cdot \nu_\Gamma \,|\, \partial_\nu \phi)_\Gamma.$$

Since

$$\| \nabla \phi \|_{H^1_{p'}(\Omega)} \leq c \| \psi \|_{L_{p'}(\Omega)}$$

for some constant $c > 0$, which depends on Ω only, we may use the regularity of the functions involved on the right hand side and apply the operator ∂_t^τ to deduce

$$\| (p \,|\, \psi)_\Omega \|_{_0 H^\tau_p(0, a)}$$

$$\leq c \left(\| u \|_{_0 \mathbb{X}_u(a, \Omega)} + \| f \|_{_0 H^\tau_p((0, a), L_p(\Omega, \mathbb{R}^n))} + \beta \| Q_\Gamma h \|_{_0 \mathbb{N}^\beta_{h, \gamma}(a, \Omega)} \right) \| \psi \|_{L_{p'}(\Omega)},$$

where the modified constant $c > 0$ is independent of $a \in (0, 1]$ thanks to Proposition 3.9 and Corollary 4.71. Thus, we obtain the asserted estimate and the proof is complete. $\quad\square$

7.4 Boundary Conditions Involving the Normal Velocity

7.15 Let us focus on the case $\beta = 0$ and $\gamma = -\infty$ and start with the derivation of the estimate (7.1a). Suppose that

$$(u, p) \in {}_0\mathbb{X}^\beta_\gamma(a, \Omega)$$

is a maximal regular solution to the Stokes equations $(\mathrm{S} \,|\, a, \Omega)^{\alpha, \beta}_{0, 0, h, 0}$ with data

$$h \in {}_\tau \mathbb{Y}^\alpha_{h, \gamma}(a, \Gamma).$$

If we set $u_k := \Phi_k u$ and $p_k := \Phi_k p$ for $k = 0, 1, \ldots, N$, then the decomposition

$$(u, p) = \sum_{k=0}^N (\phi_k u, \phi_k p) = \sum_{k=0}^N (\Psi_k \Phi_k u, \Psi_k \Phi_k p) = \sum_{k=0}^N (\Psi_k u_k, \Psi_k p_k)$$

is valid, $(u_0, p_0) \in {}_0\mathbb{X}^\beta_\gamma(a, \mathbb{R}^n)$ is a solution to the Stokes equations

$$\rho \partial_t u_0 - \mu \Delta u_0 + \nabla p_0 = \rho F_0(u, p) \quad \text{in } (0, a) \times \mathbb{R}^n,$$
$$\operatorname{div} u_0 = G_0(u) \quad \text{in } (0, a) \times \mathbb{R}^n,$$
$$u_0(0) = 0 \quad \text{in } \mathbb{R}^n$$

and $(u_k, p_k) \in {}_0\mathbb{X}_\gamma^\beta(a, \Omega_k)$ are solutions to the Stokes equations

$$\rho\partial_t u_k - \mu\Delta u_k + \nabla p_k = \rho F_k(u, p) \qquad \text{in } (0, a) \times \Omega_k,$$
$$\operatorname{div} u_k = G_k(u) \qquad \text{in } (0, a) \times \Omega_k,$$
$$P_{\Gamma_k}\mathcal{B}^\alpha(u_k) = P_{\Gamma_k}H_k^\alpha(u) + \Phi_k P_\Gamma h \qquad \text{on } (0, a) \times \partial\Omega_k,$$
$$[u_k]_{\Gamma_k} \cdot \nu_{\Gamma_k} = 0 \qquad \text{on } (0, a) \times \partial\Omega_k,$$
$$u_k(0) = 0 \qquad \text{in } \Omega_k$$

for $k = 1, 2, \ldots, N$.

7.16 Now, $u_0 = v_0 + \nabla\varphi_0$ and $p_0 = q_0 - \rho\partial_t\varphi_0 + \mu\Delta\varphi_0$, where

$$\varphi_0 \in {}_0H_p^1((0, a), H_p^2(\mathbb{R}^n)) \cap {}_0H_p^{1/2}((0, a), H_p^3(\mathbb{R}^n))$$

is a solution to the elliptic problem

$$-\Delta\varphi_0 = -G_0(u) \quad \text{in } (0, a) \times \mathbb{R}^n$$

and $(v_0, q_0) \in {}_0\mathbb{X}_\gamma^\beta(a, \mathbb{R}^n)$ is a solution to the Stokes equations

$$\rho\partial_t v_0 - \mu\Delta v_0 + \nabla q_0 = \rho F_0(u, p) \quad \text{in } (0, a) \times \mathbb{R}^n,$$
$$\operatorname{div} v_0 = 0 \qquad \text{in } (0, a) \times \mathbb{R}^n,$$
$$v_0(0) = 0 \qquad \text{in } \mathbb{R}^n.$$

Analogously, $u_k = v_k + \bar{v}_k + \nabla\varphi_k$ and $p_k = q_k + \bar{q}_k - \rho\partial_t\varphi_k + \mu\Delta\varphi_k$, where

$$\varphi_k \in {}_0H_p^1((0, a), H_p^2(\Omega_k)) \cap {}_0H_p^{1/2}((0, a), H_p^3(\Omega_k))$$

are solutions to the elliptic problems

$$-\Delta\varphi_k = -G_k(u) \quad \text{in } (0, a) \times \Omega_k,$$
$$\partial_\nu\varphi_k = 0 \qquad \text{on } (0, a) \times \partial\Omega_k$$

and $(v_k, q_k) \in {}_0\mathbb{X}_\gamma^\beta(a, \Omega_k)$ are solutions to the Stokes equations

$$\rho\partial_t v_k - \mu\Delta v_k + \nabla q_k = \rho F_k(u, p) \qquad \text{in } (0, a) \times \Omega_k,$$
$$\operatorname{div} v_k = 0 \qquad \text{in } (0, a) \times \Omega_k,$$
$$P_{\Gamma_k}\mathcal{B}^\alpha(v_k) = P_{\Gamma_k}H_k^\alpha(u) - P_{\Gamma_k}\mathcal{B}^\alpha(\nabla\varphi_k) \qquad \text{on } (0, a) \times \partial\Omega_k,$$
$$[v_k]_{\Gamma_k} \cdot \nu_{\Gamma_k} = 0 \qquad \text{on } (0, a) \times \partial\Omega_k,$$
$$v_k(0) = 0 \qquad \text{in } \Omega_k$$

and $(\bar{v}_k, \bar{q}_k) \in {}_0\mathbb{X}_\gamma^\beta(a, \Omega_k)$ are solutions to the Stokes equations

$$\rho\partial_t\bar{v}_k - \mu\Delta\bar{v}_k + \nabla\bar{q}_k = 0 \qquad \text{in } (0, a) \times \Omega_k,$$
$$\operatorname{div}\bar{v}_k = 0 \qquad \text{in } (0, a) \times \Omega_k,$$
$$P_{\Gamma_k}\mathcal{B}^\alpha(\bar{v}_k) = \Phi_k P_\Gamma h \qquad \text{on } (0, a) \times \partial\Omega_k,$$
$$[\bar{v}_k]_{\Gamma_k} \cdot \nu_{\Gamma_k} = 0 \qquad \text{on } (0, a) \times \partial\Omega_k,$$
$$\bar{v}_k(0) = 0 \qquad \text{in } \Omega_k$$

for $k = 1, 2, \ldots, N$.

7.17 To estimate (v_k, q_k) first observe that all terms on the right hand side of the corresponding Stokes equations carry additional time regularity. Indeed

$$\|\rho F_k(u, p)\|_{\mathbb{Y}_f(a, \Omega_k)} \leq ca^\eta \|\rho F_k(u, p)\|_{{}_0 H_p^\tau((0, a), L_p(\Omega_k, \mathbb{R}^n))}$$
$$\leq ca^\eta \left(\|u\|_{{}_0 \mathbb{X}_u(a, \Omega)} + \|p\|_{{}_0 H_p^\tau((0, a), L_p(\Omega))} \right)$$
$$\leq ca^\eta \|u\|_{{}_0 \mathbb{X}_u(a, \Omega)}$$

for $k = 0, 1, \ldots, N$ with some constants $c > 0$ and $\eta > 0$, which are independent of $a \in (0, 1]$ thanks to Proposition 7.14. Moreover,

$$\|P_{\Gamma_k} H_k^\alpha(u)\|_{\mathbb{T}_h^\alpha(a, \Gamma_k)} \leq ca^\eta \|u\|_{{}_0 \mathbb{X}_u(a, \Omega)}$$

for $k = 1, 2, \ldots, N$, since all commutators are of lower order. Again, the constants $c > 0$ and $\eta > 0$ are independent of $a \in (0, 1]$. Using the estimate (7.4c) for $\|\nabla \varphi_k\|_{{}_0 \mathbb{X}_u(a, \Omega_k)}$ we infer

(7.4a) $$\|(v_k, q_k)\|_{{}_0 \mathbb{X}_\gamma^\beta(a, \Omega_k)} \leq ca^\eta \|u\|_{{}_0 \mathbb{X}_u(a, \Omega)} + c' \|h\|_{\mathbb{Y}_{h, \gamma}^{\alpha, \beta}(a, \Gamma)}$$

with $c > 0$ and $\eta > 0$ still independent of $a \in (0, 1]$. Finally, Proposition 7.14 implies

$$\|q_k\|_{{}_0 H_p^\tau((0, a), L_p(\Omega_k))} \leq c \left(\|v_k\|_{{}_0 \mathbb{X}_u(a, \Omega_k)} + \|\rho F_k(u, p)\|_{{}_0 H_p^\tau((0, a), L_p(\Omega_k, \mathbb{R}^n))} \right)$$
$$\leq c \|u\|_{{}_0 \mathbb{X}_u(a, \Omega)}$$

for $k = 0, 1, \ldots, N$.

7.18 On the other hand, $(\bar{v}_k, \bar{q}_k)$ may first be estimated as

(7.4b) $$\|(\bar{v}_k, \bar{q}_k)\|_{{}_0 \mathbb{X}_\gamma^\beta(a, \Omega_k)} \leq c' \|h\|_{\mathbb{Y}_{h, \gamma}^{\alpha, \beta}(a, \Gamma)},$$

which then also implies

$$\|\bar{q}_k\|_{{}_0 H_p^\tau((0, a), L_p(\Omega_k))} \leq c' \|\bar{v}_k\|_{{}_0 \mathbb{X}_u(a, \Omega_k)} \leq c' \|h\|_{\mathbb{Y}_{h, \gamma}^{\alpha, \beta}(a, \Gamma)}$$

for $k = 1, 2, \ldots, N$ due to Proposition 7.14.

7.19 Therefore, we also obtain estimates for φ_k, since

$$\|\mu \Delta \varphi_0 - \rho \partial_t \varphi_0\|_{{}_0 H_p^\tau((0, a), L_p(\Omega_k))} = \|p_0 - q_0\|_{{}_0 H_p^\tau((0, a), L_p(\Omega_k))} \leq c \|u\|_{{}_0 \mathbb{X}_u(a, \Omega)}$$

thanks to Proposition 7.14 and, analogously,

$$\|\mu \Delta \varphi_k - \rho \partial_t \varphi_k\|_{{}_0 H_p^\tau((0, a), L_p(\Omega_k))} = \|p_k - q_k - \bar{q}_k\|_{{}_0 H_p^\tau((0, a), L_p(\Omega_k))}$$
$$\leq c \|u\|_{{}_0 \mathbb{X}_u(a, \Omega)} + c' \|h\|_{\mathbb{Y}_{h, \gamma}^{\alpha, \beta}(a, \Gamma)}$$

for $k = 1, 2, \ldots, N$. However,

$$\|\Delta \varphi_k\|_{{}_0 H_p^1((0, a), L_p(\Omega_k))} \leq c \|u\|_{{}_0 \mathbb{X}_u(a, \Omega)}$$

for $k = 0, 1, \ldots, N$ by construction and we infer

$$\|\partial_t \varphi_k\|_{0 H_p^\tau((0,a),\, L_p(\Omega_k))}, \; \|\Delta \varphi_k\|_{0 H_p^\tau((0,a),\, L_p(\Omega_k))} \leq c\|u\|_{0 X_u(a,\Omega)} + c'\|h\|_{Y_{h,\gamma}^{\alpha,\beta}(a,\Gamma)}$$

for $k = 0, 1, \ldots, N$. Thus, using

$$\|\partial_t \varphi_k\|_{L_p((0,a),\, H_p^2(\Omega_k))}, \; \|\Delta \varphi_k\|_{0 H_p^1((0,a),\, L_p(\Omega_k)) \cap L_p((0,a),\, H_p^2(\Omega_k))} \leq c\|u\|_{0 X_u(a,\Omega)}$$

we also have

$$
\begin{aligned}
\|\partial_t \nabla \varphi_k\|_{L_p((0,a)\times\Omega_k)} &\leq ca^\eta \|\partial_t \varphi_k\|_{0 H_p^{\tau/2}((0,a),\, H_p^1(\Omega_k))} \\
&\leq ca^\eta \|\partial_t \varphi_k\|_{0 H_p^\tau((0,a),\, L_p(\Omega_k)) \cap L_p((0,a),\, H_p^2(\Omega_k))} \\
&\leq ca^\eta \|u\|_{0 X_u(a,\Omega)} + c'\|h\|_{Y_{h,\gamma}^{\alpha,\beta}(a,\Gamma)}
\end{aligned}
$$

and

$$
\begin{aligned}
\|\Delta \nabla \varphi_k\|_{L_p((0,a)\times\Omega_k)} &\leq ca^\eta \|\Delta \varphi_k\|_{0 H_p^{1/2}((0,a),\, H_p^1(\Omega_k))} \\
&\leq ca^\eta \|\Delta \varphi_k\|_{0 H_p^1((0,a),\, L_p(\Omega_k)) \cap L_p((0,a),\, H_p^2(\Omega_k))} \\
&\leq ca^\eta \|u\|_{0 X_u(a,\Omega)}
\end{aligned}
$$

for $k = 0, \ldots, N$. Hence, we have

(7.4c) $$\|(\nabla \varphi_k, \, \mu \Delta \varphi_k - \rho \partial_t \varphi_k)\|_{0 X_\gamma^\beta(a,\Omega_k)} \leq ca^\eta \|u\|_{0 X_u(a,\Omega)} + c'\|h\|_{Y_{h,\gamma}^{\alpha,\beta}(a,\Gamma)}$$

for $k = 0, \ldots, N$.

7.20 Finally,

$$(u, p) = \sum_{k=0}^{N} (\Psi_k u_k, \, \Psi_k p_k)$$

and (7.4) implies (7.1a).

7.21 It remains to construct a right inverse for $_0 L_{a,\gamma}^{\alpha,\beta}$. First note that Theorem 4.60 ensures the existence of a bounded linear operator

$$S_{a,\gamma,\#}^{\alpha,\beta} : {}_0 Y_\gamma^{\alpha,\beta}(a, \Omega) \longrightarrow {}_0 X_\gamma^\beta(a, \Omega)$$

with the property that $(v, q) = S_{a,\gamma,\#}^{\alpha,\beta}(f, g, h)$ is a solution to the Stokes equations

$$
\begin{aligned}
\rho \partial_t v - \mu \Delta v + \nabla q &= \rho f && \text{in } (0, a) \times \Omega, \\
\operatorname{div} v &= g && \text{in } (0, a) \times \Omega, \\
[v]_\Gamma \cdot \nu_\Gamma &= h \cdot \nu_\Gamma && \text{on } (0, a) \times \partial\Omega, \\
v(0) &= 0 && \text{in } \Omega,
\end{aligned}
$$

whenever $(f, g, h) \in {}_0 Y_\gamma^{\alpha,\beta}(a, \Omega)$, cf. 4.63.

7.22 Note that only the desired tangential boundary condition

$$P_\Gamma \mathcal{B}^\alpha(v) = P_\Gamma h \quad \text{on } (0, a) \times \partial\Omega,$$

which will in general not be satisfied, prevents (v, q) from being a solution to the Stokes equations $(S \mid a, \Omega)_{f,g,h,0}^{\alpha,\beta}$. Therefore, we will construct a second bounded linear operator

$$S_{a,\gamma,*}^{\alpha,\beta} : {}_\tau\mathbb{Y}_{h,\gamma}^\alpha(a, \Gamma) \longrightarrow {}_0\mathbb{X}_\gamma^\beta(a, \Omega)$$

to adjust the tangential boundary condition and then set

$$S_{a,\gamma}^{\alpha,\beta}(f, g, h) := S_{a,\gamma,\#}^{\alpha,\beta}(f, g, h) + S_{a,\gamma,*}^{\alpha,\beta}(P_\Gamma h - P_\Gamma \mathcal{B}^\alpha(S_{a,\gamma,\#}^{\alpha,\beta}(f, g, h)))$$

for $(f, g, h) \in {}_0\mathbb{Y}_\gamma^{\alpha,\beta}(a, \Omega)$.

7.23 Given

$$\bar{h} \in {}_\tau\mathbb{Y}_{h,\gamma}^\alpha(a, \Gamma)$$

we first define $(\bar{v}_k, \bar{q}_k) \in {}_0\mathbb{X}_\gamma^\beta(a, \Omega_k)$ to be the unique maximal regular solution to the Stokes equations

$$\begin{aligned}
\rho\partial_t\bar{v}_k - \mu\Delta\bar{v}_k + \nabla\bar{q}_k &= 0 & &\text{in } (0, a) \times \Omega_k, \\
\operatorname{div}\bar{v}_k &= 0 & &\text{in } (0, a) \times \Omega_k, \\
P_{\Gamma_k}\mathcal{B}^\alpha(\bar{v}_k) &= \Phi_k P_\Gamma\bar{h} & &\text{on } (0, a) \times \partial\Omega_k, \\
[\bar{v}_k]_{\Gamma_k} \cdot \nu_{\Gamma_k} &= 0 & &\text{on } (0, a) \times \partial\Omega_k, \\
\bar{v}_k(0) &= 0 & &\text{in } \Omega_k
\end{aligned}$$

for $k = 1, 2, \ldots, N$. Then we define

$$\eta_k \in {}_0H_p^1((0, a), H_p^2(\Omega)) \cap {}_0H_p^{1/2}((0, a), H_p^3(\Omega))$$

to be the solutions to the elliptic problems

$$\begin{aligned}
-\Delta\eta_k &= E_\Omega(\nabla\psi_k \cdot R_k\bar{v}_k) & &\text{in } (0, a) \times \Omega, \\
\partial_\nu\eta_k &= 0 & &\text{on } (0, a) \times \partial\Omega
\end{aligned}$$

for $k = 1, 2, \ldots, N$ and set

$$S_{a,\gamma,*}^{\alpha,\beta}(\bar{h}) := \sum_{k=1}^N (\Psi_k\bar{v}_k - \nabla\eta_k, \, \Psi_k\bar{q}_k + \rho\partial_t\eta_k - \mu\Delta_k\eta_k).$$

7.24 Setting $(\bar{v}, \bar{q}) = S_{a,\gamma,*}^{\alpha,\beta}(\bar{h})$ we have

$${}_0L_{a,\gamma}^{\alpha,\beta}S_{a,\gamma,*}^{\alpha,\beta}(\bar{h}) = (0, 0, P_\Gamma\bar{h})$$

$$+ \sum_{k=1}^N \Big(-\mu[\Delta, \Psi_k]\bar{v}_k + [\nabla, \Psi_k]\bar{q}_k, \, 0, \, [P_{\Gamma_k}\mathcal{B}^\alpha, \Psi_k]\bar{v}_k - P_\Gamma\mathcal{B}^\alpha(\nabla\eta_k)\Big).$$

Therefore, if $(f, g, h) \in {}_0\mathbb{Y}_\gamma^{\alpha,\beta}(a, \Omega)$ we set

$$(v, q) := S_{a,\gamma,\#}^{\alpha,\beta}(f, g, h) \qquad \text{and} \qquad \bar{h} := P_\Gamma h - P_\Gamma\mathcal{B}^\alpha(v)$$

and infer

$$S_{a,\gamma}^{\alpha,\beta}(f, g, h) = (v + \bar{v}, q + \bar{q})$$

as well as

$$_0 L_{a,\gamma}^{\alpha,\beta} S_{a,\gamma}^{\alpha,\beta}(f, g, h) = (f, g, h) + R_{a,\gamma}^{\alpha,\beta}(f, g, h)$$

with

$$R_{a,\gamma}^{\alpha,\beta}(f, g, h) = \sum_{k=1}^{N} \Big(-\mu[\Delta, \Psi_k]\bar{v}_k + [\nabla, \Psi_k]\bar{q}_k, 0, [P_{\Gamma_k}\mathcal{B}^\alpha, \Psi_k]\bar{v}_k - P_\Gamma \mathcal{B}^\alpha(\nabla \eta_k)\Big).$$

7.25 With this definition we may obtain the desired estimate (7.3). As in 7.18 we first estimate $(\bar{v}_k, \bar{q}_k)$ as

$$\|(\bar{v}_k, \bar{q}_k)\|_{0 \mathbb{X}_\gamma^\beta(a, \Omega_k)} \le c\|\bar{h}\|_{\mathbb{Y}_{h,\gamma}^{\alpha,\beta}(a, \Gamma)},$$

which then also implies

$$\|\bar{q}_k\|_{0 H_p^\tau((0,a), L_p(\Omega_k))} \le c\|\bar{v}_k\|_{0 \mathbb{X}_u(a, \Omega_k)} \le c\|\bar{h}\|_{\mathbb{Y}_{h,\gamma}^{\alpha,\beta}(a, \Gamma)}$$

for $k = 1, 2, \ldots, N$ due to Proposition 7.14. Thus, we infer

$$(7.5a) \quad \begin{aligned} \| -\mu[\Delta, \Psi_k]\bar{v}_k &+ [\nabla, \Psi_k]\bar{q}_k\|_{\mathbb{Y}_f(a, \Omega)} \\ &\le ca^\eta \| -\mu[\Delta, \Psi_k]\bar{v}_k + [\nabla, \Psi_k]\bar{q}_k\|_{0 H_p^\tau((0,a), L_p(\Omega))} \\ &\le ca^\eta \left(\|\bar{v}_k\|_{0 \mathbb{X}_u(a, \Omega_k)} + \|\bar{q}_k\|_{0 H_p^\tau((0,a), L_p(\Omega_k))} \right) \\ &\le ca^\eta \|\bar{h}\|_{\mathbb{Y}_{h,\gamma}^{\alpha,\beta}(a, \Gamma)} \end{aligned}$$

for $k = 1, 2, \ldots, N$. Analogously, the commutator term in the tangential boundary condition is of lower order and we obtain

$$(7.5b) \qquad \|[P_{\Gamma_k}\mathcal{B}^\alpha, \Psi_k]\bar{v}_k\|_{\mathbb{Y}_{h,\gamma}^{\alpha,\beta}(a, \Gamma)} \le ca^\eta \|\bar{h}\|_{\mathbb{Y}_{h,\gamma}^{\alpha,\beta}(a, \Gamma)}$$

for $k = 1, 2, \ldots, N$.

7.26 To estimate the remaining perturbation in the tangential boundary condition we set $(\tilde{v}_k, \tilde{q}_k) := (\Psi_k \bar{v}_k - \nabla \eta_k, \Psi_k \bar{q}_k + \rho \partial_t \eta_k - \mu \Delta \eta_k)$ and observe that

$$\begin{aligned} \rho \partial_t \tilde{v}_k - \mu \Delta \tilde{v}_k + \nabla \tilde{q}_k &= -\mu[\Delta, \Psi_k]\bar{v}_k + [\nabla, \Psi_k]\bar{q}_k && \text{in } (0, a) \times \Omega, \\ \operatorname{div} \tilde{v}_k &= 0 && \text{in } (0, a) \times \Omega, \\ P_\Gamma \mathcal{B}^\alpha(\tilde{v}_k) &= \phi_k P_\Gamma \bar{h} + [P_{\Gamma_k}\mathcal{B}^\alpha, \Psi_k]\bar{v}_k - P_\Gamma \mathcal{B}^\alpha(\nabla \eta_k) && \text{on } (0, a) \times \partial\Omega, \\ [\tilde{v}_k]_\Gamma \cdot \nu_\Gamma &= 0 && \text{on } (0, a) \times \partial\Omega, \\ \tilde{v}_k(0) &= 0 && \text{in } \Omega. \end{aligned}$$

Hence,

$$\|(\tilde{v}_k, \tilde{q}_k)\|_{0 \mathbb{X}_\gamma^\beta(a, \Omega)} \le c\|\bar{h}\|_{\mathbb{Y}_{h,\gamma}^{\alpha,\beta}(a, \Gamma)}$$

and we further infer

$$\|\tilde{q}_k\|_{_0H_p^\tau((0,a),\,L_p(\Omega))}$$
$$\leq c\left(\|\tilde{v}_k\|_{_0\mathbb{X}_u(a,\Omega)} + \| -\mu[\Delta,\,\Psi_k]\bar{v}_k + [\nabla,\,\Psi_k]\bar{q}_k\|_{_0H_p^\tau((0,a),\,L_p(\Omega))}\right)$$
$$\leq c\|\bar{h}\|_{\mathbb{Y}_{h,\gamma}^{\alpha,\beta}(a,\Gamma)}$$

for $k = 1, 2, \ldots, N$ due to Proposition 7.14. Hence,

$$\|\mu\Delta\eta_k - \rho\partial_t\eta_k\|_{_0H_p^\tau((0,a),\,L_p(\Omega))} = \|\Psi_k\bar{q}_k - \tilde{q}_k\|_{_0H_p^\tau((0,a),\,L_p(\Omega))}$$
$$\leq c\|\bar{h}\|_{\mathbb{Y}_{h,\gamma}^{\alpha,\beta}(a,\Gamma)}$$

for $k = 1, 2, \ldots, N$. However,

$$\|\Delta\eta_k\|_{_0H_p^1((0,a),\,L_p(\Omega))} \leq c\|\bar{v}_k\|_{_0\mathbb{X}_u(a,\Omega_k)} \leq c\|\bar{h}\|_{\mathbb{Y}_{h,\gamma}^{\alpha,\beta}(a,\Gamma)}$$

for $k = 1, 2, \ldots, N$ and we infer

$$\|\partial_t\eta_k\|_{_0H_p^\tau((0,a),\,L_p(\Omega))}, \|\Delta\eta_k\|_{_0H_p^\tau((0,a),\,L_p(\Omega))} \leq c\|\bar{h}\|_{\mathbb{Y}_{h,\gamma}^{\alpha,\beta}(a,\Gamma)}$$

for $k = 1, 2, \ldots, N$. Thus, using

$$\|\partial_t\eta_k\|_{L_p((0,a),\,H_p^2(\Omega))}, \|\Delta\eta_k\|_{H_p^1((0,a),\,L_p(\Omega))\cap L_p((0,a),\,H_p^2(\Omega))} \leq c\|\bar{h}\|_{\mathbb{Y}_{h,\gamma}^{\alpha,\beta}(a,\Gamma)}$$

we also have

$$\|\partial_t\nabla\eta_k\|_{L_p((0,a)\times\Omega)} \leq ca^\eta\|\partial_t\eta_k\|_{_0H_p^{\tau/2}((0,a),\,H_p^1(\Omega))}$$
$$\leq ca^\eta\|\partial_t\eta_k\|_{_0H_p^\tau((0,a),\,L_p(\Omega))\cap L_p((0,a),\,H_p^2(\Omega))}$$
$$\leq ca^\eta\|\bar{h}\|_{\mathbb{Y}_{h,\gamma}^{\alpha,\beta}(a,\Gamma)}$$

as well as

$$\|\Delta\nabla\eta_k\|_{L_p((0,a)\times\Omega)} \leq ca^\eta\|\Delta\eta_k\|_{_0H_p^{1/2}((0,a),\,H_p^1(\Omega))}$$
$$\leq ca^\eta\|\Delta\eta_k\|_{_0H_p^1((0,a),\,L_p(\Omega))\cap L_p((0,a),\,H_p^2(\Omega))}$$
$$\leq ca^\eta\|\bar{h}\|_{\mathbb{Y}_{h,\gamma}^{\alpha,\beta}(a,\Gamma)}$$

for $k = 1, 2, \ldots, N$. Hence, we have

$$\|(\nabla\eta_k,\,\mu\Delta\eta_k - \rho\partial_t\eta_k)\|_{_0\mathbb{X}_\gamma^\beta(a,\Omega)} \leq ca^\eta\|\bar{h}\|_{\mathbb{Y}_{h,\gamma}^{\alpha,\beta}(a,\Gamma)}$$

for $k = 1, 2, \ldots, N$. In particular, we obtain

$$(7.5c) \qquad \|P_\Gamma\mathcal{B}^\alpha(\nabla\eta_k)\|_{\mathbb{Y}_{h,\gamma}^{\alpha,\beta}(a,\Gamma)} \leq ca^\eta\|\bar{h}\|_{\mathbb{Y}_{h,\gamma}^{\alpha,\beta}(a,\Gamma)}$$

and (7.5) implies (7.3). In summary, the proof for $\beta = 0$ is complete.

7.5 Boundary Conditions Involving the Pressure

7.27 To complete the proof of Theorem 3.30 let us finally consider the case $\beta = +1$ and $\gamma = 1/2 - 1/2p$. We start with the derivation of the estimate (7.1b). Suppose that

$$(u, p) \in {}_0\mathbb{X}_\gamma^\beta(a, \Omega)$$

is a maximal regular solution to the Stokes equations $(S \,|\, a, \Omega)_{0,0,h,0}^{\alpha,\beta}$ with data

$$h \in {}_\nu\mathbb{Y}_{h,\gamma}^{+1}(a, \Gamma).$$

If we set $u_k := \Phi_k u$ and $p_k := \Phi_k p$ for $k = 0, 1, \ldots, N$, then the decomposition

$$(u, p) = \sum_{k=0}^N (\phi_k u, \phi_k p) = \sum_{k=0}^N (\Psi_k \Phi_k u, \Psi_k \Phi_k p) = \sum_{k=0}^N (\Psi_k u_k, \Psi_k p_k)$$

is valid, $(u_0, p_0) \in {}_0\mathbb{X}_\gamma^\beta(a, \mathbb{R}^n)$ is a solution to the Stokes equations

$$
\begin{aligned}
\rho \partial_t u_0 - \mu \Delta u_0 + \nabla p_0 &= \rho F_0(u, p) && \text{in } (0, a) \times \mathbb{R}^n, \\
\operatorname{div} u_0 &= G_0(u) && \text{in } (0, a) \times \mathbb{R}^n, \\
u_0(0) &= 0 && \text{in } \mathbb{R}^n
\end{aligned}
$$

and $(u_k, p_k) \in {}_0\mathbb{X}_\gamma^\beta(a, \Omega_k)$ are solutions to the Stokes equations

$$
\begin{aligned}
\rho \partial_t u_k - \mu \Delta u_k + \nabla p_k &= \rho F_k(u, p) && \text{in } (0, a) \times \Omega_k, \\
\operatorname{div} u_k &= G_k(u) && \text{in } (0, a) \times \Omega_k, \\
P_{\Gamma_k} \mathcal{B}^\alpha(u_k) &= P_{\Gamma_k} H_k^\alpha(u) && \text{on } (0, a) \times \partial\Omega_k, \\
2\mu \, \partial_\nu u_k \cdot \nu_{\Gamma_k} - [p_k]_{\Gamma_k} &= Q_{\Gamma_k} H_k^{+1}(u) \cdot \nu_{\Gamma_k} + \Phi_k Q_\Gamma h \cdot \nu_{\Gamma_k} && \text{on } (0, a) \times \partial\Omega_k, \\
u_k(0) &= 0 && \text{in } \Omega_k
\end{aligned}
$$

for $k = 1, 2, \ldots, N$.

7.28 Now, $u_0 = v_0 + \nabla\varphi_0$ and $p_0 = q_0 - \rho\partial_t\varphi_0 + \mu\Delta\varphi_0$, where

$$\varphi_0 \in {}_0H_p^1((0, a), H_p^2(\mathbb{R}^n)) \cap {}_0H_p^{1/2}((0, a), H_p^3(\mathbb{R}^n))$$

is a solution to the elliptic problem

$$-\Delta\varphi_0 = -G_0(u) \quad \text{in } (0, a) \times \mathbb{R}^n$$

and $(v_0, q_0) \in {}_0\mathbb{X}_\gamma^\beta(a, \mathbb{R}^n)$ is a solution to the Stokes equations

$$
\begin{aligned}
\rho \partial_t v_0 - \mu \Delta v_0 + \nabla q_0 &= \rho F_0(u, p) && \text{in } (0, a) \times \mathbb{R}^n, \\
\operatorname{div} v_0 &= 0 && \text{in } (0, a) \times \mathbb{R}^n, \\
v_0(0) &= 0 && \text{in } \mathbb{R}^n.
\end{aligned}
$$

Analogously, $u_k = v_k + \bar{v}_k + \nabla\varphi_k$ and $p_k = q_k + \bar{q}_k - \rho\partial_t\varphi_k + \mu\Delta\varphi_k$, where

$$\varphi_k \in {}_0H_p^1((0,\,a),\,H_p^2(\Omega_k)) \cap {}_0H_p^{1/2}((0,\,a),\,H_p^3(\Omega_k))$$

are solutions to the elliptic problems

$$-\Delta\varphi_k = -G_k(u) \quad \text{in } (0,\,a)\times\Omega_k,$$
$$[\varphi_k]_{\Gamma_k} = 0 \qquad \text{on } (0,\,a)\times\partial\Omega_k$$

and $(v_k,\,q_k) \in {}_0\mathbb{X}_\gamma^\beta(a,\,\Omega_k)$ are solutions to the Stokes equations

$$\rho\partial_t v_k - \mu\Delta v_k + \nabla q_k = \rho F_k(u,\,p) \qquad \text{in } (0,\,a)\times\Omega_k,$$
$$\operatorname{div} v_k = 0 \qquad \text{in } (0,\,a)\times\Omega_k,$$
$$P_{\Gamma_k}\mathcal{B}^\alpha(v_k) = P_{\Gamma_k}H_k^\alpha(u) - P_{\Gamma_k}\mathcal{B}^\alpha(\nabla\varphi_k) \qquad \text{on } (0,\,a)\times\partial\Omega_k,$$
$$2\mu\,\partial_\nu v_k \cdot \nu_{\Gamma_k} - [q_k]_{\Gamma_k} = Q_{\Gamma_k}H_k^{+1}(u)\cdot\nu_{\Gamma_k} \qquad \text{on } (0,\,a)\times\partial\Omega_k,$$
$$v_k(0) = 0 \qquad \text{in } \Omega_k$$

and $(\bar{v}_k,\,\bar{q}_k) \in {}_0\mathbb{X}_\gamma^\beta(a,\,\Omega_k)$ are solutions to the Stokes equations

$$\rho\partial_t\bar{v}_k - \mu\Delta\bar{v}_k + \nabla\bar{q}_k = 0 \qquad \text{in } (0,\,a)\times\Omega_k,$$
$$\operatorname{div}\bar{v}_k = 0 \qquad \text{in } (0,\,a)\times\Omega_k,$$
$$P_{\Gamma_k}\mathcal{B}^\alpha(\bar{v}_k) = 0 \qquad \text{on } (0,\,a)\times\partial\Omega_k,$$
$$2\mu\,\partial_\nu\bar{v}_k \cdot \nu_{\Gamma_k} - [\bar{q}_k]_{\Gamma_k} = \Phi_k Q_\Gamma h \cdot \nu_{\Gamma_k} \qquad \text{on } (0,\,a)\times\partial\Omega_k,$$
$$\bar{v}_k(0) = 0 \qquad \text{in } \Omega_k$$

for $k = 1,\,2,\,\ldots,\,N$.

7.29 To estimate $(v_k,\,q_k)$ first observe that all terms on the right hand side of the corresponding Stokes equations carry additional time regularity. Indeed

$$\begin{aligned}
\|\rho F_k(u,\,p)\|_{\mathbb{Y}_f(a,\Omega_k)} &\le ca^\eta\|\rho F_k(u,\,p)\|_{{}_0H_p^\tau((0,a),\,L_p(\Omega_k,\mathbb{R}^n))} \\
&\le ca^\eta\left(\|u\|_{{}_0\mathbb{X}_u(a,\Omega)} + \|p\|_{{}_0H_p^\tau((0,a),\,L_p(\Omega))}\right) \\
&\le ca^\eta\|u\|_{{}_0\mathbb{X}_u(a,\Omega)} + c'\|h\|_{\mathbb{Y}_{h,\gamma}^{\alpha,\beta}(a,\Gamma)}
\end{aligned}$$

for $k = 0,\,1,\,\ldots,\,N$ with some constants $c > 0$ and $\eta > 0$, which are independent of $a \in (0,\,1]$ thanks to Proposition 7.14. Moreover,

$$\|P_{\Gamma_k}H_k^\alpha(u)\|_{\mathbb{T}_h^\alpha(a,\Gamma_k)},\ \ \|Q_{\Gamma_k}H_k^{+1}(u)\|_{\mathbb{N}_{h,1/2-1/2p}^{+1}(a,\Gamma_k)} \le ca^\eta\|u\|_{{}_0\mathbb{X}_u(a,\Omega)}$$

for $k = 1,\,2,\,\ldots,\,N$, since all commutators are of lower order. Again, the constants $c > 0$ and $\eta > 0$ are independent of $a \in (0,\,1]$. Using (7.6c), we infer

(7.6a) $$\qquad \|(v_k,\,q_k)\|_{{}_0\mathbb{X}_\gamma^\beta(a,\Omega_k)} \le ca^\eta\|u\|_{{}_0\mathbb{X}_u(a,\Omega)} + c'\|h\|_{\mathbb{Y}_{h,\gamma}^{\alpha,\beta}(a,\Gamma)}$$

with $c > 0$ and $\eta > 0$ still independent of $a \in (0, 1]$. Finally, Proposition 7.14 implies

$$\|q_k\|_{_0H_p^\tau((0,a),\, L_p(\Omega_k))}$$
$$\leq c \left(\|v_k\|_{_0\mathbb{X}_u(a,\Omega_k)} + \|\rho F_k(u, p)\|_{_0H_p^\tau((0,a),\, L_p(\Omega_k, \mathbb{R}^n))} + \|u\|_{_0\mathbb{X}_u(a,\Omega)} \right)$$
$$\leq c\|u\|_{_0\mathbb{X}_u(a,\Omega)} + c'\|h\|_{\mathbb{Y}_{h,\gamma}^{\alpha,\beta}(a,\Gamma)}$$

for $k = 0, 1, \ldots, N$.

7.30 On the other hand, $(\bar{v}_k, \bar{q}_k)$ may first be estimated as

$$(7.6b) \qquad\qquad \|(\bar{v}_k, \bar{q}_k)\|_{_0\mathbb{X}_\gamma^\beta(a,\Omega_k)} \leq c'\|h\|_{\mathbb{Y}_{h,\gamma}^{\alpha,\beta}(a,\Gamma)},$$

which then also implies

$$\|\bar{q}_k\|_{_0H_p^\tau((0,a),\, L_p(\Omega_k))} \leq c \left(\|\bar{v}_k\|_{_0\mathbb{X}_u(a,\Omega_k)} + \|h\|_{\mathbb{Y}_{h,\gamma}^{\alpha,\beta}(a,\Gamma)} \right) \leq c'\|h\|_{\mathbb{Y}_{h,\gamma}^{\alpha,\beta}(a,\Gamma)}$$

for $k = 1, 2, \ldots, N$ due to Proposition 7.14.

7.31 Therefore, we also obtain estimates for φ_k, since

$$\|\mu\Delta\varphi_0 - \rho\partial_t\varphi_0\|_{_0H_p^\tau((0,a),\, L_p(\Omega_k))} = \|p_0 - q_0\|_{_0H_p^\tau((0,a),\, L_p(\Omega_k))}$$
$$\leq c\|u\|_{_0\mathbb{X}_u(a,\Omega)} + c'\|h\|_{\mathbb{Y}_{h,\gamma}^{\alpha,\beta}(a,\Gamma)}$$

thanks to Proposition 7.14 and, analogously,

$$\|\mu\Delta\varphi_k - \rho\partial_t\varphi_k\|_{_0H_p^\tau((0,a),\, L_p(\Omega_k))} = \|p_k - q_k - \bar{q}_k\|_{_0H_p^\tau((0,a),\, L_p(\Omega_k))}$$
$$\leq c\|u\|_{_0\mathbb{X}_u(a,\Omega)} + c'\|h\|_{\mathbb{Y}_{h,\gamma}^{\alpha,\beta}(a,\Gamma)}$$

for $k = 1, 2, \ldots, N$. However,

$$\|\Delta\varphi_k\|_{_0H_p^1((0,a),\, L_p(\Omega_k))} \leq c\|u\|_{_0\mathbb{X}_u(a,\Omega)}$$

for $k = 0, 1, \ldots, N$ by construction and we infer

$$\|\partial_t\varphi_k\|_{_0H_p^\tau((0,a),\, L_p(\Omega_k))}, \ \|\Delta\varphi_k\|_{_0H_p^\tau((0,a),\, L_p(\Omega_k))} \leq c\|u\|_{_0\mathbb{X}_u(a,\Omega)} + c'\|h\|_{\mathbb{Y}_{h,\gamma}^{\alpha,\beta}(a,\Gamma)}$$

for $k = 0, 1, \ldots, N$. Thus, using

$$\|\partial_t\varphi_k\|_{L_p((0,a),\, H_p^2(\Omega_k))}, \ \|\Delta\varphi_k\|_{H_p^1((0,a),\, L_p(\Omega_k)) \cap L_p((0,a),\, H_p^2(\Omega_k))} \leq c\|u\|_{_0\mathbb{X}_u(a,\Omega)}$$

we also have

$$\|\partial_t\nabla\varphi_k\|_{L_p((0,a)\times\Omega_k)} \leq ca^\eta \|\partial_t\varphi_k\|_{_0H_p^{\tau/2}((0,a),\, H_p^1(\Omega_k))}$$
$$\leq ca^\eta \|\partial_t\varphi_k\|_{_0H_p^\tau((0,a),\, L_p(\Omega_k)) \cap L_p((0,a),\, H_p^2(\Omega_k))}$$
$$\leq ca^\eta \|u\|_{_0\mathbb{X}_u(a,\Omega)} + c'\|h\|_{\mathbb{Y}_{h,\gamma}^{\alpha,\beta}(a,\Gamma)}$$

and

$$\|\Delta\nabla\varphi_k\|_{L_p((0,a)\times\Omega_k)} \leq ca^\eta \|\Delta\varphi_k\|_{0 H_p^{1/2}((0,a),\, H_p^1(\Omega_k))}$$
$$\leq ca^\eta \|\Delta\varphi_k\|_{0 H_p^1((0,a),\, L_p(\Omega_k))\cap L_p((0,a),\, H_p^2(\Omega_k))}$$
$$\leq ca^\eta \|u\|_{0\mathbb{X}_u(a,\Omega)}$$

for $k = 0, \ldots, N$. Hence, we have

(7.6c) $$\|(\nabla\varphi_k, \mu\Delta\varphi_k - \rho\partial_t\varphi_k)\|_{0\mathbb{X}_\gamma^\beta(a,\Omega_k)} \leq ca^\eta \|u\|_{0\mathbb{X}_u(a,\Omega)} + c'\|h\|_{\mathbb{Y}_{h,\gamma}^{\alpha,\beta}(a,\Gamma)},$$

for $k = 0, \ldots, N$.

7.32 Finally,

$$(u, p) = \sum_{k=0}^{N} (\Psi_k u_k,\, \Psi_k p_k)$$

and (7.6) implies (7.1b).

7.33 It remains to construct a right inverse for $_0 L_{a,\gamma}^{\alpha,\beta}$. First note that Theorem 4.60 ensures the existence of a bounded linear operator

$$S_{a,\gamma,\#}^{\alpha,\beta} : {}_0\mathbb{Y}_\gamma^{\alpha,\beta}(a,\,\Omega) \longrightarrow {}_0\mathbb{X}_\gamma^\beta(a,\,\Omega)$$

with the property that $(v, q) = S_{a,\gamma,\#}^{\alpha,\beta}(f, g, h)$ is a solution to the Stokes equations

$$\rho\partial_t v - \mu\Delta v + \nabla q = \rho f \quad \text{in } (0,a)\times\Omega,$$
$$\operatorname{div} v = g \quad \text{in } (0,a)\times\Omega,$$
$$P_\Gamma \mathcal{B}^\alpha(v) = P_\Gamma h \quad \text{on } (0,a)\times\partial\Omega,$$
$$v(0) = 0 \quad \text{in } \Omega,$$

whenever $(f, g, h) \in {}_0\mathbb{Y}_\gamma^{\alpha,\beta}(a,\,\Omega)$, cf. 4.61.

7.34 Note that only the desired normal boundary condition

$$2\mu\, \partial_\nu v \cdot \nu_\Gamma - [p]_\Gamma = Q_\Gamma h \cdot \nu_\Gamma \quad \text{on } (0,a)\times\partial\Omega,$$

which will in general not be satisfied, prevents (v, q) from being a solution to the Stokes equations $(\mathrm{S}\,|\,a,\,\Omega)_{f,g,h,0}^{\alpha,\beta}$. Therefore, we will construct a second bounded linear operator

$$S_{a,\gamma,*}^{\alpha,\beta} : {}_\nu\mathbb{Y}_{h,\gamma}^{+1}(a,\,\Gamma) \longrightarrow {}_0\mathbb{X}_\gamma^\beta(a,\,\Omega)$$

to adjust the normal boundary condition and then set

$$S_{a,\gamma}^{\alpha,\beta}(f, g, h) = S_{a,\gamma,\#}^{\alpha,\beta}(f, g, h) + S_{a,\gamma,*}^{\alpha,\beta}(Q_\Gamma h - Q_\Gamma \mathcal{B}^\alpha(S_{a,\gamma,\#}^{\alpha,\beta}(f, g, h)))$$

for $(f, g, h) \in {}_0\mathbb{Y}_\gamma^{\alpha,\beta}(a,\,\Omega)$.

7.35 Given

$$\bar{h} \in {}_\nu\mathbb{Y}_{h,\gamma}^{+1}(a,\,\Gamma)$$

we first define $(\bar{v}_k, \bar{q}_k) \in {}_0\mathbb{X}_\gamma^\beta(a, \Omega_k)$ to be the unique maximal regular solution to the Stokes equations

$$
\begin{aligned}
\rho\partial_t\bar{v}_k - \mu\Delta\bar{v}_k + \nabla\bar{q}_k &= 0 && \text{in } (0, a) \times \Omega_k, \\
\operatorname{div}\bar{v}_k &= 0 && \text{in } (0, a) \times \Omega_k, \\
P_{\Gamma_k}\mathcal{B}^\alpha(\bar{v}_k) &= 0 && \text{on } (0, a) \times \partial\Omega_k, \\
2\mu\,\partial_\nu\bar{v}_k \cdot \nu_{\Gamma_k} - [\bar{q}_k]_{\Gamma_k} \cdot \nu_{\Gamma_k} &= \Phi_k Q_\Gamma\bar{h} \cdot \nu_{\Gamma_k} && \text{on } (0, a) \times \partial\Omega_k, \\
\bar{v}_k(0) &= 0 && \text{in } \Omega_k
\end{aligned}
$$

for $k = 1, 2, \ldots, N$. Then, we define

$$
\eta_k \in {}_0H_p^1((0, a), H_p^2(\Omega)) \cap {}_0H_p^{1/2}((0, a), H_p^3(\Omega))
$$

to be the solutions to the elliptic problems

$$
\begin{aligned}
-\Delta\eta_k &= E_\Omega(\nabla\psi_k \cdot R_k\bar{v}_k) && \text{in } (0, a) \times \Omega, \\
[\eta_k]_\Gamma &= 0 && \text{on } (0, a) \times \partial\Omega
\end{aligned}
$$

for $k = 1, 2, \ldots, N$ and set

$$
S_{a,\gamma,*}^{\alpha,\beta}(\bar{h}) := \sum_{k=1}^N (\Psi_k\bar{v}_k - \nabla\eta_k, \Psi_k\bar{q}_k + \rho\partial_t\eta_k - \mu\Delta_k\eta_k).
$$

7.36 Setting $(\bar{v}, \bar{q}) = S_{a,\gamma,*}^{\alpha,\beta}(\bar{h})$ we have

$$
{}_0L_{a,\gamma}^{\alpha,\beta}S_{a,\gamma,*}^{\alpha,\beta}(\bar{h}) = (0, 0, Q_\Gamma\bar{h})
$$

$$
+ \sum_{k=1}^N \Big(-\mu[\Delta, \Psi_k]\bar{v}_k + [\nabla, \Psi_k]\bar{q}_k, 0, [P_{\Gamma_k}\mathcal{B}^\alpha, \Psi_k]\bar{v}_k - P_\Gamma\mathcal{B}^\alpha(\nabla\eta_k) \Big)
$$

$$
+ \sum_{k=1}^N \Big(0, 0, [Q_{\Gamma_k}\mathcal{B}^{+1}, \Psi_k]\bar{v}_k - (2\mu\,Q_\Gamma[\nabla^2\eta_k]_\Gamma - \mu[\Delta\eta_k]_\Gamma\,\nu_\Gamma) \Big).
$$

Therefore, if $(f, g, h) \in {}_0\mathbb{Y}_\gamma^{\alpha,\beta}(a, \Omega)$ we set

$$
(v, q) := S_{a,\gamma,\#}^{\alpha,\beta}(f, g, h) \qquad \text{and} \qquad \bar{h} := Q_\Gamma h - Q_\Gamma\mathcal{B}^{+1}(v)
$$

and infer

$$
S_{a,\gamma}^{\alpha,\beta}(f, g, h) = (v + \bar{v}, q + \bar{q})
$$

as well as

$$
{}_0L_{a,\gamma}^{\alpha,\beta}S_{a,\gamma}^{\alpha,\beta}(f, g, h) = (f, g, h) + R_{a,\gamma}^{\alpha,\beta}(f, g, h)
$$

with

$$
R_{a,\gamma}^{\alpha,\beta}(f, g, h) = \sum_{k=1}^N \Big(-\mu[\Delta, \Psi_k]\bar{v}_k + [\nabla, \Psi_k]\bar{q}_k, 0, [P_{\Gamma_k}\mathcal{B}^\alpha, \Psi_k]\bar{v}_k - P_\Gamma\mathcal{B}^\alpha(\nabla\eta_k) \Big)
$$

$$
+ \sum_{k=1}^N \Big(0, 0, [Q_{\Gamma_k}\mathcal{B}^{+1}, \Psi_k]\bar{v}_k - (2\mu\,Q_\Gamma[\nabla^2\eta_k]_\Gamma - \mu[\Delta\eta_k]_\Gamma\,\nu_\Gamma) \Big).
$$

7.37 With this definition, we may obtain the desired estimate (7.3). As in 7.30 we first estimate $(\bar{v}_k, \bar{q}_k)$ as

$$\|(\bar{v}_k, \bar{q}_k)\|_{0\mathbb{X}^\beta_\gamma(a,\Omega_k)} \le c\|\bar{h}\|_{\mathbb{Y}^{\alpha,\beta}_{h,\gamma}(a,\Gamma)},$$

which then also implies

$$\|\bar{q}_k\|_{0H^\sigma_p((0,a),L_p(\Omega_k))} \le c\|\bar{v}_k\|_{0\mathbb{X}_u(a,\Omega_k)} + c\|\bar{h}\|_{\mathbb{Y}^{\alpha,\beta}_{h,\gamma}(a,\Gamma)} \le c\|\bar{h}\|_{\mathbb{Y}^{\alpha,\beta}_{h,\gamma}(a,\Gamma)}$$

for $k = 1, 2, \ldots, N$ due to Proposition 7.14. Thus, we infer

$$\begin{aligned}
\| &-\mu[\Delta, \Psi_k]\bar{v}_k + [\nabla, \Psi_k]\bar{q}_k\|_{\mathbb{Y}_f(a,\Omega)} \\
&\le ca^\eta\|-\mu[\Delta, \Psi_k]\bar{v}_k + [\nabla, \Psi_k]\bar{q}_k\|_{0H^\tau_p((0,a),L_p(\Omega))} \\
&\le ca^\eta\left(\|\bar{v}_k\|_{0\mathbb{X}_u(a,\Omega_k)} + \|\bar{q}_k\|_{0H^\tau_p((0,a),L_p(\Omega_k))} + \|\bar{h}\|_{\mathbb{Y}^{\alpha,\beta}_{h,\gamma}(a,\Gamma)}\right) \\
&\le ca^\eta\|\bar{h}\|_{\mathbb{Y}^{\alpha,\beta}_{h,\gamma}(a,\Gamma)}
\end{aligned}$$
(7.7a)

for $k = 1, 2, \ldots, N$. Analogously, the commutator terms in the boundary conditions are of lower order and we obtain

(7.7b) $\qquad \|[P_{\Gamma_k}\mathcal{B}^\alpha, \Psi_k]\bar{v}_k\|_{\mathbb{Y}^{\alpha,\beta}_{h,\gamma}(a,\Gamma)}, \; \|[Q_{\Gamma_k}\mathcal{B}^{+1}, \Psi_k]\bar{v}_k\|_{\mathbb{Y}^{\alpha,\beta}_{h,\gamma}(a,\Gamma)} \le ca^\eta\|\bar{h}\|_{\mathbb{Y}^{\alpha,\beta}_{h,\gamma}(a,\Gamma)}$

for $k = 1, 2, \ldots, N$. Furthermore, we have

$$\begin{aligned}
\|\nabla^2\eta_k\|_{0H^{1/2}_p((0,a),L_p(\Omega))\cap L_p((0,a),H^1_p(\Omega))} \\
\le ca^\eta\|\nabla^2\eta_k\|_{0H^1_p((0,a),L_p(\Omega))\cap 0H^{1/2}_p((0,a),H^1_p(\Omega))} \\
\le ca^\eta\|\bar{v}_k\|_{0\mathbb{X}_u(a,\Omega)} \\
\le ca^\eta\|\bar{h}\|_{\mathbb{Y}^{\alpha,\beta}_{h,\gamma}(a,\Gamma)},
\end{aligned}$$

which implies

(7.7c) $\qquad \|2\mu Q_\Gamma[\nabla^2\eta_k]_\Gamma - [\Delta\eta_k]_\Gamma\,\nu_\Gamma\|_{\mathbb{Y}^{\alpha,\beta}_{h,\gamma}(a,\Gamma)} \le ca^\eta\|\bar{h}\|_{\mathbb{Y}^{\alpha,\beta}_{h,\gamma}(a,\Gamma)}.$

7.38 To estimate the remaining perturbation in the boundary conditions we now set $(\tilde{v}_k, \tilde{q}_k) := (\Psi_k\bar{v}_k - \nabla\eta_k, \Psi_k\bar{q}_k + \rho\partial_t\eta_k - \mu\Delta\eta_k)$ and observe that

$$\begin{aligned}
\rho\partial_t\tilde{v}_k - \mu\Delta\tilde{v}_k + \nabla\tilde{q}_k &= -\mu[\Delta, \Psi_k]\bar{v}_k + [\nabla, \Psi_k]\bar{q}_k && \text{in } (0,a)\times\Omega, \\
\operatorname{div}\tilde{v}_k &= 0 && \text{in } (0,a)\times\Omega, \\
P_\Gamma\mathcal{B}^\alpha(\tilde{v}_k) &= [P_{\Gamma_k}\mathcal{B}^\alpha, \Psi_k]\bar{v}_k - P_\Gamma\mathcal{B}^\alpha(\nabla\eta_k) && \text{on } (0,a)\times\partial\Omega, \\
2\mu\,\partial_\nu\tilde{v}_k\cdot\nu_\Gamma - [\tilde{q}_k]_\Gamma &= \phi_k Q_\Gamma\bar{h}\cdot\nu_\Gamma + [Q_{\Gamma_k}\mathcal{B}^{+1}, \Psi_k]\bar{v}_k\cdot\nu_\Gamma \\
&\quad -(2\mu Q_\Gamma[\nabla^2\eta_k]_\Gamma\cdot\nu_\Gamma - \mu[\Delta\eta_k]_\Gamma) && \text{on } (0,a)\times\partial\Omega, \\
\tilde{v}_k(0) &= 0 && \text{in } \Omega.
\end{aligned}$$

Hence,

$$\|(\tilde{v}_k, \tilde{q}_k)\|_{0\mathbb{X}^\beta_\gamma(a,\Omega)} \le c\|\bar{h}\|_{\mathbb{Y}^{\alpha,\beta}_{h,\gamma}(a,\Gamma)}$$

and we further infer

$$\|\tilde{q}_k\|_{0H_p^\tau((0,a),\,L_p(\Omega))}$$
$$\leq c\left(\|\tilde{v}_k\|_{0X_u(a,\,\Omega)} + \|-\mu[\Delta,\,\Psi_k]\bar{v}_k + [\nabla,\,\Psi_k]\bar{q}_k\|_{0H_p^\tau((0,a),\,L_p(\Omega))}\right)$$
$$+ c\|2\mu\,Q_\Gamma[\nabla^2\eta_k]_\Gamma - \mu[\Delta\eta_k]_\Gamma\,\nu_\Gamma\|_{Y_{h,\gamma}^{\alpha,\beta}(a,\,\Gamma)}$$
$$\leq c\|\bar{h}\|_{Y_{h,\gamma}^{\alpha,\beta}(a,\,\Gamma)}$$

for $k = 1, 2, \ldots, N$ due to Proposition 7.14. Hence,

$$\|\mu\Delta\eta_k - \rho\partial_t\eta_k\|_{0H_p^\tau((0,a),\,L_p(\Omega))} = \|\Psi_k\bar{q}_k - \tilde{q}_k\|_{0H_p^\tau((0,a),\,L_p(\Omega))}$$
$$\leq c\|\bar{h}\|_{Y_{h,\gamma}^{\alpha,\beta}(a,\,\Gamma)}$$

for $k = 1, 2, \ldots, N$. However,

$$\|\Delta\eta_k\|_{0H_p^1((0,a),\,L_p(\Omega))} \leq c\|\bar{v}_k\|_{0X_u(a,\,\Omega_k)} \leq c\|\bar{h}\|_{Y_{h,\gamma}^{\alpha,\beta}(a,\,\Gamma)}$$

for $k = 1, 2, \ldots, N$ and we infer

$$\|\partial_t\eta_k\|_{0H_p^\tau((0,a),\,L_p(\Omega))},\ \|\Delta\eta_k\|_{0H_p^\tau((0,a),\,L_p(\Omega))} \leq c\|\bar{h}\|_{Y_{h,\gamma}^{\alpha,\beta}(a,\,\Gamma)}$$

for $k = 1, 2, \ldots, N$. Thus, using

$$\|\partial_t\eta_k\|_{L_p((0,a),\,H_p^2(\Omega))},\ \|\Delta\eta_k\|_{H_p^1((0,a),\,L_p(\Omega))\cap L_p((0,a),\,H_p^2(\Omega))} \leq c\|\bar{h}\|_{Y_{h,\gamma}^{\alpha,\beta}(a,\,\Gamma)}$$

we also have

$$\|\partial_t\nabla\eta_k\|_{L_p((0,a)\times\Omega)} \leq ca^\eta\|\partial_t\eta_k\|_{0H_p^{\tau/2}((0,a),\,H_p^1(\Omega))}$$
$$\leq ca^\eta\|\partial_t\eta_k\|_{0H_p^\tau((0,a),\,L_p(\Omega))\cap L_p((0,a),\,H_p^2(\Omega))}$$
$$\leq ca^\eta\|\bar{h}\|_{Y_{h,\gamma}^{\alpha,\beta}(a,\,\Gamma)}$$

as well as

$$\|\Delta\nabla\eta_k\|_{L_p((0,a)\times\Omega)} \leq ca^\eta\|\Delta\eta_k\|_{0H_p^{1/2}((0,a),\,H_p^1(\Omega))}$$
$$\leq ca^\eta\|\Delta\eta_k\|_{0H_p^1((0,a),\,L_p(\Omega))\cap L_p((0,a),\,H_p^2(\Omega))}$$
$$\leq ca^\eta\|\bar{h}\|_{Y_{h,\gamma}^{\alpha,\beta}(a,\,\Gamma)}$$

for $k = 1, 2, \ldots, N$. Hence, we have

$$\|\nabla\eta_k,\ \mu\Delta\eta_k - \rho\partial_t\eta_k\|_{0X_\gamma^\beta(a,\,\Omega)} \leq ca^\eta\|\bar{h}\|_{Y_{h,\gamma}^{\alpha,\beta}(a,\,\Gamma)}$$

for $k = 1, 2, \ldots, N$. In particular, we have

$$(7.7\text{d}) \qquad \|P_\Gamma\mathcal{B}^\alpha(\nabla\eta_k)\|_{Y_{h,\gamma}^{\alpha,\beta}(a,\,\Gamma)} \leq ca^\tau\|\bar{h}\|_{Y_{h,\gamma}^{\alpha,\beta}(a,\,\Gamma)}$$

and (7.7) implies (7.3). In summary, the proof for $\beta = +1$ is complete.

Remarks

7.39 The localisation presented in this section to reduce the problem in a bounded domain to finitely many problems in bent halfspaces is well-known for parabolic problems, see e. g. the monograph [DHP03] by R. DENK, M. HIEBER and J. PRÜSS and the article [DHP07], where a complete theory for parabolic problems subject to inhomogeneous boundary conditions is developed. However, in case of the Stokes equations the pressure and the divergence constraint demand a more complex approach. To the best of our knowledge, the construction of a right inverse for the Stokes equations as presented in this chapter may not be found in the literature.

References

[DHP03] R. DENK, M. HIEBER, and J. PRÜSS: $\mathcal{R}$-Boundedness, Fourier-Multipliers and Problems of Elliptic and Parabolic Type, *Mem. Amer. Math. Soc.*, vol. 166. American Mathematical Society, 2003.

[DHP07] R. DENK, M. HIEBER, and J. PRÜSS: *Optimal L_p-L_q-Regularity for Parabolic Problems with Inhomogeneous Boundary Data.* Math. Z., 257, 193–224, 2007.

Bounded Weakly Singular Domains

Chapter 8

L_p-Theory in Weakly Singular Domains

The aim of the this is to extend the L_p-Theory for incompressible Newtonian flows subject to one of the energy preserving respectively artificial boundary conditions introduced in Chapter 2, which has been developed for bounded smooth domains in Chapters 3 – 7, to a certain class of domains having a non-smooth boundary. As has been mentioned in the introduction, domains like the tube in the figure on page 3 occur frequently as model problems in computational fluid dynamics. However, these domains are not covered by the theory developed so far.

We focus on bounded domains $\Omega \subseteq \mathbb{R}^n$, whose boundary $\partial\Omega$ may be decomposed as $\partial\Omega = \Gamma_1 \cup \Gamma_2 \cup \cdots \cup \Gamma_m \cup S$ into m smooth submanifolds of $\mathbb{R}^n$, denoted by Γ_1, Γ_2, ..., Γ_m, and a remaining *singular part* S, which satisfies $\mathcal{H}^{n-1}(S) = 0$. Moreover, we assume the smooth parts to be mutually disjoint and to obey certain geometrical constraints.

Given $a > 0$, a constant density $\rho > 0$ and a constant viscosity $\mu > 0$ as well as parameter vectors α, $\beta \in \{-1, 0, +1\}^m$, which define the boundary conditions to be imposed, we will study the Navier-Stokes equations

$$
\begin{aligned}
\rho\partial_t u + \rho(u \cdot \nabla)u - \mu\Delta u + \nabla p &= \rho f &&\text{in } (0, a) \times \Omega, \\
\operatorname{div} u &= g &&\text{in } (0, a) \times \Omega, \\
\mathcal{B}^{\alpha_j, \beta_j}(u, p) &= h_j &&\text{on } (0, a) \times \Gamma_j, \\
u(0) &= u_0 &&\text{in } \Omega
\end{aligned}
$$

$(\mathrm{N}\,|\,a,\,\Omega)^{\alpha,\beta}_{f,g,h,u_0}$

as well as the Stokes equations

$$
\begin{aligned}
\rho\partial_t u - \mu\Delta u + \nabla p &= \rho f &&\text{in } (0, a) \times \Omega, \\
\operatorname{div} u &= g &&\text{in } (0, a) \times \Omega, \\
\mathcal{B}^{\alpha_j, \beta_j}(u, p) &= h_j &&\text{on } (0, a) \times \Gamma_j, \\
u(0) &= u_0 &&\text{in } \Omega,
\end{aligned}
$$

$(\mathrm{S}\,|\,a,\,\Omega)^{\alpha,\beta}_{f,g,h,u_0}$

where $j = 1, 2, \ldots, m$, in a functional analytic framework based on L_p-spaces.

The boundary operators $\mathcal{B}^{\alpha_j,\beta_j}$ realise in each case one of the energy preserving respectively artificial boundary conditions $(\mathrm{B}\,|\,a,\,\Omega)^{\alpha_j,\beta_j}$ introduced in 2.19, 2.22 and 2.23 with parameters $\alpha_j,\,\beta_j \in \{-1,\,0,\,+1\}$. We adapt the definitions from Chapter 3 and set

$$P_\Gamma \mathcal{B}^{0,\beta_j}(u,\,p) := P_\Gamma[u]_{\Gamma_j}, \qquad P_\Gamma \mathcal{B}^{\pm 1,\beta_j}(u,\,p) := \mu P_\Gamma[\nabla u \pm \nabla u^{\mathsf{T}}]_{\Gamma_j}\,\nu_\Gamma$$

for $\beta_j \in \{-1,\,0,\,+1\}$ and

$$Q_\Gamma \mathcal{B}^{\alpha_j,0}(u,\,p) \cdot \nu_\Gamma := [u]_{\Gamma_j} \cdot \nu_\Gamma,$$

$$Q_\Gamma \mathcal{B}^{\alpha_j,+1}(u,\,p) \cdot \nu_\Gamma := 2\mu\,\partial_\nu u \cdot \nu_\Gamma - [p]_{\Gamma_j}, \qquad Q_\Gamma \mathcal{B}^{\alpha_j,-1}(u,\,p) \cdot \nu_\Gamma := -[p]_{\Gamma_j}$$

for $\alpha_j \in \{-1,\,0,\,+1\}$. Here $[\,\cdot\,]_{\Gamma_j}$ denotes the trace of a function defined in Ω on the smooth boundary component Γ_j. Consequently, the normal derivative has to be understood as $\partial_\nu = [\nabla \cdot \,^{\mathsf{T}}]_{\Gamma_j}\,\nu_\Gamma$. Moreover, $P_\Gamma = P_\Gamma(y) := 1 - \nu_\Gamma(y) \otimes \nu_\Gamma(y)$ denotes the projection onto the tangent space of Γ at a point $y \in \Gamma$. Here and in the following sections we denote by $\Gamma := \partial\Omega \setminus S$ the smooth part of the boundary $\partial\Omega$. Furthermore, $Q_\Gamma := 1 - P_\Gamma$ denotes the projection onto the normal bundle of Γ and $\nu_\Gamma : \Gamma \longrightarrow \mathbb{R}^n$ denotes the outer unit normal field of Ω, which we assume to be well-defined on each smooth part Γ_j.

Since we want to adapt the techniques used in Chapters 3 – 7, in particular the localisation procedure, we restrict our considerations to a class of domains, which we call *weakly singular*.

8.1 Weakly Singular Domains

8.1 We define the notion of a *weakly singular domain* based on a larger class of domains, which have a *singular boundary*, which, however, has a sufficiently large *regular part* to allow for the definition of an outer unit normal field and to allow for a reasonable definition of a surface integral.

8.2 DEFINITION. Let $\mu \in \mathbb{N}$ with $\mu \geq 2$. A Lipschitz domain $\Omega \subseteq \mathbb{R}^n$ is said to be a singular $C^{\mu-}$-domain, if there exists an $(n-1)$-dimensional $C^{\mu-}$-submanifold M of $\mathbb{R}^n$, such that

(1) $M \subseteq \partial\Omega$;

(2) $\mathcal{H}^{n-1}(\partial\Omega \setminus M) = 0$;

(3) M is open and dense in $\partial\Omega$;

(4) every point $y \in M$ satisfies the constraints 7.7.

The largest $(n-1)$-dimensional $C^{\mu-}$-submanifold M of $\mathbb{R}^n$ with the properties *(1)–(4)* is called the regular boundary of Ω and denoted by $\partial_r\Omega$. The remaining part $\partial_s\Omega = \partial\Omega \setminus \partial_r\Omega$ is called the singular boundary of Ω.

8.3 Note that for a singular $C^{\mu-}$-domain $\Omega \subseteq \mathbb{R}^n$ we have $\partial_r\Omega,\,\partial_s\Omega \subseteq \partial\Omega$ by definition. Moreover, $\partial\Omega = \partial_r\Omega \cup \partial_s\Omega$ and $\partial_r\Omega \cap \partial_s\Omega = \varnothing$. The second constraint ensures that the singular boundary may be neglected concerning integration. The third constraint ensures that domains like $B_1(0) \setminus \{0\}$ are excluded in the above definition. Furthermore, the fourth constraint ensures that there exists a well-defined outer unit normal field

$$\nu_\Gamma : \Gamma = \partial_r\Omega \longrightarrow \mathbb{R}^n.$$

In particular, locally *the domain is located on one side of its regular boundary.*

8.4 To adapt the localisation procedure, which has been presented in Chapters 5, 6 and 7 for bounded smooth domains, the class of singular $C^{\mu-}$-domains is not restrictive enough, since a *local flattening* of the boundary may not be performed for the singular boundary. However, we want to allow for the presence of a singular boundary to include domains like the tube shown in the figure on page 3. This gives raise to the following definition.

8.5 DEFINITION. Let $\mu \in \mathbb{N}$ with $\mu \geq 2$. A singular $C^{\mu-}$-domain $\Omega \subseteq \mathbb{R}^n$ is said to be weakly singular of order one, if

(1) $\Gamma = \partial_r \Omega$ consists of finitely many connected components Γ_1, Γ_2, ..., Γ_m;

(2) the outer unit normal field
$$\nu_{\Gamma_j} : \Gamma_j \longrightarrow \mathbb{R}^n$$
extends continuously to $\bar{\Gamma}_j$ for all $j = 1, 2, \ldots, m$;

(3) $\bar{\Gamma}_j \cap \bar{\Gamma}_k \cap \bar{\Gamma}_l = \varnothing$, if $j, k, l \in \{1, 2, \ldots, m\}$ with $j \neq k$, $j \neq l$, $k \neq l$;

(4) $\nu_{\Gamma_k}(y) \perp \nu_{\Gamma_l}(y)$ for all $y \in \bar{\Gamma}_k \cap \bar{\Gamma}_l \cap \partial_s \Omega$ and $k, l \in \{1, 2, \ldots, m\}$.

Moreover, for $k, l \in \{1, 2, \ldots, m\}$ with $k \neq l$ and $y \in \bar{\Gamma}_k \cap \bar{\Gamma}_l$ there has to exist an $\bar{r} = \bar{r}(y) > 0$ with the following properties:

(i) for all $0 < r < \bar{r}$ and $j \in \{k, l\}$ there exist $BUC^{\mu-}$-functions $\bar{\omega}_j : \Sigma_j \longrightarrow \mathbb{R}$, where $\Sigma_j \subseteq \mathcal{T}_y \Gamma_j$ denotes the projection of $\Gamma_j \cap B_r(y)$ onto the tangent plane $\mathcal{T}_y \Gamma_j$, such that
$$\Gamma_j \cap B_r(y) = \Big\{ \eta - \bar{\omega}_j(\eta) \nu_{\Gamma_j}(y) \, : \, \eta \in \Sigma_j \Big\},$$
i. e. $\Gamma_j \cap B_r(y)$ may be parametrised over its projection onto the tangent plane;

(ii) the above functions $\bar{\omega}_j$ may be extended to $BUC^{\mu-}$-functions $\omega_j : \mathcal{T}_y \Gamma_j \longrightarrow \mathbb{R}$ with compact support, which by definition satisfy
$$\omega_j(y) = y, \qquad \nabla_{\mathcal{T}_y \Gamma_j} \omega_j(y) = 0;$$

(iii) either the acute bent wedges
$$\mathfrak{V} := \Big\{ x \in \mathbb{R}^n \, : \, -(x - \Pi_j(x)) \cdot \nu_{\Gamma_j}(y) > \omega_j(\Pi_j(x)), \; j \in \{k, l\} \Big\},$$
where $\Pi_j : \mathbb{R}^n \longrightarrow \mathcal{T}_y \Gamma_j$ denotes the metric projection onto $\mathcal{T}_y \Gamma_j$, or the obtuse bent wedges
$$\mathfrak{V} := \left\{ x \in \mathbb{R}^n \, : \, \begin{array}{c} -(x - \Pi_k(x)) \cdot \nu_{\Gamma_k}(y) > \omega_k(\Pi_k(x)) \\ \text{or } -(x - \Pi_l(x)) \cdot \nu_{\Gamma_l}(y) > \omega_l(\Pi_l(x)) \end{array} \right\}$$
satisfy
$$\Omega \cap B_r(y) = \mathfrak{V} \cap B_r(y).$$

Finally, Ω is said to be acute, if for all $y \in \partial_s \Omega$ constraint *(iii)* is satisfied by the acute bent wedges.

8.6 A weakly singular $C^{\mu-}$-domain $\Omega \subseteq \mathbb{R}^n$ of order one enjoys many nice properties. By definition $\Gamma = \partial_r \Omega$ is open in $\partial \Omega$ and, hence, Γ_j is open in $\partial \Omega$ for $j = 1, 2, \ldots, m$ and the smooth boundary components are mutually disjoint. Due to the third constraint no more than two smooth components may meet at any singular point $y \in \partial_s \Omega$. On the other hand, a smooth part Γ_j is not allowed to *meet itself* at such a singular point, since this would violate at least one of the constraints two or four. For illustration consider the domain $(-1, 0)^2 \cup B_1(0) \subseteq \mathbb{R}^2$. This domain satisfies the constraints one, three and four with $m = 1$ and its boundary has a single singular point at $y = (-1, -1)$. However, the outer unit normal field may not be continuously extended into y. On the other hand, a domain that satisfies the constraints one and two may not satisfy the orthogonality constraint four at a point $y \in \partial_s \Omega$, if a smooth component of the boundary meets itself in y. Finally, the singular boundary $\partial_s \Omega$ is locally a subset of the intersection $\bar{\Gamma}_k \cap \bar{\Gamma}_l$ for some $k, l \in \{1, 2, \ldots, m\}$ with $k \neq l$. Hence, the singular boundary is locally an $(n-2)$-dimensional submanifold of $\mathbb{R}^n$, which is the reason for calling such domains weakly singular of order one. Of course, this definition may be generalised to weakly singular domains of order N, if we allow at most $N + 1$ smooth components of the boundary to meet at a single singular point $y \in \partial_s \Omega$. However, in this case the singular boundary is locally an $(n - k - 1)$-dimensional submanifold of $\mathbb{R}^n$ for some $k \in \{1, 2, \ldots, N\}$.

8.7 The most valuable property of a weakly singular $C^{\mu-}$-domain $\Omega \subseteq \mathbb{R}^n$ of order one is certainly the possibility to generalise the localisation procedure employed in Chapters 5, 6 and 7. Indeed, after a suitable rotation and translation of the coordinate system we may consider any singular point as $0 \in \bar{\Gamma}_k \cap \bar{\Gamma}_l$ for some $k, l \in \{1, 2, \ldots, m\}$ with $k \neq l$. Moreover, we may assume

$$\mathcal{T}_0 \Gamma_k = \left\{ (x, 0, z) : x \in \mathbb{R}^{n-2}, z \in \mathbb{R} \right\}, \quad \mathcal{T}_0 \Gamma_l = \left\{ (x, y, 0) : x \in \mathbb{R}^{n-2}, y \in \mathbb{R} \right\}$$

due to the orthogonality constraint four. Thus, by definition there exist two $BUC^{\mu-}$-functions $\omega_y, \omega_z : \mathbb{R}^{n-1} \longrightarrow \mathbb{R}$ such that

$$(8.1) \qquad\qquad \omega_y(0) = 0 = \omega_z(0), \qquad \nabla \omega_y(0) = 0 = \nabla \omega_z(0)$$

and, if Ω is acute, the acute bent wedge

$$\mathfrak{V} = \left\{ (x, y, z) \in \mathbb{R}^{n-2} \times \mathbb{R} \times \mathbb{R} : y > \omega_y(x, z), \ z > \omega_z(x, y) \right\}$$

satisfies

$$\Omega \cap B_r(0) = \mathfrak{V} \cap B_r(0).$$

Now, we may define an isomorphism

$$\Theta_{\omega_y, \omega_z} : \mathbb{R}^n_{++} := \left\{ (x, y, z) \in \mathbb{R}^{n-2} \times \mathbb{R} \times \mathbb{R} : y, z > 0 \right\} \longrightarrow \mathfrak{V}$$

via

$$\Theta_{\omega_y, \omega_z}(x, y, z) = (\bar{x}, \bar{y}, \bar{z}) \qquad \Leftrightarrow \qquad \begin{aligned} \bar{x} &= x, \\ \bar{y} &= y + \omega_y(\bar{x}, \bar{z}), \\ \bar{z} &= z + \omega_z(\bar{x}, \bar{y}). \end{aligned}$$

Indeed, for fixed $(x, y, z) \in \mathbb{R}^n_{++}$ the mapping

$$M : \mathbb{R}^n \longrightarrow \mathbb{R}^n, \qquad M(\bar{x}, \bar{y}, \bar{z}) = (x, y + \omega_y(\bar{x}, \bar{z}), z + \omega_z(\bar{x}, \bar{y}))$$

admits a unique fixed point $(\bar{x}, \bar{y}, \bar{z}) \in \mathbb{R}^n$, provided

(8.2) $\|\nabla \omega_y\|_{L_\infty(\mathbb{R}^{n-1})}, \ \|\nabla \omega_z\|_{L_\infty(\mathbb{R}^{n-1})} < \delta$

with $\delta > 0$ sufficiently small. However, due to (8.1) we may arrange (8.2) to be satisfied for an arbitrary $\delta > 0$, provided we choose $r > 0$ to be sufficiently small.

8.8 Now, $\partial_s \Omega$ is closed in $\partial \Omega$ and, therefore, compact in $\mathbb{R}^n$, if Ω is bounded. Hence, we may adapt the partition of the domain presented in Section 7.2 and start with a finite covering of $\partial_s \Omega$ with balls $U_1, U_2, \ldots, U_M$ such that the cut-off technique leads to problems in acute bent wedges. This covering may then be completed by finitely many balls $U_{M+1}, U_{M+2}, \ldots, U_N$ to a covering of $\partial \Omega$ such that the cut-off technique leads to problems in bent halfspaces exactly as in Section 7.2. Finally, we complete this covering by an open set $U_0 \subseteq \Omega$ to a covering of Ω and construct a partition of unity subordinate to $U_0, U_1, \ldots, U_N$. This way the parabolic problem introduced in Section 4.2 as well as the Stokes equations may be treated by exactly the same arguments as used in Chapter 7, where we may impose different boundary conditions on the different smooth components of the boundary, provided a maximal L_p-regularity result is available for the corresponding problem in an acute bent wedge.

8.9 However, a parabolic problem or the Stokes equations in an acute bent wedge may be reduced to a corresponding problem in the acute wedge $\mathbb{R}^n_{++}$ using the techniques presented in Chapter 6. Indeed, the transformation $\Theta_{\omega_y, \omega_z}$ introduced in 8.7 satisfies

$$\Theta_{\omega_y, \omega_z}^{-1}(x, y, z) = (x, y - \omega_y(x, z), z - \omega_z(x, y)), \qquad (x, y, z) \in \mho$$

and we obtain

$$\partial_t u = (\partial_t \bar{u}) \circ \Theta_{\omega_y, \omega_z}^{-1},$$

$$\partial_k u = (\partial_k \bar{u}) \circ \Theta_{\omega_y, \omega_z}^{-1} - (\partial_k \omega_y) \left\{ (\partial_y \bar{u}) \circ \Theta_{\omega_y, \omega_z}^{-1} \right\} - (\partial_k \omega_z) \left\{ (\partial_z \bar{u}) \circ \Theta_{\omega_y, \omega_z}^{-1} \right\},$$

$$\partial_y u = (\partial_y \bar{u}) \circ \Theta_{\omega_y, \omega_z}^{-1} - (\partial_y \omega_z) \left\{ (\partial_z \bar{u}) \circ \Theta_{\omega_y, \omega_z}^{-1} \right\},$$

$$\partial_z u = (\partial_y \bar{u}) \circ \Theta_{\omega_y, \omega_z}^{-1} - (\partial_z \omega_y) \left\{ (\partial_y \bar{u}) \circ \Theta_{\omega_y, \omega_z}^{-1} \right\}$$

for the functions $\bar{u} = u \circ \Theta_{\omega_y, \omega_z}$, where $k = 1, 2, \ldots, n - 2$. Therefore, the perturbation terms introduced by a transformation of an acute bent wedge into an acute wedge are of exactly the same type as those, which were introduced by the transformation of a bent halfspace into a halfspace as presented in Chapter 6. Hence, we may reuse the very same estimates to obtain a maximal L_p-regularity result for the parabolic problem introduced in Section 4.2 as well as the Stokes equations in an acute bent wedge, where we may impose different boundary conditions on the two smooth parts of the boundary, provided a maximal L_p-regularity result is at hand for the corresponding problem in an acute wedge.

8.10 Hence, by considering the prototype domains $\mathbb{R}^n$ and $\mathbb{R}^n_+$, which has been treated in Chapter 5, as well as the prototype domain $\mathbb{R}^n_{++}$ we may generalise the maximal L_p-regularity results obtained for the parabolic problem introduced in Section 4.2 and the Stokes equations to the situation of a bounded, acute weakly singular C^{3-}-domain of order one with different boundary conditions imposed on the different smooth components of the boundary.

8.2 Necessary Regularity Conditions

8.11 In order to establish a suitable functional analytic framework for the Stokes equations $(S \mid a, \Omega)$ in a bounded, acute weakly singular C^{3-}-domain $\Omega \subseteq \mathbb{R}^n$ of order one we proceed as in Chapter 3 and first derive the necessary regularity and compatibility conditions on the data. To save some notation, we always denote by $\Gamma := \partial_r \Omega$ the regular boundary of Ω in this and the following sections. Since Ω is by definition in particular a Lipschitz domain, there are generalisations of Propositions 3.9 and 3.14 available, cf. Remark 8.39.

8.12 PROPOSITION. *Let $a > 0$ and let $\Omega \subseteq \mathbb{R}^n$ be a bounded, acute weakly singular C^{3-}-domain of order one. Let $1 < p < \infty$, $\tau \in (0, 1]$ and $\sigma \in (0, 2]$. Then the embeddings*

$$H_p^\tau((0, a), L_p(\Omega)) \cap L_p((0, a), H_p^\sigma(\Omega))$$
$$\hookrightarrow H_p^{(1-\theta)\tau}((0, a), H_p^{\theta\sigma}(\Omega)), \quad \theta \in [0, 1]$$

are valid. Moreover, the embeddings

$$_0H_p^\tau((0, \bar{a}), L_p(\Omega)) \cap L_p((0, \bar{a}), H_p^\sigma(\Omega))$$
$$\hookrightarrow {}_0H_p^{(1-\theta)\tau}((0, \bar{a}), H_p^{\theta\sigma}(\Omega)), \quad \theta \in [0, 1], \ \bar{a} \in (0, a]$$

are valid, where the embedding constants are independent of $\bar{a} \in (0, a]$. □

8.13 PROPOSITION. *Let $\Omega \subseteq \mathbb{R}^n$ be a bounded, acute weakly singular C^{3-}-domain of order one with smooth boundary components $\Gamma_1, \Gamma_2, \ldots, \Gamma_m$. Let $1 < p < \infty$. Moreover, let $J \subseteq \{1, 2, \ldots, m\}$ and set*

$$\Gamma_0 := int_{\partial\Omega} \bigcup_{j \in J} \bar{\Gamma}_j.$$

Then the following assertions are valid.

(i) Let $\sigma \in (1/p, 1 + 1/p) \cup (1 + 1/p, 2]$. Then the trace operator

$$[\cdot]_\Gamma : W_p^\sigma(\Omega) \longrightarrow W_p^{\sigma-1/p}(\Gamma_0)$$

is bounded and surjective.

(ii) The trace operator

$$[\cdot]_\Gamma : \dot{H}_p^1(\Omega) \longrightarrow \dot{W}_p^{1-1/p}(\Gamma_0)$$

is bounded and surjective. □

8.14 Here, $int_{\partial\Omega} A$ denotes the interior of a set $A \subseteq \partial\Omega$ with respect to the trace topology on $\partial\Omega$. This way the above boundary part Γ_0 does not only contain the smooth components Γ_j with $j \in J$ but also those edges of the singular boundary $\partial_s\Omega$ that connect two smooth components Γ_k and Γ_l with $k, l \in J$. Hence, if Γ_0 happens to include two smooth boundary components Γ_k and Γ_l with $k, l \in J$ that meet at the singular boundary, then the trace space $W_p^{\sigma-1/p}(\Gamma_0)$ does not only encode regularity constraints but also compatibility constraints. As an illustration, consider the domain $\Omega = (-1, 1)^2 \subseteq \mathbb{R}^2$ with smooth boundary components

$$\Gamma_1 = \{-1\} \times (-1, 1), \quad \Gamma_2 = \{1\} \times (-1, 1),$$
$$\Gamma_3 = (-1, 1) \times \{-1\}, \quad \Gamma_4 = (-1, 1) \times \{1\}.$$

For $\sigma = 1$ and $J = \{1, 2\}$ the space $W_p^{\sigma-1/p}(\Gamma_0)$ simply encodes the corresponding regularity constraints on Γ_1 and Γ_2. However, for $J = \{1, 3\}$ it encodes in addition the constraint of continuity across the singular point $(-1, -1)$, provided $p > 2$. These additional compatibility conditions are necessary to obtain the surjectivity of the trace operator. Thus, Proposition 8.13 establishes the possibility to simultaneously extend functions defined on different smooth components of the boundary to functions defined in Ω, provided the necessary compatibility conditions on the joining edges are satisfied.

8.15 Note that Proposition 8.13 especially covers the case of a particular smooth boundary component $\Gamma_0 = \Gamma_j$ with $J = \{j\}$ and the case of the whole boundary $\Gamma_0 = \partial\Omega$ with $J = \{1, 2, \ldots, m\}$.

8.16 Hence, given vectors of parameters α, $\beta \in \{-1, 0, +1\}^m$ and regularity parameters $\gamma_1, \gamma_2, \ldots, \gamma_m$ with $\gamma_j = -\infty$, if $\beta_j = 0$, $\gamma_j \in \{-\infty\} \cup [0, 1/2 - 1/2p]$, if $\beta_j = +1$, respectively $\gamma_j \in \{-\infty\} \cup [0, \infty)$, if $\beta_j = -1$, the same considerations as in Sections 3.1 and 3.2 lead to the solution spaces

$$\mathbb{X}_u(a, \Omega) := H_p^1((0, a), L_p(\Omega, \mathbb{R}^n)) \cap L_p((0, a), H_p^2(\Omega, \mathbb{R}^n)),$$

$$\mathbb{X}_{p,\gamma}^0(a, \Omega) := \left\{ q \in L_p((0, a), \dot{H}_p^1(\Omega)) : (q)_\Omega = 0 \right\}$$

and

$$\mathbb{X}_{p,\gamma}^\beta(a, \Omega) := \left\{ q \in L_p((0, a), \dot{H}_p^1(\Omega)) : \begin{array}{l} [q]_{\Gamma_j} \in W_p^{\gamma_j}((0, a), L_p(\Gamma_j)) \\ \cap\, L_p((0, a), W_p^{1-1/p}(\Gamma_j)), \text{ if } \gamma_j \geq 0 \end{array} \right\}$$

for $\beta \neq (0, 0, \ldots, 0)$ and to the data spaces

$$\mathbb{Y}_f(a, \Omega) := L_p((0, a) \times \Omega, \mathbb{R}^n),$$

$$\mathbb{Y}_g(a, \Omega) := H_p^{1/2}((0, a), H_p^1(\Omega)) \cap L_p((0, a), H_p^1(\Omega)),$$

$$\mathbb{Y}_0(\Omega) := W_p^{2-2/p}(\Omega, \mathbb{R}^n)$$

and

$$\mathbb{Y}_{h,\gamma_j}^{\alpha_j,\beta_j}(a, \Gamma_j) = \left\{ \eta \in \mathbb{Y}_h(a, \Gamma_j) : P_\Gamma \eta \in \mathbb{T}_h^{\alpha_j}(a, \Gamma_j), \ Q_\Gamma \eta \in \mathbb{N}_{h,\gamma_j}^{\beta_j}(a, \Gamma_j) \right\},$$

where $\mathbb{Y}_h(a, \Gamma_j) := L_p((0, a), L_{p,loc}(\Gamma_j, \mathbb{R}^n))$ and

$$\mathbb{T}_h^0(a, \Gamma_j) := W_p^{1-1/2p}((0, a), L_p(\Gamma_j, T\Gamma)) \cap L_p((0, a), W_p^{2-1/p}(\Gamma_j, T\Gamma)),$$

$$\mathbb{T}_h^{\pm 1}(a, \Gamma_j) := W_p^{1/2-1/2p}((0, a), L_p(\Gamma_j, T\Gamma)) \cap L_p((0, a), W_p^{1-1/p}(\Gamma_j, T\Gamma)),$$

$$\mathbb{N}_{h,-\infty}^0(a, \Gamma_j) := W_p^{1-1/2p}((0, a), L_p(\Gamma_j, N\Gamma)) \cap L_p((0, a), W_p^{2-1/p}(\Gamma_j, N\Gamma)),$$

$$\mathbb{N}_{h,-\infty}^{\pm 1}(a, \Gamma_j) := L_p((0, a), \dot{W}_p^{1-1/p}(\Gamma_j, N\Gamma)),$$

$$\mathbb{N}_{h,\gamma_j}^{\pm 1}(a, \Gamma_j) := W_p^{\gamma_j}((0, a), L_p(\Gamma_j, N\Gamma)) \cap L_p((0, a), W_p^{1-1/p}(\Gamma_j, N\Gamma)) \quad \text{for } \gamma_j \geq 0$$

for $j = 1, 2, \ldots, m$.

8.3 Necessary Compatibility Conditions

8.17 Analogously to the considerations in Section 3.3 there are several compatibility conditions, which have to be satisfied by the data. First of all, the compatibility condition

$$(C_1)_{g,u_0} \qquad\qquad\qquad \operatorname{div} u_0 = g(0)$$

is necessary, i. e. the right hand side of the divergence equation has to be compatible with the initial data, regardless of the particular boundary conditions.

8.18 Moreover, the boundary conditions imply the compatibility conditions

$$(C_2)_{h,u_0}^{\alpha} \qquad \begin{aligned} P_\Gamma[u_0]_{\Gamma_j} &= P_\Gamma h_j(0), \text{ if } \alpha_j = 0 \quad\text{and } p > \tfrac{3}{2}, \\ \mu P_\Gamma[\nabla u_0 \pm \nabla u_0^\mathsf{T}]_{\Gamma_j}\, \nu_\Gamma &= P_\Gamma h_j(0), \text{ if } \alpha_j = \pm 1 \text{ and } p > 3 \end{aligned}$$

to be necessary for $(S\,|\,a,\,\Omega)_{f,g,h,u_0}^{\alpha,\beta}$ to admit a maximal regular solution.

8.19 Furthermore, there exists a compatibility condition between the right hand side of the divergence equation, the boundary data and the initial velocity. Indeed, there necessarily exist

$$\eta_j \in W_p^{1-1/2p}((0,a),\, L_p(\Gamma_j)) \cap L_p((0,a),\, W_p^{2-1/p}(\Gamma_j))$$

for $j = 1,\, 2,\, \ldots,\, m$, such that

$$(C_3)_{g,h,u_0}^{\beta} \qquad \begin{aligned} [u_0]_{\Gamma_j}\cdot \nu_\Gamma &= \eta_j(0), \\ \eta_j &= h_j\cdot \nu_\Gamma, \qquad \text{if } \beta_j = 0, \\ (\,g,\eta_1,\eta_2,\ldots,\eta_m\,) &\in H_p^1((0,a),\, {}_0\dot{H}_p^{-1}(\Omega)), \end{aligned}$$

where the functional $(\,\cdot\,,\,\cdot\,,\,\ldots,\,\cdot\,)$ is defined as

$$\langle\,\phi\,|\,(\,g,\eta_1,\eta_2,\ldots,\eta_m\,)\,\rangle = \sum_{j=1}^{m}\int_{\Gamma_j} [\phi]_{\Gamma_j}\,\eta_j\, \mathrm{d}\mathcal{H}^{n-1} - \int_\Omega \phi g\, \mathrm{d}\mathcal{H}^n, \quad \phi \in H_{p'}^1(\Omega).$$

As usual, $1/p + 1/p' = 1$. This is the analog to the compatibility condition derived in Section 3.3 between the right hand side g of the divergence equation, the boundary data prescribing the normal velocity and the initial velocity u_0. Note that this condition may be obtained by partial integration as in Section 3.3, since the singular boundary satisfies $\mathcal{H}^{n-1}(\partial_s\Omega) = 0$.

8.20 Now, in addition to these three compatibility conditions, which are the canonical analogs to the necessary compatibility conditions in case of a bounded, smooth domain there are in addition necessary compatibility conditions on the data of the boundary conditions on the joining edges of two adjacent smooth boundary components. In short notation, these necessary conditions may be written as: There exist $v \in \mathbb{X}_u(a,\Omega)$ and $q \in \mathbb{X}_{p,\gamma}^{\beta}(a,\Omega)$, such that

$$\mathcal{B}^{\alpha_j,\beta_j}(v,q) = h_j \quad \text{on } (0,a)\times\Gamma_j \qquad \text{for } j = 1,\, 2,\, \ldots,\, m.$$

However, this description is not very instructive. On the other hand, a more concrete formulation of the necessary compatibility conditions is only possible, if we treat each of the 45 possible combinations of the 9 energy preserving respectively artificial boundary conditions separately. Therefore, we want to focus on a small subset of the admissible combinations of different boundary conditions and assume

(8.3)
$$\text{for } k,\, l \in \{\, 1,\, 2,\, \dots,\, m \,\} \text{ with } \bar{\Gamma}_k \cap \bar{\Gamma}_l \neq \varnothing \text{ and } k \neq l \text{ we have}$$
$$(\alpha_k,\, \beta_k) = (+1,\, 0) \text{ and } \alpha_l = 0 \qquad \text{or} \qquad \alpha_k = 0 \text{ and } (\alpha_l,\, \beta_l) = (+1,\, 0),$$

i. e. one of the imposed boundary conditions is always the perfect slip condition and the other one prescribes the tangential velocity. On one hand, this is not a severe restriction, if we consider model problems like the tube in the figure on page 3, where the perfect slip condition may be assumed on the shell of the tube, a homogeneous Dirichlet condition may be assumed on the surface of the obstacle, an inhomogeneous Dirichlet condition may be assumed on the inflow boundary Γ_{in} and an artificial outflow boundary condition prescribing the tangential velocity may be assumed on the outflow boundary Γ_{out}. On the other hand, this constraint on the boundary conditions ensures, that all arising model problems in an acute wedge may be solved via reflection methods.

8.21 Now, the restriction (8.3) allows for a concrete derivation of the necessary compatibility conditions for the boundary data. In fact, if $k,\, l \in \{\, 1,\, 2,\, \dots,\, m \,\}$ with $k \neq l$, $\mathcal{E} := \bar{\Gamma}_k \cap \bar{\Gamma}_l \neq \varnothing$ and $(\alpha_k,\, \beta_k) = (+1,\, 0)$ as well as $\alpha_l = 0$, then the following compatibility conditions are necessary:

(8.4)
$$Q_{\Gamma_k}[h_l]_{\mathcal{E}} = Q_{\Gamma_k}[h_k]_{\mathcal{E}},$$
$$\mu P_{\Gamma_k} P_{\Gamma_l}[\partial_{\nu_k} h_l]_{\mathcal{E}} + \mu P_{\Gamma_l}[\nabla_{\Gamma_k} h_k]_{\mathcal{E}} \nu_k = P_{\Gamma_k} P_{\Gamma_l}[h_k]_{\mathcal{E}},$$
$$\mu P_{\Gamma_k} Q_{\Gamma_l}[\partial_{\nu_k} h_l]_{\mathcal{E}} + \mu Q_{\Gamma_l}[\nabla_{\Gamma_k} h_k]_{\mathcal{E}} \nu_k = P_{\Gamma_k} Q_{\Gamma_l}[h_k]_{\mathcal{E}}, \quad \text{if } \beta_l = 0.$$

8.22 As our analysis will reveal, the regularity and compatibility conditions derived in Section 8.2 and above are also sufficient to construct a unique maximal regular solution

$$(u,\, p) \in \mathbb{X}_{\gamma}^{\beta}(a,\, \Omega) := \mathbb{X}_u(a,\, \Omega) \times \mathbb{X}_{p,\gamma}^{\beta}(a,\, \Omega)$$

to $(S \,|\, a,\, \Omega)_{f,g,h,u_0}^{\alpha,\beta}$ for all $(f,\, g,\, h,\, u_0) \in \mathbb{Y}_{\gamma}^{\alpha,\beta}(a,\, \Omega)$, where the data space $\mathbb{Y}_{\gamma}^{\alpha,\beta}(a,\, \Omega)$ is defined to incorporate all necessary regularity and compatibility, i. e. to consist of all

$$(f,\, g,\, h,\, u_0) \in \mathbb{Y}_f(a,\, \Omega) \times \mathbb{Y}_g(a,\, \Omega) \times \prod_{j=1}^{m} \mathbb{Y}_{h,\gamma_j}^{\alpha_j,\beta_j}(a,\, \Gamma_j) \times \mathbb{Y}_0(\Omega)$$

that satisfy the compatibility conditions $(C_1)_{g,u_0}$, $(C_2)_{h,u_0}^{\alpha}$ and $(C_3)_{g,h,u_0}^{\beta}$ as well as the additional constraint (8.4). The detailed results will be stated as Theorem 8.24 below.

8.4 Maximal L_p-Regularity of the Stokes Equations

8.23 With the above preparations at hand we are able to formulate our final main theorem, which postulates the maximal L_p-regularity of the Stokes equations $(S \,|\, a,\, \Omega)$ for a bounded, acute weakly singular C^{3-}-domain of order one.

8.24 THEOREM. *Let $a > 0$, let $\Omega \subseteq \mathbb{R}^n$ be a bounded, acute weakly singular C^{3-}-domain of order one with smooth boundary components $\Gamma_1, \Gamma_2, \ldots, \Gamma_m$ and let $1 < p < \infty$ with $p \neq \frac{3}{2}, 3$. Let $\rho, \mu > 0$ and let $\alpha, \beta \in \{-1, 0, +1\}^m$ such that (8.3) is satisfied. Moreover, for $j = 1, 2, \ldots, m$*

- *let $\gamma_j = -\infty$, if $\beta_j = 0$;*
- *let $\gamma_j \in \{-\infty\} \cup [0, 1/2 - 1/2p]$, if $\beta_j = +1$;*
- *let $\gamma_j \in \{-\infty\} \cup [0, \infty)$, if $\beta_j = -1$.*

Then there exists a unique maximal regular solution

$$(u, p) \in \mathbb{X}_\gamma^\beta(a, \Omega)$$

to the Stokes equations $(S \mid a, \Omega)_{f,g,h,u_0}^{\alpha,\beta}$, if and only if the data satisfies

$$(f, g, h, u_0) \in \mathbb{Y}_\gamma^{\alpha,\beta}(a, \Omega).$$

Furthermore, the solutions depend continuously on the data.

8.25 As has been shown in Section 8.1, the proof of Theorem 8.24 may be carried out with the exactly same arguments as the proof of Theorem 3.30. The missing necessary ingredients are a *splitting scheme* as formulated in Theorem 4.60 for weakly singular domains, which makes a reduction to nearly homogeneous data possible, and a maximal L_p-regularity result for an acute wedge. However, Theorem 4.60 relies on the theory of (weak) elliptic problems subject to Dirichlet or Neumann boundary conditions, which has been presented in Section 4.1, and on the theory of parabolic problems subject to boundary conditions involving the divergence, which has been presented in Section 4.2. A suitable generalisation to the case of a weakly singular domain will be presented in Sections 8.5 and 8.6. In particular a generalised splitting scheme, which is applicable to the situation considered in Theorem 8.24, is available for acute weakly singular domains. Moreover, the case of an acute wedge will be treated in Section 8.7, which completes the proof of Theorem 8.24.

8.5 L_p-Theory for Elliptic Problems

8.26 The generalisation of the localisation method employed in Chapters 5 – 7 requires an L_p-theory for elliptic problems in weakly singular domains subject to *mixed boundary conditions*. It has already been mentioned by C. G. SIMADER and H. SOHR, that their methods to treat the Dirichlet respectively Neumann boundary conditions for the Laplacian, which is presented in [SS96] respectively [SS92], extends to problems with mixed boundary conditions, if the domain allows for a localisation to an acute wedge, cf. [SS96, page 17].

8.27 To formulate a suitable generalisation, we assume $\Omega \subseteq \mathbb{R}^n$ to be a bounded, acute weakly singular C^{2-}-domain of order one or a (bent) acute wedge of type BUC^{2-}. Furthermore, we assume the boundary to consist of the smooth components $\Gamma_1, \Gamma_2, \ldots, \Gamma_m$, we choose $J \subseteq \{1, 2, \ldots, m\}$ and set

$$\Sigma := \mathrm{int}_{\partial\Omega} \bigcup_{j \in J} \bar{\Gamma}_j \qquad \text{and} \qquad \Gamma := \mathrm{int}_{\partial\Omega} \bigcup_{j \notin J} \bar{\Gamma}_j.$$

In this setting, the mixed boundary value problem

$$-\Delta q = \mathrm{div}\, f \qquad \text{in } \Omega,$$
$$[q]_\Sigma = h_\Sigma \qquad \text{on } \Sigma,$$
$$\partial_\nu q = h_\Gamma - [f]_\Gamma \cdot \nu_\Gamma \qquad \text{on } \Gamma$$

admits a unique solution $q \in \dot{H}_p^1(\Omega)$, whenever $f \in L_p(\Omega, \mathbb{R}^n)$, $h_\Sigma \in \dot{W}_p^{1-1/p}(\Sigma)$ and

$$h_\Gamma \in \dot{W}_p^{-1/p}(\Gamma) := {}_0\dot{W}_{p'}^{1-1/p'}(\Gamma)'.$$

Note that uniqueness has to be understood as uniqueness up to a constant, if $\Sigma = \varnothing$.

8.28 As in Section 4.1 it is convenient to have a weak formulation available, which avoids the notion of a generalised normal trace as employed above. Therefore, we set

$$\Sigma\dot{H}_p^1(\Omega) := \left\{ \phi \in \dot{H}_p^1(\Omega) : [\phi]_\Sigma = 0 \right\}$$

and observe, that for every $\psi \in {}_0\dot{W}_{p'}^{1-1/p'}(\Gamma)$, there exists a $\Psi \in \Sigma\dot{H}_{p'}^1(\Omega)$, such that $[\Psi]_\Gamma = \psi$. Indeed, the extension of ψ by zero to all of $\partial\Omega$ may be extended to Ω by the surjectivity of the trace operator, cf. Proposition 8.13. Hence, we obtain a generalised normal trace

$$[\,\cdot\,]_{\Gamma,\nu} : L_{p,s}(\Omega) \longrightarrow \dot{W}_p^{-1/p}(\Gamma)$$

via

$$\langle \psi \,|\, [v]_{\Gamma,\nu} \rangle := (\nabla\Psi \,|\, v)_\Omega, \qquad \begin{array}{l} \psi \in {}_0\dot{W}_{p'}^{1-1/p'}(\Gamma),\ \Psi \in \Sigma\dot{H}_p^1(\Omega), \\ [\Psi]_\Gamma = \psi,\ v \in L_{p,s}(\Omega) \end{array}$$

and the mixed boundary value problem is equivalent to its weak formulation

$$-(\nabla q \,|\, \nabla\phi)_\Omega = (f \,|\, \nabla\phi)_\Omega - \langle [\phi]_\Gamma \,|\, h_\Gamma \rangle, \qquad \phi \in \Sigma\dot{H}_{p'}^1(\Omega),$$
$$[q]_\Sigma = h_\Sigma \qquad \text{on } \Sigma.$$

Note the similarity to the weak formulations of the Dirichlet respectively Neumann problem derived in Section 4.1.

8.29 As explained in 8.26, the construction of a unique solution to this (weak) elliptic problem subject to mixed boundary conditions may be performed by the methods employed in [SS92, SS96], since the prototype geometry of an acute wedge may be treated by reflection methods. However, we want to note two immediate consequences. On one hand, we obtain the direct topological decomposition

$$L_p(\Omega, \mathbb{R}^n) = {}_\Gamma L_{p,s}(\Omega) \oplus \nabla_\Sigma \dot{H}_p^1(\Omega),$$

which is a decomposition of Helmholtz-Weyl type. Here,

$${}_\Gamma L_{p,s}(\Omega) := \left\{ v \in L_{p,s}(\Omega) : [v]_{\Gamma,\nu} = 0 \right\}$$

and the above decomposition coincides with the Weyl decomposition for $\Gamma = \varnothing$ and with the Helmholtz decomposition for $\Sigma = \varnothing$. On the other hand, we obtain generalisations of Propositions 4.58 and 4.60 to the setting of acute weakly singular domains. Indeed, if we define $J \subseteq \{1, 2, \ldots, m\}$ by requiring $j \in J$, if and only if $\beta_j \in \{-1, +1\}$, then we obtain a splitting scheme, which reduces the Stokes equations in an acute weakly singular domain to the case, where only the data of the normal boundary conditions is non-vanishing.

8.6 L_p-Theory for Parabolic Problems

8.30 The maximal L_p-regularity results for the parabolic problems subject to boundary conditions involving the divergence, which have been obtained in Section 4.2, may also be generalised to the setting of an acute weakly singular domain. Indeed, if we assume $\Omega \subseteq \mathbb{R}^n$ to be a bounded, acute weakly singular C^{3-}-domain of order one, then the localisation procedure presented in 5 – 7 for the Stokes equations may be adapted based on the observations in Section 8.1. The only missing ingredient is a maximal L_p-regularity result for an acute wedge.

8.31 However, if we restrict our considerations to a set of boundary conditions given by a vector $\alpha \in \{-1, 0, +1\}^m$, which satisfies

for $k, l \in \{1, 2, \ldots, m\}$ with $\bar{\Gamma}_k \cap \bar{\Gamma}_l \neq \varnothing$ and $k \neq l$ we have

$$\alpha_k = 0 \qquad \text{or} \qquad \alpha_l = 0,$$

where $\Gamma_1, \Gamma_2, \ldots, \Gamma_m$ denote the smooth components of the boundary, then all model problems in an acute wedge, which occur during the localisation procedure, may be solved by reflection methods. Note that a result with the above restrictions on the combination of different boundary conditions covers all cases of Theorem 8.24.

8.32 The technique of reflections in an acute wedge will be presented in Section 8.7 for the Stokes equations and may be adapted to the parabolic problems considered in Section 4.2. Hence, all necessary ingredients for a generalisation of the splitting scheme presented in Theorem 4.60 to weakly singular C^{3-}-domains are available for the cases of Theorem 8.24. Thus, the treatment of the Stokes equations in an acute wedge as presented in Section 8.7 completes the proof of Theorem 8.24.

8.7 The Reflection Method

8.33 Finally, we consider the model problems in an acute wedge, which form the basis of the proof of Theorem 8.24, i. e. we establish a maximal L_p-regularity result for $1 < p < \infty$ with $p \neq \frac{3}{2}$, 3, provided the imposed boundary conditions satisfy the constraint (8.3), which restricts the possible combinations of boundary conditions on adjacent smooth components of the boundary.

8.34 To be precise, the first model problem reads

$$\rho\partial_t\bar{u} - \mu\Delta\bar{u} + \nabla\bar{p} = \rho f \quad \text{in } (0, a) \times \mathbb{R}^n_+,$$

$$\operatorname{div}\bar{u} = g \quad \text{in } (0, a) \times \mathbb{R}^n_+,$$

$$-\mu[\partial_y u]_y - \mu\nabla_x[v]_y = h_{y,u} \quad \text{on } (0, a) \times \partial_y\mathbb{R}^n_+,$$

$$[v]_y = h_{y,v} \quad \text{on } (0, a) \times \partial_y\mathbb{R}^n_+,$$

(8.5a) $$\quad -\mu[\partial_y w]_y - \mu\partial_z[v]_y = h_{y,w} \quad \text{on } (0, a) \times \partial_y\mathbb{R}^n_+,$$

$$[u]_z = h_{z,u} \quad \text{on } (0, a) \times \partial_z\mathbb{R}^n_+,$$

$$[v]_z = h_{z,v} \quad \text{on } (0, a) \times \partial_z\mathbb{R}^n_+,$$

$$[w]_z = h_{z,w} \quad \text{on } (0, a) \times \partial_z\mathbb{R}^n_+,$$

$$\bar{u}(0) = u_0 \quad \text{in } \mathbb{R}^n_+,$$

where

$$\partial_y\mathbb{R}^n_+ := \mathbb{R}^{n-2} \times \{0\} \times (0, \infty) \quad \text{and} \quad \partial_z\mathbb{R}^n_+ := \mathbb{R}^{n-2} \times (0, \infty) \times \{0\}$$

denote the two smooth components of the boundary. Moreover, we decomposed the spatial variable into a tangential part $x \in \mathbb{R}^{n-2}$ and two normal parts $y, z > 0$ and the desired maximal regular solution

$$(\bar{u}, \bar{p}) \in \mathbb{X}^\beta_\gamma(a, \mathbb{R}^n_+)$$

as $\bar{u} = (u, v, w)$ into a tangential velocity field $u : \mathbb{R}^n_+ \longrightarrow \mathbb{R}^{n-2}$ and two normal velocities $v, w : \mathbb{R}^n_+ \longrightarrow \mathbb{R}$. Finally, $[\,\cdot\,]_y$ respectively $[\,\cdot\,]_z$ denotes the trace on the boundary component $\partial_y\mathbb{R}^n_+$ respectively $\partial_z\mathbb{R}^n_+$ and the boundary data has been decomposed as $h_y = (h_{y,u}, h_{y,v}, h_{y,w})$ respectively $h_z = (h_{z,u}, h_{z,v}, h_{z,w})$. Note that in this case the boundary data has to satisfy the compatibility conditions

(8.5b) $$\begin{aligned} -\mu[\partial_y h_{z,u}]_y - \mu\nabla_x[h_{y,v}]_z &= [h_{y,u}]_z \quad \text{on } (0, a) \times \mathbb{R}^{n-2}, \\ [h_{z,v}]_y &= [h_{y,v}]_z \quad \text{on } (0, a) \times \mathbb{R}^{n-2}, \\ -\mu[\partial_y h_{z,w}]_y - \mu[\partial_z h_{y,v}]_z &= [h_{y,w}]_z \quad \text{on } (0, a) \times \mathbb{R}^{n-2}. \end{aligned}$$

8.35 The second and third model problems read

$$\rho\partial_t\bar{u} - \mu\Delta\bar{u} + \nabla\bar{p} = \rho f \quad \text{in } (0, a) \times \mathbb{R}^n_+,$$

$$\operatorname{div}\bar{u} = g \quad \text{in } (0, a) \times \mathbb{R}^n_+,$$

$$-\mu[\partial_y u]_y - \mu\nabla_x[v]_y = h_{y,u} \quad \text{on } (0, a) \times \partial_y\mathbb{R}^n_+,$$

$$[v]_y = h_{y,v} \quad \text{on } (0, a) \times \partial_y\mathbb{R}^n_+,$$

(8.6a) $$\quad -\mu[\partial_y w]_y - \mu\partial_z[v]_y = h_{y,w} \quad \text{on } (0, a) \times \partial_y\mathbb{R}^n_+,$$

$$[u]_z = h_{z,u} \quad \text{on } (0, a) \times \partial_z\mathbb{R}^n_+,$$

$$[v]_z = h_{z,v} \quad \text{on } (0, a) \times \partial_z\mathbb{R}^n_+,$$

$$-2\mu\tau[\partial_z w]_z + [\bar{p}]_z = h_{z,w} \quad \text{on } (0, a) \times \partial_z\mathbb{R}^n_+,$$

$$\bar{u}(0) = u_0 \quad \text{in } \mathbb{R}^n_+$$

with $\tau \in \{0, 1\}$. Note that in this case the boundary data has to satisfy the compatibility conditions

(8.6b)
$$-\mu[\partial_y h_{z,u}]_y - \mu\nabla_x[h_{y,v}]_z = [h_{y,u}]_z \quad \text{on } (0, a) \times \mathbb{R}^{n-2},$$
$$[h_{z,v}]_y = [h_{y,v}]_z \quad \text{on } (0, a) \times \mathbb{R}^{n-2}.$$

8.36 Employing the elliptic and parabolic problems treated in Sections 8.5 and 8.6 and the thereby induced splitting scheme, we may assume all data to vanish except $h_{y,v}$ and $h_{z,w}$. In this situation we may extend $h_{y,v}$ to a function on all of $\mathbb{R}^{n-2} \times \{0\} \times \mathbb{R}$ and solve the corresponding halfspace problem. In summary, this leads to the simplification $f = 0$, $g = 0$, $h_y = 0$ and $u_0 = 0$. However, the components of h_z may be non-trivial but satisfy the compatibility conditions

$$[\partial_y h_{z,u}]_y = 0, \quad [h_{z,v}]_y = 0, \quad [\partial_y h_{z,w}]_y = 0$$

in case of system (8.5a), which stem from (8.5b), respectively

$$[\partial_y h_{z,u}]_y = 0, \quad [h_{z,v}]_y = 0$$

in case of system (8.6a), which stem from (8.6b).

8.37 In this situation the system (8.5a) may be solved as follows: We extend $h_{z,u}$ and $h_{z,w}$ even in y and $h_{z,v}$ odd in y to all of $\mathbb{R}^{n-2} \times \mathbb{R} \times \{0\}$. This leads to functions in the right regularity classes thanks to the above compatibility conditions. Now, the solution to the corresponding halfspace problem has the property that u, w and $\bar{p}$ are odd functions of y and v is an even function in y. Hence, the restriction of this solution to the acute wedge $\mathbb{R}^n_{++}$ satisfies the homogeneous perfect slip condition on the boundary $\partial_y \mathbb{R}^n_{++}$.

8.38 The system (8.6a) may be solved by exactly the same procedure. However, in this case we have

$$h_{z,w} \in {}_0W_p^{1/2-1/2p}((0, a), L_p(\partial_z\mathbb{R}^n_{++})) \cap L_p((0, a), W_p^{1-1/p}(\partial_z\mathbb{R}^n_{++}))$$

and the even extension in y belongs to the same regularity class without the need of any condition at $y = 0$. Thus, the proof of Theorem 8.24 is complete.

Remarks

8.39 The mixed derivative theorem, Proposition 8.12, for an acute wedge may be proved as Proposition 3.9, cf. Remark 3.60. This result then carries over to an acute bent wedge as considered in Section 8.1 and finally to an acute, weakly singular domain via localisation. The trace theorem, Proposition 8.13 is a consequence of [Mar87, Theorem 2], which is applicable even to Lipschitz domains, see also Remarks 3.60, 3.62, 3.63 and 3.64.

8.40 Not all combinations of the 9 energy preserving respectively artificial boundary conditions lead to model problems in an acute wedge that may be solved by reflection methods. In these cases, a more refined approach is necessary. Concerning e.g. the combination of two Dirichlet conditions, maximal L_2-regularity may be proved employing Laplace and Fourier transformations. However, it is not known up to now, whether these results extend to L_p for $p \neq 2$.

References

[Mar87] J. MARSCHALL: *The Trace of Sobolev-Slobodeckij Spaces on Lipschitz Domains.* Manuscripta Math., 58, 47–65, 1987.

[SS92] C. G. SIMADER and H. SOHR: *A New Approach to the Helmholtz Decomposition and the Neumann Problem in L_q-Spaces for Bounded and Exterior Domains.* In: Mathematical Problems Relating to the Navier-Stokes Equations, *Ser. Adv. Math. Appl. Sci.*, vol. 11 (G. P. GALDI, ed.), (1–35), World Scientific Publishing, 1992.

[SS96] C. G. SIMADER and H. SOHR: The Dirichlet Problem for the Laplacian in Bounded and Unbounded Domains, *Pitman Research Notes in Mathematics*, vol. 360. Longman, 1996.

Bibliography

[AF03]   R. A. ADAMS and J. J. F. FOURNIER: Sobolev Spaces, *Pure and Applied Mathematics*, vol. 140. Academic Press, 2nd ed., 2003.

[Ama90]   H. AMANN: Ordinary Differential Equations. de Gruyter, 1990.

[Ama95]   H. AMANN: Linear and Quasilinear Parabolic Problems. Volume I. Abstract Linear Theory, *Monographs in Mathematics*, vol. 89. Birkhäuser, 1995.

[Ama09]   H. AMANN: Anisotropic Function Spaces and Maximal Regularity for Parabolic Problems. Part 1: Function Spaces, *Jindřich Nečas Center for Mathematical Modeling Lecture Notes*, vol. 6. MATFYZPRESS, 2009.

[And98]   J. D. ANDERSON, JR.: *Some Reflections on the History of Fluid Dynamics*. In: The Handbook of Fluid Dynamics (R. W. JOHNSON, ed.), chap. 2, Springer, 1998.

[Ari90]   R. ARIS: Vectors, Tensors, and the Basic Equations of Fluid Mechanics. Dover Publications, 1990.

[Bat00]   G. K. BATCHELOR: An Introduction to Fluid Dynamics. Cambridge University Press, 3rd ed., 2000.

[BF07]   F. BOYER and P. FABRIE: *Outflow Boundary Conditions for the Incompressible Non-Homogeneous Navier-Stokes Equations*. Discrete Contin. Dyn. Syst. Ser. B, 7 (2), 219–250, 2007.

[BFG$^+$12] D. BOTHE, R. FARWIG, M. GEISSERT, H. HECK, M. HIEBER, W. STANNAT, C. TROPEA, and S. ULBRICH: Mathematical Fluid Dynamics. Preprint, 2012.

[BKP12]   D. BOTHE, M. KÖHNE, and J. PRÜSS: *On a Class of Energy Preserving Boundary Conditions for Incompressible Newtonian Flows*. Preprint at arXiv.org, URL http://arxiv.org/abs/1207.0707, 2012.

[BLS07]   M. P. BRENNER, E. LAUGA, and H. A. STONE: *Microfluidics: The No-Slip Boundary Condition*. In: Handbook of Experimental Fluid Dynamics (J. FOSS, C. TROPEA, and A. YARIN, eds.), (1219–1240), Springer, 2007.

[BM88]   W. BORCHERS and T. MIYAKAWA: *L^2 Decay for the Navier-Stokes Flow in Halfspaces*. Math. Ann., 282, 139–155, 1988.

[BNP04]   H. BELLOUT, J. NEUSTUPA, and P. PENEL: *On the Navier-Stokes Equation with Boundary Conditions Based on Vorticity*. Math. Nachr., 269-270, 59–72, 2004.

[BP07]   D. BOTHE and J. PRÜSS: *L_p-Theory for a Class of Non-Newtonian Fluids*. SIAM J. Math. Anal., 39, 379–421, 2007.

[CD98]   S. CHEN and G. D. DOOLEN: *Lattice Boltzmann Method for Fluid Flows*. Annual Review of Fluid Mechanics, 30, 329–364, 1998.

[CM00]   A. J. CHORIN and J. E. MARSDEN: A Mathematical Introduction to Fluid Mechanics. Springer, 3rd ed., 2000.

[CMP94]   C. CONCA, F. MURAT, and O. PIRONNEAU: The Stokes and Navier-Stokes Equations with Boundary Conditions Involing the Pressure. Japan J. Math., 20 (2), 279–318, 1994.

[CPPT95] C. CONCA, C. PARÈS, O. PIRONNEAU, and M. THIRIET: Navier-Stokes Equations with Imposed Pressure and Velocity Fluxes. Int. J. Numer. Meth. Fluids, 20, 267–287, 1995.

[DHP01]   W. DESCH, M. HIEBER, and J. PRÜSS: L^p-Theory of the Stokes Equation in a Half-Space. J. Evol. Equ., 1, 115–142, 2001.

[DHP03]   R. DENK, M. HIEBER, and J. PRÜSS: $\mathcal{R}$-Boundedness, Fourier-Multipliers and Problems of Elliptic and Parabolic Type, Mem. Amer. Math. Soc., vol. 166. American Mathematical Society, 2003.

[DHP07]   R. DENK, M. HIEBER, and J. PRÜSS: Optimal L_p-L_q-Regularity for Parabolic Problems with Inhomogeneous Boundary Data. Math. Z., 257, 193–224, 2007.

[DS11]   R. DENK and J. SEILER: On the Maximal L_p-Regularity of Parabolic Mixed Order Systems. J. Evol. Equ., 11, 371–404, 2011.

[DSS08]   R. DENK, J. SAAL, and J. SEILER: Inhomogeneous Symbols, the Newton Polygon, and Maximal L_p-Regularity. Russian J. Math. Phys., 15 (2), 171–192, 2008.

[DV87]   G. DORE and A. VENNI: On the Closedness of the Sum of Two Closed Operators. Math. Z., 196, 189–201, 1987.

[Eva98]   L. C. EVANS: Partial Differential Equations. American Mathematical Society, 1998.

[Fef07]   C. L. FEFFERMAN: Existence & Smoothness of the Navier-Stokes Equation. Official Problem Description of the Clay Mathematics Institute, 2007.

[FK62]   H. FUJITA and T. KATO: On the Nonstationary Navier-Stokes System. Rend. Sem. Mat. Univ. Padova, 32, 234–260, 1962.

[FK64]   H. FUJITA and T. KATO: On the Navier-Stokes Initial Value Problem. I. Arch. Ration. Mech. Anal., 16 (4), 269–315, 1964.

[FV89]   R. FOSDICK and E. VIRGA: A Variational Proof of the Stress Theorem of Cauchy. Arch. Rational Mech. Anal., 105, 95–103, 1989.

[Gal94a] G. P. GALDI: An Introduction to the Mathematical Theory of the Navier-Stokes Equations, Volume 1: Linearized Steady Problems. Springer, 1994.

[Gal94b] G. P. GALDI: An Introduction to the Mathematical Theory of the Navier-Stokes Equations, Volume 2: Nonlinear Steady Problems. Springer, 1994.

[Gig85]   Y. GIGA: Domains of Fractional Powers of the Stokes Operator in L_r Spaces. Arch. Ration. Mech. Anal., 89, 251–265, 1985.

[GM85]   Y. GIGA and T. MIYAKAWA: Solutions in L_r of the Navier-Stokes Initial Value Problem. Arch. Ration. Mech. Anal., 89, 269–281, 1985.

[Gre95]   P. M. GRESHO: Incompressible Fluid Mechanics: Some Fundamental Formulation Issues. Annual Review of Fluid Mechanics, 23, 413–453, 1995.

[Gri97]   D. F. GRIFFITHS: *The 'No Boundary Condition' Outflow Boundary Condition.* Int. J. Numer. Meth. Fluids, 24 (4), 393–411, 1997.

[Gru95a]   G. GRUBB: *Nonhomogeneous Time-Dependent Navier-Stokes Problems in L_p-Sobolev Spaces.* Differential and Integral Equations, 8 (5), 1013–1046, 1995.

[Gru95b]   G. GRUBB: *Parameter-Elliptic and Parabolic Pseudodifferential Boundary Problems in Global L_p-Sobolev Spaces.* Math. Zeitschr., 218 (1), 43–90, 1995.

[Gru98]   G. GRUBB: *Nonhomogeneous Navier-Stokes Problems in L_p Sobolev Spaces over Exterior and Interior Domains.* In: Theory of the Navier-Stokes Equations, *Ser. Adv. Math. Appl. Sci.*, vol. 47 (J. G. HEYWOOD, K. MASUDA, R. RAUTMANN, and V. A. SOLONNIKOV, eds.), (46–63), World Scientific Publishing, 1998.

[GS90a]   G. GRUBB and V. A. SOLONNIKOV: *Reduction of Basic Initial-Boundary Value Problems for the Stokes Equation to Initial-Boundary Value Problems for Systems of Pseudodifferential Equations.* J. Sov. Math., 49 (5), 1140–1147, 1990.

[GS90b]   G. GRUBB and V. A. SOLONNIKOV: *Solution of Parabolic Pseudo-Differential Initial Boundary Value Problems.* J. Diff. Equ., 87 (2), 256–304, 1990.

[GS91a]   G. GRUBB and V. A. SOLONNIKOV: *Boundary Value Problems for the Nonstationary Navier-Stokes Equations Treated by Pseudo-Differential Methods.* Math. Scand., 69 (2), 217–290, 1991.

[GS91b]   G. GRUBB and V. A. SOLONNIKOV: *Reduction of Basic Initial-Boundary Value Problems for the Navier-Stokes Equations to Nonlinear Parabolic Systems of Pseudodifferential Equations.* J. Sov. Math., 56 (2), 2300–2308, 1991.

[Hop51]   E. HOPF: *Über die Anfangswertaufgabe für die hydrodynamischen Grundgleichungen.* Math. Nachr., 4, 213–231, 1951.

[HP98]   M. HIEBER and J. PRÜSS: *Functional Calculi for Linear Operators in Vector-Valued L^p-Spaces via the Transference Principle.* Adv. Differential Equations, 3, 847–872, 1998.

[HRT92]   J. G. HEYWOOD, R. RANACHER, and S. TUREK: *Artificial Boundaries and Flux and Pressure Conditions for Incompressible Navier-Stokes Equations.* Int. J. Numer. Meth. Fluids, 22, 325–352, 1992.

[Köh07]   M. KÖHNE: Zur Analysis und Numerik der Navier-Stokes-Gleichungen in Gebieten mit künstlichen Rändern. Diplomarbeit, Universität Paderborn, 2007.

[KS04]   T. KUBO and Y. SHIBATA: *On some Properties of Solutions to the Stokes Equation in the Half-Space and Perturbed Half-Space.* Quad. Math., 15, 149–220, 2004.

[KS05a]   T. KUBO and Y. SHIBATA: *On the Stokes and Navier-Stokes Equations in a Perturbed Half-Space.* Adv. Differential Equations, 10, 695–720, 2005.

[KS05b]   T. KUBO and Y. SHIBATA: *On the Stokes and Navier-Stokes Flows in a Perturbed Half-Space.* Banach Center Publ., 70, 157–167, 2005.

[Kud68a]   L. D. KUDRJAVCEV: *Imbedding Theorem for a Class of Functions Defined on the Entire Space or on a Halfspace. I.* Amer. Math. Soc. Transl., 74, 199–225, 1968.

[Kud68b]   L. D. KUDRJAVCEV: *Imbedding Theorem for a Class of Functions Defined on the Entire Space or on a Halfspace. II.* Amer. Math. Soc. Transl., 74, 227–260, 1968.

[KW01]   N. J. KALTON and L. WEIS: *The H^∞-Calculus and Sums of Closed Operators*. Math. Ann., 321, 319–345, 2001.

[Lad69]   O. A. LADYZHENSKAYA: The Mathematical Theory of Viscous Incompressible Flow. Gordon and Breach, 1969.

[Lam32]   H. LAMB: Hydrodynamics. Cambridge University Press, 6th ed., 1932.

[Ler34a]   J. LERAY: *Essai sur les Mouvements Plans d'un Liquide Visqueux que Limitent des Parois*. J. Math. Pures Appl., 13, 331–418, 1934.

[Ler34b]   J. LERAY: *Sur le Mouvement d'un Liquide Visqueux Emplissant l'Espace*. Acta Math., 63, 193–248, 1934.

[LS03]   E. LAUGA and H. A. STONE: *Effective Slip in Pressure Driven Stokes Flow*. J. Fluid Mech., 489, 55–77, 2003.

[Mar87]   J. MARSCHALL: *The Trace of Sobolev-Slobodeckij Spaces on Lipschitz Domains*. Manuscripta Math., 58, 47–65, 1987.

[McC81]   M. MCCRACKEN: *The Resolvent Problem for the Stokes Equation on Halfspaces in L_p*. SIAM J. Math. Anal., 12, 201–228, 1981.

[MM08]   M. MITREA and S. MONNIAUX: *The Regularity of the Stokes Operator and the Fujita-Kato Approach to the Navier-Stokes Initial Value Problem in Lipschitz Domains*. J. Funct. Anal., 254, 1522–1574, 2008.

[MMW11]   M. MITREA, S. MONNIAUX, and M. WRIGHT: *The Stokes Operator with Neumann Boundary Conditions in Lipschitz Domains*. J. Math. Sci. (N. Y.), 176 (3), 2011.

[Nav23]   C. L. M. H. NAVIER: *Mémoir sur les Lois du Mouvement des Fluides*. Mem. Acad. Inst. Sci. Fr., 6, 389–440, 1823.

[Neč67]   J. NEČAS: Les Méthodes Directes en Théorie des Équations Elliptiques. Academia, 1967.

[NP07]   J. NEUSTUPA and P. PENEL: *On Regularity of a Weak Solution to the Navier-Stokes Equation with Generalized Impermeability Boundary Conditions*. Nonlinear Anal., 66 (8), 1753–1769, 2007.

[NP08]   J. NEUSTUPA and P. PENEL: *The Navier-Stokes Equation with Inhomogeneous Boundary Conditions Based on Vorticity*. Banach Center Publ., 81, 321–335, 2008.

[NP10]   J. NEUSTUPA and P. PENEL: *Local in Time Strong Solvability of the Non-Steady Navier-Stokes Equations with Navier's Boundary Condition and the Question of the Inviscid Limit*. C. R., Math., Acad. Sci. Paris, 348 (19-20), 1093–1097, 2010.

[NT03]   W. NOLL and C. TRUESDELL: The Non-Linear Field Theories of Mechanics. Springer, 2003.

[PG75]   G. DA PRATO and P. GRISVARD: *Sommes d'Opératours Linéaires et équations Différentielles Opérationelles*. J. Math. Pures Appl., 54, 305–387, 1975.

[Prü03]   J. PRÜSS: *Maximal Regularity for Evolution Equations in L_p-Spaces*. Conf. Sem. Mat. Univ. Bari, 285, 1–39, 2003.

[PS90]   J. PRÜSS and H. SOHR: *On Operators with Bounded Imaginary Powers in Banach Spaces*. Math. Z., 203, 429–452, 1990.

[PS07]   J. Prüss and G. Simonett: *H∞-Calculus for the Sum of Non-Commuting Operators*. Trans. Amer. Math. Soc., 359, 3549–3565, 2007.

[PSS07]   J. Prüss, J. Saal, and G. Simonett: *Existence of Analytic Solutions for the Classical Stefan Problem*. Math. Ann., 338, 703–755, 2007.

[Rha60]   G. de Rham: Variétés Différentiables. Formes, Courants, Formes Harmoniques. III, *Actualites Scientifiques et Industrielles*, vol. 1222. Publications de lInstitut de Mathématique de lUniversité de Nancago. Hermann & Cie., 2nd ed., 1960.

[RT00]   K. R. Rajagopal and C. Truesdell: An Introduction to the Mechanics of Fluids. Birkhäuser, 2000.

[Saa03]   J. Saal: Robin Boundary Conditions and Bounded $\mathcal{H}^\infty$-Calculus for the Stokes Operator. Logos, 2003.

[Saa06]   J. Saal: *Stokes and Navier-stokes Equations with Robin Boundary Conditions in a Half-Space*. J. Math. Fluid Mech., 8, 211–241, 2006.

[Saa07]   J. Saal: *The Stokes Operator with Robin Boundary Conditions in Solenoidal Subspaces of $L^1(\mathbb{R}^n_+)$ and $L^\infty(\mathbb{R}^n_+)$*. Commun. Partial Differ. Equations, 32 (3), 343–373, 2007.

[See71]   R. T. Seeley: *Norms and Domains of the Complex Powers A_b^z*. Amer. J. Math., 93, 299–309, 1971.

[Sob64]   P. E. Sobolevskii: *Study of Navier-Stokes Equations by Methods of the Theory of Parabolic Equations*. Soviet Math. Dokl., 5, 720–723, 1964.

[Sob75]   P. E. Sobolevskii: *Fractional Powers of Coercively Positive Sums of Operators*. Soviet Math. Dokl., 16, 1638–1641, 1975.

[Soh01]   H. Sohr: The Navier-Stokes Equations. Birkhäuser, 2001.

[Sol77a] V. A. Solonnikov: *Estimates for Solutions of Nonstationary Navier-Stokes Equations*. J. Sov. Math., 8, 467–529, 1977.

[Sol77b] V. A. Solonnikov: *Solvability of a Problem on the Motion of a Viscous Incompressible Fluid Bounded by a Free Surface*. Math. USSR Izv., 11, 1323–1358, 1977.

[Sol78]   V. A. Solonnikov: *On the Solvability of the Second Initial-Boundary Value Problem for the Linear Nonstationary Navier-stokes System*. J. Sov. Math., 10, 141–193, 1978.

[Sol88]   V. A. Solonnikov: *Unsteady Motion of a finite Mass of Fluid, Bounded by a Free Surface*. J. Sov. Math., 40, 672–686, 1988.

[Sol91]   V. A. Solonnikov: *On an Initial-Boundary Value Problem for the Stokes System Arising in the Study of a Problem with a Free Surface*. Proc. Steklov Inst. Math., 3, 191–239, 1991.

[SS92]   C. G. Simader and H. Sohr: *A New Approach to the Helmholtz Decomposition and the Neumann Problem in L_q-Spaces for Bounded and Exterior Domains*. In: Mathematical Problems Relating to the Navier-Stokes Equations, Ser. Adv. Math. Appl. Sci., vol. 11 (G. P. Galdi, ed.), (1–35), World Scientific Publishing, 1992.

[SS96]   C. G. SIMADER and H. SOHR: The Dirichlet Problem for the Laplacian in Bounded
           and Unbounded Domains, *Pitman Research Notes in Mathematics*, vol. 360. Long-
           man, 1996.

[SS03]   Y. SHIBATA and S. SHIMIZU: *On a Resolvent Estimate for the Stokes Sytem with
           Neumann Boundary Condition*. Differential Integral Equations, 16, 385–426, 2003.

[SS05]   Y. SHIBATA and S. SHIMIZU: L_p-L_q *Maximal Regularity and Viscous Incompressible
           Flows with Free Surface*. Proc. Japan Acad. Ser. A Math. Sci., 81 (9), 151–155, 2005.

[SS07a]   Y. SHIBATA and R. SHIMADA: *On the Stokes Equation with Robin Boundary Con-
           dition*. Adv. Stud. Pure Math., 47-1, 341–348, 2007.

[SS07b]   Y. SHIBATA and S. SHIMIZU: *Decay Properties of the Stokes Semigroup in Exterior
           Domains with Neumann Boundary Condition*. J. Math. Soc. Japan, 59 (1), 1–34,
           2007.

[SS08]   Y. SHIBATA and S. SHIMIZU: *On the L_p-L_q Maximal Regularity of the Neumann
           Problem for the Stokes Equations in a Bounded Domain*. J. reine angew. Math., 615,
           157–209, 2008.

[SS11]   Y. SHIBATA and S. SHIMIZU: *Maximal L_p-L_q Regularity for the Stokes Equations;
           Model Problems*. Differential Equations, 251 (2), 373–419, 2011.

[Sto45]   G. G. STOKES: *On the Theories of the Internal Friction of Fluids in Motion, and
           of the Equilibrium and Motion of Elastic Solids*. Trans. Cambridge Phil. Soc., 8,
           287–319, 1845.

[Tem77]   R. TEMAM: Navier-Stokes Equations. Noth-Holland, 1977.

[Tem05]   R. TEMAM: Mathematical Modeling in Continuum Mechanics. Cambridge University
           Press, 2005.

[Tri98]   H. TRIEBEL: Interpolation Theory. Function Spaces. Differential Operators. Wiley-
           VCH, 2nd ed., 1998.

[Uka87]   S. UKAI: *A Solution Formula for the Stokes Equation in* $\mathbb{R}_+^n$. Comm. Pure Appl.
           Math., 40 (5), 611–621, 1987.

[Wei80]   F. B. WEISSLER: *The Navier-Stokes Initial Value Problem in* L^p. Arch. Ration.
           Mech. Anal., 74, 219–230, 1980.

Table of Symbols

Fluid and Flow Quantities

D	the rate of deformation tensor, $D = \frac{1}{2}(\nabla u + \nabla u^{\mathsf{T}})$, see 1.12
μ	the (dynamic) viscosity of the fluid, a positive constant, see 1.13
ω	the vorticity of the flow, $\omega = \operatorname{rot} u$, see 1.12
p	the pressure in the flow, an *unknown* scalar function, see 1.11 and 1.12
R	the rate of rotation tensor, $R = \frac{1}{2}(\nabla u - \nabla u^{\mathsf{T}})$, see 1.12
ρ	the density of the fluid, a positive constant, see 1.6 and
S	the stress tensor, $S = 2\mu D - p$, see 1.10, 1.11, 1.12 and 1.13
u	the velocity field of the flow, an *unknown* vector field, see 1.5
V	the asymmetric counterpart of the stress tensor, $V = 2\mu R - p$, see 2.17

Basic Operations

$a \cdot b$	the inner product of two vectors $a, b \in \mathbb{R}^n$, i.e. $a \cdot b = \sum_{k=1}^{n} a_k b_k$
$A : B$	the inner product of two matrices $A, B \in \mathbb{R}^{n \times n}$, i.e. $A : B = \sum_{k,l=1}^{n} A_{k,l} B_{k,l}$
$(\cdot)_\Omega$	the mean value functional, see 3.6
$(\cdot \mid \cdot)_\Omega$	the $L_2(\Omega)$-inner product, see 3.41

Domain Boundaries

The following quantities and operators are used for a domain $\Omega \subseteq \mathbb{R}^n$ with boundary $\Gamma := \partial\Omega$ of class C^1. Note that this implies the outer unit normal field $\nu_\Gamma : \Gamma \longrightarrow \mathbb{R}^n$ to be continuous.

ν_Γ the outer unit normal field,
$\nu_\Gamma : \Gamma \longrightarrow \mathbb{R}^n$

$[\cdot]_\Gamma$ the trace of a function defined in Ω on the boundary Γ,
$[\phi]_\Gamma = \phi|_\Gamma$ for $\phi \in C(\bar{\Omega})$

P_Γ the projection onto the tangential bundle $T\Gamma$ of Γ,
$P_\Gamma(y) = 1 - \nu_\Gamma(y) \otimes \nu_\Gamma(y)$ for $y \in \Gamma$

Q_Γ the projection onto the normal bundle $N\Gamma$ of Γ,
$Q_\Gamma(y) = \nu_\Gamma(y) \otimes \nu_\Gamma(y)$ for $y \in \Gamma$, i.e. $Q_\Gamma = 1 - P_\Gamma$

Boundary Conditions

The following boundary conditions are realised by an operator $\mathcal{B}^{\alpha,\beta}$ as introduced in Chapter 3. The parameters $\alpha,\ \beta \in \{-1,\ 0,\ +1\}$ determine a particular energy preserving respectively artificial boundary condition as introduced in 2.19, 2.22 and 2.23.

$\mathcal{B}^{0,\beta}(u,\,p)$ boundary condition prescribing the tangential velocity $P_\Gamma[u]_\Gamma$,
i.e. $P_\Gamma\mathcal{B}^{0,\beta}(u,\,p) = P_\Gamma[u]_\Gamma$ with $\beta \in \{-1,\ 0,\ +1\}$,
see Chapter 3, see also 2.19, 2.22 and 2.23

$\mathcal{B}^{+1,\beta}(u,\,p)$ boundary condition prescribing the tangential stress $P_\Gamma[S]_\Gamma\,\nu_\Gamma$,
i.e. $P_\Gamma\mathcal{B}^{+1,\beta}(u,\,p) := \mu P_\Gamma[\nabla u + \nabla u^\mathsf{T}]_\Gamma\,\nu_\Gamma$ with $\beta \in \{-1,\ 0,\ +1\}$,
see Chapter 3, see also 2.19, 2.22 and 2.23

$\mathcal{B}^{-1,\beta}(u,\,p)$ boundary condition prescribing the tangential vorticity $P_\Gamma[V]_\Gamma\,\nu_\Gamma$,
i.e. $P_\Gamma\mathcal{B}^{-1,\beta}(u,\,p) := \mu P_\Gamma[\nabla u - \nabla u^\mathsf{T}]_\Gamma\,\nu_\Gamma$ with $\beta \in \{-1,\ 0,\ +1\}$,
see Chapter 3, see also 2.19, 2.22 and 2.23

$\mathcal{B}^{\alpha,0}(u,\,p)$ boundary condition prescribing the normal velocity $Q_\Gamma[u]_\Gamma$,
i.e. $Q_\Gamma\mathcal{B}^{\alpha,0}(u,\,p) \cdot \nu_\Gamma := [u]_\Gamma \cdot \nu_\Gamma$ with $\alpha \in \{-1,\ 0,\ +1\}$,
see Chapter 3, see also 2.19

$\mathcal{B}^{\alpha,+1}(u,\,p)$ boundary condition prescribing the normal stress $Q_\Gamma[S]_\Gamma\,\nu_\Gamma$,
i.e. $Q_\Gamma\mathcal{B}^{\alpha,+1}(u,\,p) \cdot \nu_\Gamma := 2\mu\,\partial_\nu u \cdot \nu_\Gamma - [p]_\Gamma$ with $\alpha \in \{-1,\ 0,\ +1\}$,
see Chapter 3, see also 2.22 and 2.23

$\mathcal{B}^{\alpha,-1}(u,\,p)$ boundary condition prescribing the pressure $Q_\Gamma[V]_\Gamma\,\nu_\Gamma$,
i.e. $Q_\Gamma\mathcal{B}^{\alpha,-1}(u,\,p) \cdot \nu_\Gamma := -[p]_\Gamma$ with $\alpha \in \{-1,\ 0,\ +1\}$,
see Chapter 3, see also 2.22 and 2.23

Equations

$(B \mid a,\,\Omega)_h^{\alpha,\beta}$	energy preserving respectively artificial boundary condition on the time-space cylinder $(0,\,a) \times \Omega$, see 2.19, 2.22 and 2.22
$(C_1)_{g,u_0}$	necessary compatibility condition for the Stokes equations, see 3.25 and 8.17
$(C_2)_{h,u_0}^{\alpha}$	necessary compatibility condition for the Stokes equations, see 3.26 and 8.18
$(C_3)_{g,h,u_0}^{\beta}$	necessary compatibility condition for the Stokes equations, see 3.27 and 8.19
$(C)_{h,u_0}^{\alpha,\mathrm{div}}$	necessary compatibility condition for parabolic problems with divergence type boundary condition, see 4.19
$(L_D \mid \Omega)_{f,h}$	the Dirichlet problem for the Laplacian on the domain Ω, see 4.1
$(L_D \mid \infty,\,\Omega)_{f,h}$	the Dirichlet problem for the Laplacian lifted to the time-space cylinder $(0,\,\infty) \times \Omega$, see 4.28
$(\mathcal{L}_D \mid \Omega)_{f,h}$	the weak Dirichlet problem for the Laplacian on the domain Ω, see 4.3
$(L_N \mid \Omega)_{f,h}$	the Neumann problem for the Laplacian on the domain Ω, see 4.1
$(L_N \mid \infty,\,\Omega)_{f,h}$	the Neumann problem for the Laplacian lifted to the time-space cylinder $(0,\,\infty) \times \Omega$, see 4.28
$(\mathcal{L}_N \mid \Omega)_{f,h}$	the weak Neumann problem for the Laplacian on the domain Ω, see 4.9
$(S \mid a,\,\Omega)_{f,g,h,u_0}^{\alpha,\beta}$	the Stokes equations on the time-space cylinder $(0,\,a) \times \Omega$ subject to a boundary condition given by the operator $\mathcal{B}^{\alpha,\beta}$, see Chapters 3 and 8, see also Chapters 6 and 7
$(S \mid \infty,\,\Omega)_{f,g,h,u_0}^{\alpha,\beta,\varepsilon}$	the shifted Stokes equations on the time-space cylinder $(0,\,\infty) \times \Omega$ subject to a boundary condition given by the operator $\mathcal{B}^{\alpha,\beta}$, see Chapter 5
$(N \mid a,\,\Omega)_{f,g,h,u_0}^{\alpha,\beta}$	the Navier-Stokes equations on the time-space cylinder $(0,\,a) \times \Omega$ subject to a boundary condition given by the operator $\mathcal{B}^{\alpha,\beta}$, see Chapters 3 and 8
$(P \mid a,\,\Omega)_{f,h,u_0}^{\alpha}$	a parabolic problem on the time-space cylinder $(0,\,a) \times \Omega$ subject to a boundary condition given by the operator $\mathcal{B}^{\alpha,\mathrm{div}}$, see 4.16
$(P \mid \infty,\,\Omega)_{f,h,u_0}^{\alpha,\varepsilon}$	a shifted parabolic problem on the time-space cylinder $(0,\,\infty) \times \Omega$ subject to a boundary condition given by the operator $\mathcal{B}^{\alpha,\mathrm{div}}$, see 4.28

Function Spaces

BUC^m	the spaces of bounded, uniformly continuous functions, $m \in \mathbb{N}$, see 3.19
BUC^{m-}	the subspaces of the spaces BUC^{m-1} consisting of the functions with Lipschitz continuous partial derivatives of order $m-1$, $m \in \mathbb{N}$, see 6.2
$_0BUC$	the spaces of bounded, uniformly continuous functions with homogeneous initial value, $m \in \mathbb{N}$, see 3.50
H_p^s	the Bessel potential spaces, $0 \le s < \infty$ and $1 < p < \infty$, see 3.2, 3.3, 3.10, 3.57 and 3.58
$_0H_p^s$	the Bessel potential spaces with zero initial value, $0 \le s \le 1$ and $1 < p < \infty$, see 3.2, 3.10, 3.57 and 3.58
$\dot{H}_p^1$	the homogeneous Sobolev spaces, $1 < p < \infty$, see 3.5 and 3.57
$_0\dot{H}_p^1$	the homogeneous Sobolev spaces with zero boundary value, $1 < p < \infty$, see 3.41, 3.57 and 4.3
$_0\dot{H}_p^{-1}$	the dual spaces of the homogeneous Sobolev spaces, $1 < p < \infty$, see 3.27 and 3.57
$\hat{H}_p^1$	the homogeneous Sobolev spaces factored by the polynomials, $1 < p < \infty$, see 3.57, 3.59, 4.9
L_p	the Lebesgue spaces, $1 \le p \le \infty$, see 3.2 and 3.57
$L_{p,s}$	the Lebesgue spaces of vector fields with zero divergence, $1 < p < \infty$, see 3.37 and 4.7
$L_{p,\sigma}$	the Lebesgue spaces of solenoidal vector fields, $1 < p < \infty$, see 3.37 and 4.13
W_p^s	the Sobolev resp. Sobolev-Slobodeckij spaces, $0 \le s < \infty$ and $1 < p < \infty$, see 3.3, 3.4, 3.11, 3.57 and 3.58
$_0W_p^s$	the Sobolev resp. Sobolev-Slobodeckij spaces with zero initial value, $0 \le s \le 1$ and $1 < p < \infty$, see 3.11, 3.57 and 3.58
$\dot{W}_p^{1-1/p}$	the homogeneous Sobolev-Slobodeckij spaces, $1 < p < \infty$, see 3.15 and 3.57
$_0\dot{W}_p^{1-1/p}$	the homogeneous Sobolev-Slobodeckij spaces with zero boundary value, $1 < p < \infty$, see 8.27
$\dot{W}_p^{-1/p}$	the dual spaces of the homogeneous Sobolev-Slobodeckij spaces with zero boundary value, $1 < p < \infty$, see 4.1 and 8.27

Maximal Regularity Spaces

For the following spaces the parameters α, $\beta \in \{-1, 0, +1\}$ prescribe an energy preserving respectively artificial boundary condition as introduced in 2.19, 2.22 and 2.23. The parameter $\gamma \in \{-\infty\} \cup [0, \infty)$ prescribes the regularity of the pressure trace with

- $\gamma = -\infty$, if $\beta = 0$;
- $\gamma \in \{-\infty\} \cup [0, 1/2 - 1/2p]$, if $\beta = +1$;
- $\gamma \in \{-\infty\} \cup [0, \infty)$, if $\beta = -1$.

$\mathbb{X}_\gamma^\beta(a, \Omega)$	the solution space, see 3.28 and 3.59
$_0\mathbb{X}_\gamma^\beta(a, \Omega)$	the solution space with homogeneous initial value, see 3.31 and 3.59
$\mathbb{X}_u(a, \Omega)$	the velocity space, see 3.1 and 3.59
$_0\mathbb{X}_u(a, \Omega)$	the velocity space with homogeneous initial value, see 3.31 and 3.59
$\mathbb{X}_{p,\gamma}^\beta(a, \Omega)$	the pressure space, see 3.21, 3.22, 3.23 and 3.59
$\mathbb{Y}_\gamma^{\alpha,\beta}(a, \Omega)$	the data space, see 3.28 and 3.59
$_0\mathbb{Y}_\gamma^{\alpha,\beta}(a, \Omega)$	the data space with homogeneous initial condition, see 3.31 and 3.59
$\mathbb{Y}^{\alpha,\mathrm{div}}(a, \Omega)$	the data space for parabolic problems with divergence type boundary condition, see 4.20
$\mathbb{Y}_f(a, \Omega)$	the data space of the momentum equation, see 3.1 and 3.59
$\mathbb{Y}_g(a, \Omega)$	the data space of the divergence equation, see 3.12 and 3.59
$\mathbb{Y}_{h,\gamma}^{\alpha,\beta}(a, \Gamma)$	the data space of the boundary condition, see 3.23 and 3.59
$_\tau\mathbb{Y}_{h,\gamma}^\alpha(a, \Gamma)$	the data space of the boundary condition with vanishing normal component and homogeneous initial value, see 7.2
$_\nu\mathbb{Y}_{h,\gamma}^\beta(a, \Gamma)$	the data space of the boundary condition with vanishing tangential component and homogeneous initial value, see 6.5 and 7.3
$\mathbb{Y}_h^{\alpha,\mathrm{div}}(a, \Gamma)$	the data space of the divergence type boundary condition, see 4.18
$\mathbb{Y}_0(\Omega)$	the data space of the initial condition, see 3.19 and 3.59

Index